Edited by
Laurens D. A. Siebbeles and
Ferdinand C. Grozema

Charge and Exciton Transport through Molecular Wires

Related Titles

Chujo, Y. (ed.)

Conjugated Polymer Synthesis
Methods and Reactions

2011
ISBN: 978-3-527-32267-1

Cosnier, S., Karyakin, A. (eds.)

Electropolymerization
Concepts, Materials and Applications

2010
ISBN: 978-3-527-32414-9

Leclerc, M., Morin, J.-F. (eds.)

Design and Synthesis of Conjugated Polymers

2010
ISBN: 978-3-527-32474-3

Guldi, D. M., Martín, N. (eds.)

Carbon Nanotubes and Related Structures
Synthesis, Characterization, Functionalization, and Applications

2010
ISBN: 978-3-527-32406-4

Brabec, C., Scherf, U., Dyakonov, V. (eds.)

Organic Photovoltaics
Materials, Device Physics, and Manufacturing Technologies

2008
ISBN: 978-3-527-31675-5

Balzani, V., Credi, A., Venturi, M.

Molecular Devices and Machines
Concepts and Perspectives for the Nanoworld

2008
ISBN: 978-3-527-31800-1

Freund, M. S., Deore, B. A.

Self-Doped Conducting Polymers

2007
ISBN: 978-0-470-02969-5

Hadziioannou, G., Malliaras, G. G. (eds.)

Semiconducting Polymers
Chemistry, Physics and Engineering

2007
ISBN: 978-3-527-31271-9

Edited by Laurens D. A. Siebbeles and Ferdinand C. Grozema

Charge and Exciton Transport through Molecular Wires

WILEY-VCH Verlag GmbH & Co. KGaA

The Editors

Prof. Dr. L. D. A. Siebbeles
Delft ChemTech
Techn. University Delft
Julianalaan 136
2628 BL Delft
The Netherlands

Dr. F. C. Grozema
Delft ChemTech
Techn. University of Delft
Julianalaan 136
2628 BL Delft
The Netherlands

Library of Congress Card No.: applied for

British Library Cataloguing-in-Publication Data
A catalogue record for this book is available from the British Library.

Bibliographic information published by the Deutsche Nationalbibliothek
The Deutsche Nationalbibliothek lists this publication in the Deutsche Nationalbibliografie; detailed bibliographic data are available on the Internet at <http://dnb.d-nb.de>.

Typesetting Laserwords Private Limited, Chennai, India
Printing and Binding Strauss GmbH, Mörlenbach
Cover Design Formgeber, Eppelheim

Printed in the Federal Republic of Germany.
Printed on acid-free paper

ISBN: 978-3-527-32501-6

Contents

Charge and Exciton Transport through Molecular Wires. Edited by L.D.A. Siebbeles and F.C. Grozema

ISBN: 978-3-527-32501-6

List of Contributors

Bo Albinsson
Chalmers University of Technology
Department of Chemical and Biological Engineering/Physical and Organic Chemistry
Kemigarden 4
412 96 Göteborg
Sweden

David Q. Andrews
Environmental Working Group
Washington, D.C. 20009
USA

Trisha L. Andrew
Massachusetts Institute of Technology
Department of Chemistry
77 Massachusetts Avenue
Cambridge, MA 02139
USA

Timothy Clark
Friedrich-Alexander-Universität-Erlangen-Nürnberg
Department of Chemistry and Pharmacy & Interdisciplinary Center for Molecular Materials (ICMM)
Egerlandstr. 3
91058 Erlangen
Germany

Seong Ho Choi
University of Minnesota
Department of Chemistry and Department of Chemical Engineering and Materials Science
421, Washinton Ave. SE
Minneapolis, MN 55455
USA

Andrew R. Cook
Brookhaven National Laboratory
Chemistry Department
Upton, NY 11973
USA

Charge and Exciton Transport through Molecular Wires. Edited by L.D.A. Siebbeles and F.C. Grozema

ISBN: 978-3-527-32501-6

Mattias P. Eng
Chalmers University of Technology
Department of Chemical and Biological Engineering/Physical and Organic Chemistry
Kemigarden 4
412 96 Göteborg
Sweden

C. Daniel Frisbie
University of Minnesota
Department of Chemistry and Department of Chemical Engineering and Materials Science
421, Washinton Ave. SE
Minneapolis, MN 55455
USA

Ferdinand C. Grozema
Delft University of Technology
Department of Chemical Engineering
Optoelectronic Materials Section
Julianalaan 136
2628 BL Delft
The Netherlands

Dirk M. Guldi
Friedrich-Alexander-Universität-Erlangen-Nürnberg
Department of Chemistry and Pharmacy & Interdisciplinary Center for Molecular Materials (ICMM)
Egerlandstr. 3
91058 Erlangen
Germany

Magnus Hultell
Linköping University
Department of Physics
Chemistry and Biology
IFM Bldg F
Room G405
58183 Linköping
Sweden

Frederick D. Lewis
Argonne/Northwestern Solar Energy Research (ANSER) Center
Department of Chemistry
2145 Sheridan Road
Evanston, IL 60208-3113
USA

Jerker Mårtensson
Chalmers University of Technology
Department of Chemical and Biological Engineering/Physical and Organic Chemistry
Kemigarden 4
412 96 Göteborg
Sweden

Nazario Martín
Universidad Complutense
Departamento de Química Orgánica
Facultad de Química
28040 Madrid
Spain

John R. Miller
Brookhaven National Laboratory
Chemistry Department
Upton, NY 11973
USA

Mark A. Ratner
Argonne/Northwestern Solar
Energy Research (ANSER) Center
Department of Chemistry
2145 Sheridan Road
Evanston, IL 60208-3113
USA

Kirk S. Schanze
University of Florida
Chemistry Department
Gainesville, FL 32611
USA

Laurens D. A. Siebbeles
Delft University of Technology
Department of Chemical
Engineering
Optoelectronic Materials Section
Julianalaan 136
2628 BL Delft
The Netherlands

Gemma C. Solomon
Nano-Science Center and
Department of Chemistry
University of Copenhagen
Universitetparken 5
Copenhagen Ø, 2100
Denmark

Paiboon Sreearunothai
Brookhaven National Laboratory
Chemistry Department
Upton, NY 11973
USA

and

Thammasat University
Sirindhorn International
Institute of Technology
Pathum Thani, 12121
Thailand

Sven Stafström
Linköping University
Department of Physics
Chemistry and Biology
IFM Bldg F
Room G405
58183 Linköping
Sweden

Timothy M. Swager
Massachusetts Institute of
Technology
Department of Chemistry
77 Massachusetts Avenue
Cambridge, MA 02139
USA

Josh Vura-Weis
Argonne/Northwestern Solar
Energy Research (ANSER) Center
Department of Chemistry
2145 Sheridan Road
Evanston, IL 60208-3113
USA

Michael R. Wasielewski
Argonne/Northwestern Solar
Energy Research (ANSER) Center
Department of Chemistry
2145 Sheridan Road
Evanston, IL 60208-3113
USA

Mateusz Wielopolski
Friedrich-Alexander-Universität-
Erlangen-Nürnberg
Department of Chemistry and
Pharmacy & Interdisciplinary
Center for Molecular
Materials (ICMM)
Egerlandstr. 3
91058 Erlangen
Germany

1
Introduction: Molecular Electronics and Molecular Wires

Ferdinand C. Grozema and Laurens D. A. Siebbeles

1.1
Introduction

According to the predictions of Gordon Moore in 1965, the number of transistors per square centimeter of silicon doubles every 18 months [1]. This requires that the size of transistors and the interconnecting wires between them decrease at the same rate. Up until now, this miniaturization has been realized by improvements in photolithographic techniques. These techniques will reach their fundamental limit in the near future, as the dimensions of the components drop below tens of nanometers. Therefore, it is of considerable practical and fundamental interest to study the smallest components that are likely to be functional, that is, components consisting of single molecules or groups of molecules.

Already in 1959 the eminent physicist Richard Feynman discussed the possibilities of devices of extremely small dimensions in his lecture entitled "There's plenty of room at the bottom" [2]:

> I don't know how to do this on a small scale in a practical way, but I do know that computing machines are very large; they fill rooms. Why can't we make them very small, make them of little wires, little elements – and by little I mean little. For instance, the wires should be 10 or 100 atoms in diameter, and the circuits should be a few thousand angstroms across. [...] There is plenty of room to make them smaller. There is nothing that I can see in the laws of physics that says the computer elements cannot be made enormously smaller than they are now.

In 1959, Feynman and the rest of the world did not know how to manipulate electronic components on a molecular scale; however, more than 30 years after that, in the 1990s several breakthroughs were achieved and now, 50 years later, a large community of scientists is working on the use of single molecules as electronic components. Among the pioneers in single-molecule conduction studies were Gimzewski and Joachim who measured the electrical conductance of a single fullerene C_{60} molecule [3]. Other seminal experimental advances were the measurement of the electrical resistance of a single benzenedithiol

Charge and Exciton Transport through Molecular Wires. Edited by L.D.A. Siebbeles and F.C. Grozema

ISBN: 978-3-527-32501-6

bonded between two Au electrodes by the group of Reed *et al.* [4] and the experimental demonstration of single-molecule rectification in an Aviram–Ratner type molecule by Metzger and coworkers [5]. Since the 1990s, a lot of progress has been made, both in the practical problem of manipulating single molecules and doing measurements on them and in the fundamental understanding of the electrical processes on this small scale. As a result of this research, a variety of single-molecule electronic components have been proposed and demonstrated.

A field, that is, very much related to molecular electronics, and has inspired it to some extent, is that of electron transfer in donor–acceptor systems. This area of science started long before the first ideas of using molecules in electronics with the work of Mulliken in the late 1940s, from which the theory of binding and charge transfer spectra emerged [6]. A theory for electron transfer with a classical description of nuclear degrees of freedom was developed in the 1950s by Marcus [7–9] and later extended by Hush [10, 11]. Jortner and others later extended this theory by including a quantum mechanical description of the nuclear degrees of freedom [12, 13]. These theoretical predictions were confirmed experimentally over the following decades by (among many others) Verhoeven [14, 15], Paddon-Row [16, 17], and Miller [18, 19]. Most of the initial groundbreaking experiments were done for donor–bridge–acceptor systems in which the bridge consisted of a nonconjugated rigid spacer, most notably the norbornyl derivatives. These donor–bridge–acceptor molecules show strong resemblance to the initial molecular diode proposed by Aviram and Ratner [20]. More recently the study of electron transfer has been extended to conjugated bridges, with particular focus on the properties of conjugated chains as molecular wires [21–27].

In this chapter, we will not give a thorough review of the enormous progress that has been made in the field of single-molecule conductance. Excellent reviews on molecular electronics are available for a deeper background [28–35]. We aim to give an impression of some of the different molecular electronic components and discuss the importance of molecular wires that should serve as interconnects between these devices. We also discuss the different approaches that are used for studying charge transport through molecular wires. These approaches, both theoretically and experimentally, vary considerably between the fields of molecular electronics where conductance measurements are most common, and electron transfer where charge transfer is often determined by spectroscopic techniques. In the following chapters in this book, these different methods are discussed in detail and applied to actual systems.

1.2
Single-Molecule Devices

1.2.1
Molecular Rectifiers

The first concrete idea for an electronic component consisting of a single molecule was the molecular rectifier described by Aviram and Ratner [20]. The

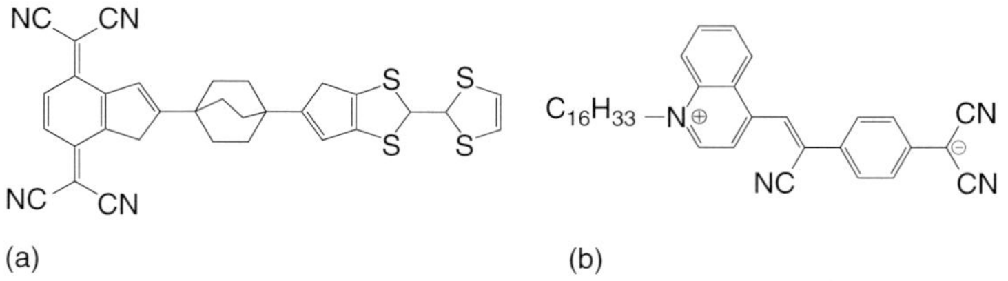

Figure 1.1 Single-molecule transistors. (a) Aviram–Ratner proposal for a single-molecule rectifier. (b) Molecular rectifier realized by Metzger.

molecular rectifier that they considered consisted of an electron donating moiety, tetrathiafulvalene, which was connected to an electron-accepting group, tetracyanoquinodimethane, by an "insulating" σ-bonded spacer, see Figure 1.1(a). This molecule can be considered as an analog of p–n junctions common to the design of traditional solid-state rectifiers. Quantum chemical calculations suggested that this molecule should indeed exhibit rectifying behavior. After this landmark proposal, it took another 25 years until such behavior was experimentally confirmed for the related donor–acceptor molecule shown in Figure 1.1(b) by Metzger *et al.* [5, 36, 37].

A more recent approach to realize a single-molecule rectifier, reported by the group of Dekker [38], is more akin to its macroscopic equivalent. It consists of single-walled nanotubes that can be either metallic or semiconducting depending on their diameter and helicity. An intramolecular junction between a metallic and a semiconducting nanotube section can be realized by introducing a pentagon and a heptagon into the hexagonal carbon lattice. Electrical transport measurements on a single carbon nanotube intramolecular metal–semiconductor junction have been performed [38]. It was shown that the transport characteristics were strongly asymmetric with respect to the bias polarity, thus exhibiting the behavior of a rectifying diode. The disadvantage of using carbon nanotubes is that there is no synthetic control over the construction of the molecules and the realization relies on coincidence during the synthesis of carbon nanotubes.

1.2.2
Molecular Switches

The basic control element in electronic architecture is the switch, which allows the control of current flow. Switches can be used in isolated form but can also be connected in arrays of multiple switches to implement logic operations [39, 40]. One example of a switch on a molecular scale is the photochromic switch consisting of a dithienylethene molecule; see Figure 1.2(a) [41]. The connection between the thienyl rings can be opened or closed by illuminating with different wavelengths of light. In the open form, the thienylene rings are not connected and, therefore, the conjugation across the molecule is broken. If the molecule is illuminated with ultra-violet (UV) light, the closed form is obtained. The molecule can be switched back to its open form by irradiation with visible light. Such a light switchable

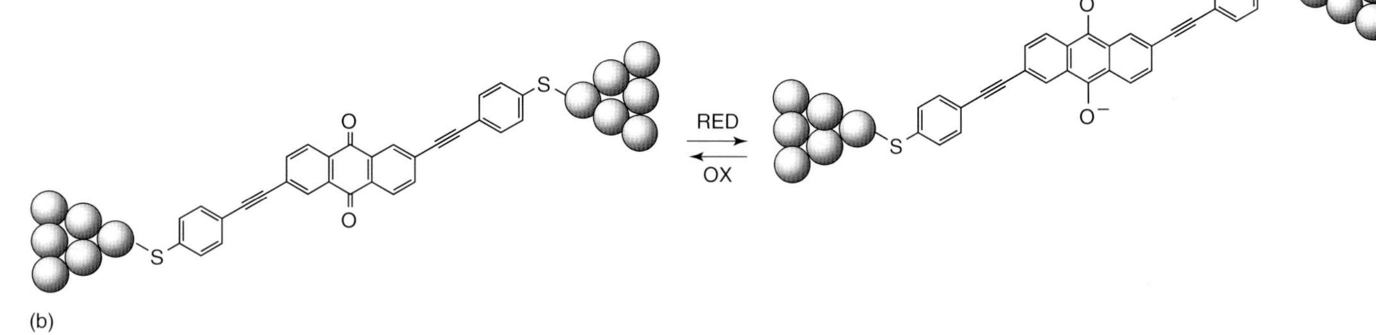

Figure 1.2 Single-molecule switches. (a) Dithienylethene switch that can be opened and closed by illumination with visible and UV light, respectively. (b) Redox switch that can be made conducting by reduction of the anthraquinone moiety.

molecule can be used as a memory element, using the open and closed form as "on" and "off" bits. The photochromic switch can also, in principle, be used for switching currents "on" or "off" on a molecular level when it is incorporated into a molecular wire or, as has been shown recently, by trapping it between two metal electrodes. The operation of this switch was demonstrated by chemisorbing it inside a mechanically controllable break junction between two gold electrodes. It was found that the resistance increased by 3 orders of magnitude upon opening of the switch by irradiation with visible light [42].

Another example is the anthraquinone-based switch reported by the group of Hummelen; see Figure 1.2(b) [43]. In this molecule, the π-electron pathway can be switched from cross-conjugated to conjugated by reduction of the anthraquinone moiety. In general, charge transfer through a cross-conjugated π-system is much less efficient than charge transfer through a conjugated pathway [44, 45]. Therefore this molecule can be considered a redox switch.

1.2.3 Molecular Transistors

The examples of single-molecule switches discussed above rely on conformational changes in the molecule. This limits the possible switching speed to a few kilohertz since usually the reverse conformational transition is relatively slow in a molecular system [46]. An approach that should in principle allow much faster switching speeds is a switch (or transistor) that relies on a single electron transfer. In 1988, 14 years after the proposal of the molecular rectifier, Aviram proposed a field-effect transistor that consists of a single molecule; see Figure 1.3(a) [47]. This transistor consists of a semiconducting piece of polythiophene, connected in such a way to a doped (oxidized) piece of polythiophene that charge transfer between these two parts of the molecule is inefficient. The oxidized polythiophene is conducting and the nondoped polythiophene will be nonconducting up to a certain threshold voltage, but the application of an electric field can result in tunneling of an electron between the two parts. In this way, the conduction of both polythiophene channels can be switched by application of an electric field [47].

Another example of a single-molecule transistor that relies on single-electron tunneling is shown in Figure 1.3(b). In this three-terminal design, described by Wada [46, 48], a central "quantum dot" unit consisting of a single thiopene ring is connected to three conjugated arms by saturated linker units. In the case when two arms are connected to electrodes, the central part with the saturated linkages acts as a tunneling barrier. The tunneling rate through this barrier can be modified by applying a potential to the third terminal, resulting in an increase or decrease in the energy levels in the quantum dot part. Therefore, by applying a potential to the "gate" electrode, the tunneling current between the source and drain can be controlled. It has been estimated that switching speeds of more than 10 THz could be reached [49].

Single-molecule transistors that consist of a single semiconducting single-walled nanotube have been proposed by the group of Dekker [50]. The nanotubes are

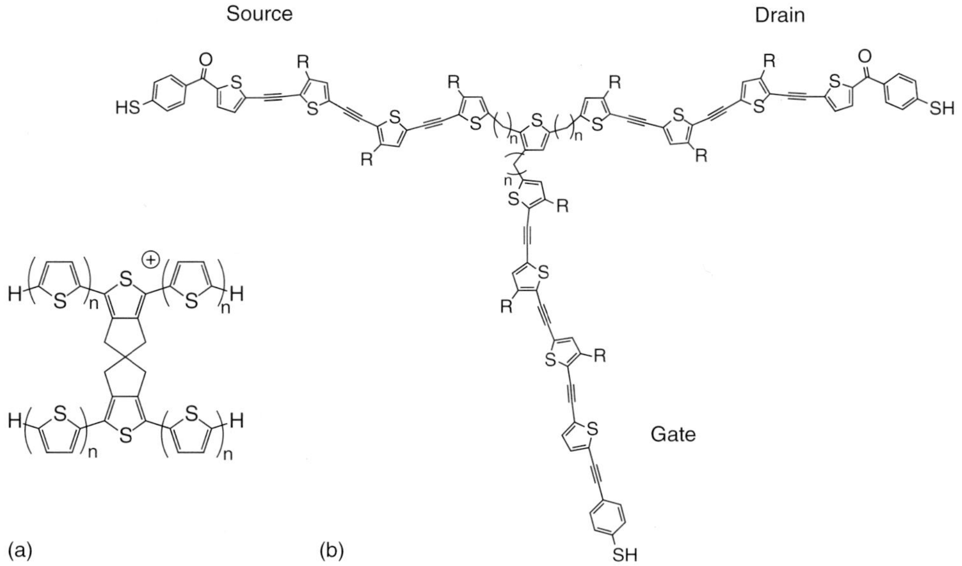

Figure 1.3 Single-molecule transistors. (a) Aviram proposal for a molecular transistor consisting of a photoconductor coupled to a conductor. (b) A central "quantum dot" unit connected to three electrodes by conjugated chains.

positioned across two Pt electrodes on a silicon oxide substrate with doped silicon as the back gate. The current through the nanotube can be manipulated by changing the voltage applied at the gate electrode. It has also been demonstrated that these devices can be assembled into one-, two-, and three-transistor circuits that perform a range of digital logic operations such as an inverter or a memory cell [51].

1.2.4 Molecular Wires: Connecting the Devices

In order to use the single-molecule electronic components described above in a functional way while preserving the small scale, they have to be connected by conducting wires of the same (molecular) dimensions. One of the first to coin the term "molecular wire" was the 1988 Nobel prize winner Lehn who described a caroviologen molecule that could be incorporated into vesicle membranes, see Figure 1.4 [52]. The charge in such a chain can transfer easily through the conjugated pathway between the two terminal groups of the molecule.

Similar conjugated molecular wires are the "simple" conjugated polymer-derived wires shown in Figures 1.5(a)–(c). These wires consist of a piece of conjugated polymer analogous to the polymers used for organic electronics. In such conjugated polymers, generally there is a considerable amount of conformational freedom, most notably the rotational freedom around the (formally) single bonds in the chain [53]. Therefore, more rigid alternatives have also been proposed as show in

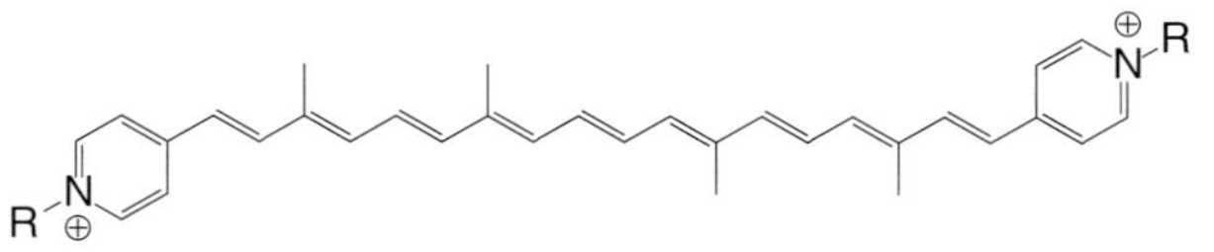

Figure 1.4 Caroviologen proposed by Lehn *et al.* as a trans-membrane molecular wire.

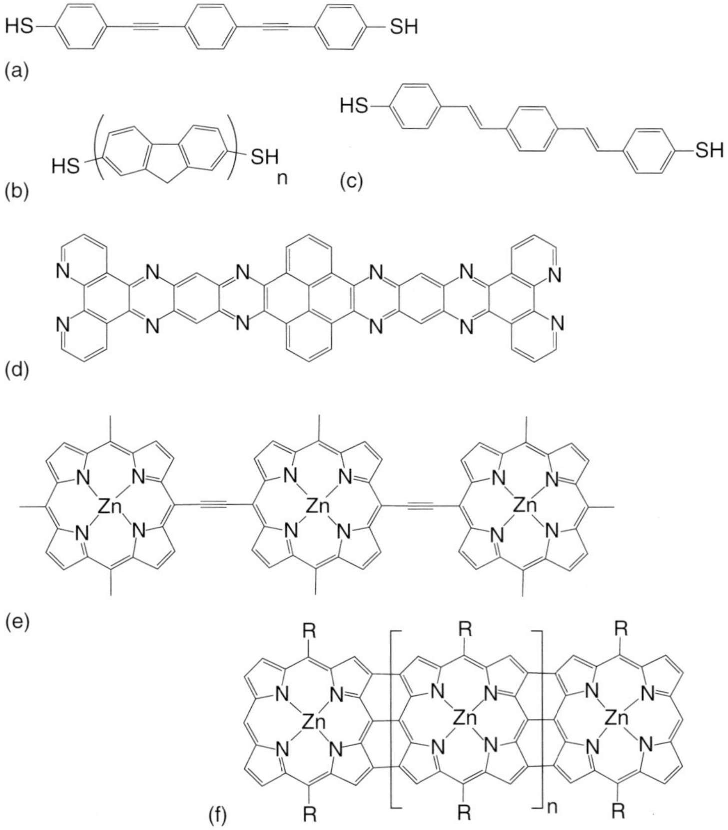

Figure 1.5 Examples of conjugated molecular wires, ranging from simple conjugated wires (a–c) to fully rigid oligo(quinoxaline) (d) and nonrigid (e) and rigid (f) porphyrin-based molecular wires.

Figure 1.5(d) [54]. One of the advantages of using organic molecules as molecular wires is the level of control over the structural and electronic properties of these wires. The conjugated wires can be designed to meet the required rigidity as, for example, in porphyrin-based molecular wires, see Figures 1.5(e) and (f). Butadiyne-linked porphyrin wires have interesting charge transfer properties but also exhibit a considerable degree of torsional disorder [25, 55, 56].

These porphyrin wires can be made much more rigid by directly coupling the porphyrin units in a ladder-type structure as has been shown by Osuka *et al.* These porphyrin "tapes" should function as exceptionally efficient pathways for charge transport [57].

Apart from the synthetic control over the structure and properties inside a molecule, organic materials also offer advantages due to their self-assembling properties. Conjugated molecules can be designed so that they self-assemble into supra-molecular structures suitable for charge transport [58]. An excellent example of this is the incorporation of specific quadruple hydrogen-bonding units in conjugated molecules as shown by Meijer and coworkers [59–61]. Such designed self-assembly can possibly be used to assemble molecular devices and wires into electronic circuits that perform a specific function. It has already been shown that supramolecular interactions can be used to control the optical and charge transport properties of conjugated molecular wires [62–64].

1.3 Transport of Charges and Excitons in Molecular Wires

In the context of the emergence of molecular electronics, the study of charge transport through molecular wires has become an important research topic. Charge transport phenomena have been studied using a variety of techniques. These techniques can largely be divided in three categories. In the first category, the molecules are positioned between electrodes in some way [65, 66]. Sometimes single molecules are trapped between electrodes but often their properties are also studied in so-called self-assembled monolayers. In the latter case, the substrate functions as one electrode, while a scanning-tunneling microscopy tip is the other [35]. An example of such measurements is described in Chapter 3 of this book.

The second approach to measuring the charge transport in conjugated molecular wires comes from the area of photo-induced electron transfer [21, 22]. In this case, an electron donor and acceptor are attached to a conjugated bridge and the rate of charge transfer upon excitation is measured by time-resolved spectroscopy (Figure 1.6 c and d). This is extensively discussed in Chapters 4 and 6 for conjugated bridges and in Chapter 5 for π-stacked DNA bridges.

The basic mechanism of charge transfer involved in these two methods is very similar, even if the experimental methods differ considerably. In both cases, charge transfer is generally due to a single-step tunneling process in which a charge, either a hole or electron, tunnels between the donor and acceptor or between the electrodes without becoming localized on the bridge. In donor–bridge–acceptor systems, the rate of charge transfer can then be described in terms of the Marcus–Hush model, which involves coupling to the vibrational states in the molecule and its surroundings. It is then typically found that the rate of charge transfer decays exponentially with distance, since the charge transfer integral exhibits an exponential distance dependence [67, 68].

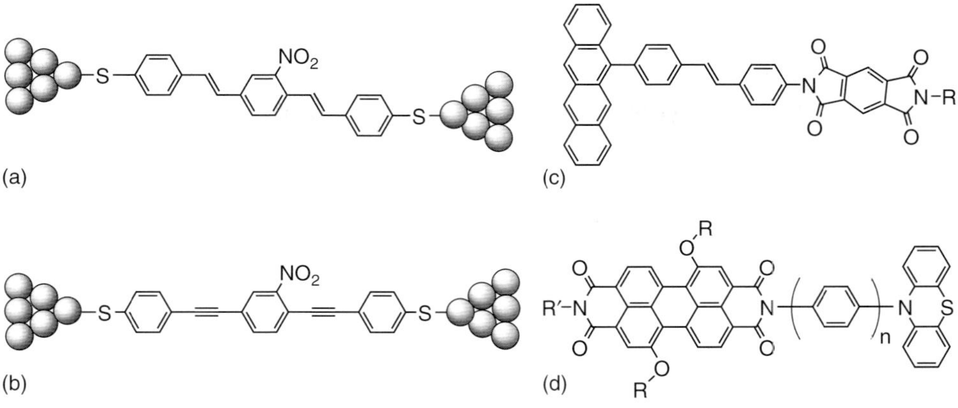

(a)

(b)

(c)

(d)

Figure 1.6 Examples of conjugated molecular wires between electrodes (a, b) and in donor–bridge–acceptor systems (c, d).

In the theoretical description of charge transport through molecules between electrodes, the Landauer approach has been used [69], see, for instance, Chapter 2 of this book. Although the two approaches may appear very different at a first glance, they are very much related. The relation between the Landauer approach [69] for molecular conductance and the Marcus charge transfer rate has been demonstrated by Nitzan [70]. Similar to the case of single-step electron transfer, the conductance through the molecule typically decays exponentially with increasing distance between the electrodes. The groups of Joachim and Grill demonstrated this in 2009 in an experiment using a scanning tunneling microscope (STM). With the STM tip, a conjugated polymer was lifted off a conducting surface, while the current between the tip and surface was measured at the same time. Lifting the polymer off the surface increases the length of the conjugated chain between surface and tip, leading to an exponential decay of the conductance [71].

1.3.1 Deviations from Exponential Distance Dependence: Transfer to Hopping

In both the single-molecule conductance approach and spectroscopic measurements on donor–bridge–acceptor systems, interesting deviations from the exponential distance dependence have been observed. For donor–bridge–acceptor systems, Wasielewski and coworkers have found that after a certain bridge-length in conjugated molecules, the distance dependence of charge transfer becomes much weaker and is in fact nonexponential [22, 23]. The same trend has been observed for charge transfer through π-stacked DNA bases by the groups of Giese [72] and Lewis *et al.* [73, 74].

Interestingly, in single-molecule conductance experiments very similar observations were reported by the group of Frisbie for a series of conjugated chains of increasing length. Although the conductance was exponential for short chains, at a certain length the decrease with distance became much weaker, see also Chapter 3 [75].

The fact that in both types of experiments the same deviations were observed confirms the strong similarity in the charge transfer process that is probed by the two approaches. The crossover from a strong exponential distance dependence to almost distance independence has in both cases been sought in a change of the mechanism by which the charge transfers. In the exponential regime, charge transfer occurs via a single-step (super-exchange) tunneling mechanism in which the charge is never localized on the bridge. For longer bridges, the superexchange tunneling process is very slow and actual population of the bridge by the charge becomes a competing process. In such cases, the charge can transfer from donor to acceptor or between electrodes by a multistep hopping mechanism. Theories describing the crossover between tunneling and incoherent hopping have been postulated [76–78]. In this respect, it is interesting to note that the presence of the charge on the bridge has not been observed experimentally, although in the case of charge transfer in DNA it was found that the charge leaves the donor faster than it arrives at the acceptor site [73, 74]. This indicates that the charge is at least temporarily localized on the bridge.

1.3.2 Charges Localized on Conjugated Chains

In both the spectroscopic studies on donor–bridge–acceptor systems and single-molecule conductance the charge does not become localized on the bridge in the majority of cases. As a consequence the rate of charge transfer or the conductance is determined to a large extent by the properties of the donor and acceptor or the molecule-electrode coupling. This means that, for instance, the charge transfer rate does not provide direct information on the motion of the charge when it is moving on the conjugated bridge. A convenient way of studying charges that are actually moving on conjugated chains is to generate the charges initially on the conjugated chains. This is possible by creating ionizations by irradiation with short pulses of high-energy electrons [79–82]. The charges generated in this way can move along conjugated polymer chains. This motion can be probed by optical spectroscopy, for instance, detecting the motion of charges toward appended traps at the chain ends, see Chapter 7 [79]. Alternatively, it is possible to directly determine the mobility of the charges along the chains using the time-resolved microwave conductivity technique, as described in Chapter 9 [80–82]. The dynamics of charges on conjugated polymers chains has also been studied theoretically as discussed in Chapters 8 and 9.

The interesting feature of both these ways of probing charge transport is that the charge is actually localized on the chain and the motion along the chain is probed. It is hard to compare the data obtained from such measurements directly to the results from single-molecule conductance experiments or spectroscopy on donor–bridge–acceptor systems. However, in the limit of very long bridges, there should be similarities. In this limit, the single-step tunneling is by definition negligible and the only pathway for transport would be motion along the chain. Although it is experimentally quite challenging to go to this limit because of the small charge transfer rates (or low currents), it is of considerable fundamental interest to enter the regime where charges are moving on the chains.

1.3.3 Motion of Excited States

Although much of the focus in molecular electronics is on charge transport and electronic functionality, the molecules that are considered often also have interesting optical properties. In fact the combination of light and charges is one of considerable interest and, as illustrated above, in principle a current can be switched on and off by illumination of molecular switches with light of different wavelengths. In this context, excited states and, in particular, the motion of excited states along the molecular wires are of interest [83, 84]. Among the possible applications is the possibility to construct chemical sensors based on specific interactions that regulate motion of excited states [85]. The motion of excitons along conjugated chains is related to charge migration, and some of the techniques to probe the motion are the same. For instance, the motion of excitons to appended traps at the ends of conjugated chains is very similar to the trapping of charges on these traps. This is discussed in Chapter 7. Other ways of probing exciton motion along conjugated chains use the specific properties, for instance, fluorescence. In Chapter 10 the depolarization of the fluorescence is used as a probe of the motion of excited states along conjugated chains.

References

1. Moore, G.E. (1965) Cramming more components onto integrated circuits. *Electronics*, **38**, 114–117.
2. Feynman, R.P. (1960) There's plenty of room at the bottom. *Eng. Sci.*, **23**, 22–36.
3. Joachim, C., Gimzewski, J.K., Schlitter, P.R., and Chavy, C. (1995) Electronic transparence of a single C_{60} molecule. *Phys. Rev. Lett.*, **74**, 2102–2105.
4. Reed, M.A., Zhou, C., Muller, C.J., Burgin, T.P., and Tour, J.M. (1997) Conductance of a molecular junction. *Science*, **278**, 252–254.
5. Metzger, R.M., Chen, B., Hopfner, U., Lakshmikantham, M.V., Vuillaume, D., Kawai, T., Wu, X., Tachibana, H., Hughes, T.V., Sakurai, H., Baldwin, J.W., Hosch, C., Cava, M.P., Brehmer, L., and Ashwell, G.J. (1997) Unimolecular electrical rectification in hexadecylquinolinium tricyanoquinodimethnide. *J. Am. Chem. Soc.*, **119**, 10455–10466.
6. Mulliken, R.S. (1952) Molecular compounds and their spectra II. *J. Am. Chem. Soc.*, **74**, 811–824.
7. Marcus, R.A. (1956) On the theory of oxidation–reduction reactions involving electron transfer. *J. Chem. Phys.*, **24**, 966–978.
8. Marcus, R.A. and Sutin, N. (1985) Electron transfers in chemistry and biology. *Biochim. Biophys. Acta*, **811**, 265–322.
9. Marcus, R.A. (1964) Chemical and electrochemical electron-transfer theory. *Annu. Rev. Phys. Chem.*, **15**, 155–196.
10. Hush, N.S. (1967) Intervalence-transfer absorption. Part 2. Theoretical considerations and spectroscopic data. *Prog. Inorg. Chem.*, **8**, 391.

11. Hush, N.S. (1968) Homogeneous and heterogeneous optical and thermal electron transfer. *Electrochim. Acta*, **13**, 1005–1023.
12. Ulstrup, J. and Jortner, J. (1976) The effect of intramolecular quantum modes on free energy relationships for electron transfer reactions. *J. Chem. Phys.*, **63**, 4358–4368.
13. Bixon, M. and Jortner, J. (1999) Electron transfer – from isolated molecules to biomolecules. *Adv. Chem. Phys.*, **106**, 35–202.
14. Oevering, H., Paddon-Row, M.N., Heppener, M., Oliver, A.M., Cotsaris, E., Verhoeven, J.W., and Hush, J.W. (1987) Long-range photoinduced through-bond electron transfer and radiative recombination via rigid nonconjugated bridges: distance and solvent dependence. *J. Am. Chem. Soc.*, **109**, 3258–3269.
15. Warman, J.M., de Haas, M.P., Paddon-Row, M.N., Cotsaris, E., Hush, N.S., Oevering, H., and Verhoeven, J.W. (1986) Light-induced giant dipoles in simple model compounds for photosynthesis. *Nature*, **320**, 615–616.
16. Hush, N.S., Paddon-Row, M.N., Cotsaris, E., Oevering, H., Verhoeven, J.W., and Heppener, M. (1985) Distance dependence of photoinduced electron transfer through non-conjugated bridges. *Chem. Phys. Lett.*, **117**, 8–11.
17. Paddon-Row, M.N. (1994) Investigating long-range electron-transfer processes with rigid covalently linked donor-(norbornylogous bridge)-acceptor systems. *Acc. Chem. Res.*, **27**, 18–25.
18. Closs, G.L. and Miller, J.R. (1988) Intramolecular long-distance electron transfer in organic molecules. *Science*, **240**, 440–447.
19. Penfield, K.W., Miller, J.R., Paddon-Row, M.N., Cotsaris, E., Oliver, A.M., and Hush, N.S. (1987) Optical and thermal electron transfer in rigid difunctional molecules of fixed distance and orientation. *J. Am. Chem. Soc.*, **109**, 5061–5065.
20. Aviram, A. and Ratner, M.A. (1974) Molecular rectifiers. *Chem. Phys. Lett.*, **29**, 277–283.
21. Davis, W.B., Svec, W.A., Ratner, M.A., and Wasielewski, M.R. (1998) Molecular-wire behavior in *p*-phenylenevinylene oligomers. *Nature*, **396**, 60–63.
22. Weiss, E.A., Ahrens, M.J., Sinks, L.E., Gusev, A.V., Ratner, M.A., and Wasielewski, M.R. (2004) Making a molecular wire: charge and spin transport through para-phenylene oligomers. *J. Am. Chem. Soc.*, **126**, 5577–5584.
23. Goldsmith, R.H., Sinks, L.E., Kelley, R.F., Betzen, L.J., Liu, W., Weiss, E.A., Ratner, M.A., and Wasielewski, M.R. (2005) Wire-like charge transport at near constant bridge energy through fluorene oligomers. *Proc. Natl. Acad. Sci. U.S.A.*, **102**, 3540–3545.
24. Wiberg, J., Guo, L., Pettersson, K., Nilsson, D., Ljungdahl, T., Martensson, J., and Albinsson, B. (2006) Charge recombination versus charge separation in donor–bridge–acceptor systems. *J. Am. Chem. Soc.*, **129**, 155–163.
25. Winters, M.U., Dahlstedt, E., Blades, H.E., Wilson, C.J., Frampton, M.J., Anderson, H.L., and Albinsson, B. (2007) Probing the efficiency of electron transfer through porphyrin-based molecular wires. *J. Am. Chem. Soc.*, **129**, 4291–4297.
26. Kilså, K., Kajanus, J., Macpherson, A.N., Mårtensson, J., and Albinsson, B. (2001) Bridge dependent electron transfer in porphyrin-based donor–bridge–acceptor systems. *J. Am. Chem. Soc.*, **123**, 3069–3080.
27. Giacalone, F., Segura, J.L., Martin, N., and Guldi, D.M. (2004) Exceptionally small attenuation factors in molecular wires. *J. Am. Chem. Soc.*, **126**, 5340–5341.
28. Muller, C.J. and Reed, M.A. (1996) There is plenty of room between two atom contacts. *Science*, **272**, 1901–1902.
29. Joachim, C., Gimzewski, J.K., and Aviram, A. (2000) Electronics using hybrid-molecular and mono-molecular devices. *Nature*, **408**, 541–548.
30. Carroll, R.L. and Gorman, C.B. (2002) The genesis of molecular electronics. *Angew. Chem. Int. Ed. Engl.*, **41**, 4378–4400.
31. Heath, J.R. and Ratner, M.A. (2003) Molecular electronics. *Phys. Today*, **5**, 43–49.

32. McCreery, R.L. (2004) Molecular electronic junctions. *Chem. Mater.*, **16**, 4477–4496.
33. Heath, J.R. (2009) Molecular electronics. *Annu. Rev. Mater. Res.*, **39**, 1–23.
34. Moth-Poulsen, K. and Bjornholm, T. (2009) Molecular electronics with single molecules in solid-state devices. *Nat. Nanotechnol.*, **4**, 551–556.
35. McCreery, R.L. and Berggren, A.J. (2009) Progress with molecular electronic junctions: meeting experimental challenges in design and fabrication. *Adv. Mater.*, **21**, 4303–4322.
36. Metzger, R.M. (2001) The quest for unimolecular rectification from Oxford to Waltham to Exeter to Tuscaloosa. *J. Macromol. Sci.*, **A38**, 1499–1517.
37. Metzger, R.M. (2003) Unimolecular electrical rectifiers. *Chem. Rev.*, **103**, 3803–3834.
38. Yao, Z., Postma, H.W.C., Balents, L., and Dekker, C. (1999) Carbon nanotube intramolecular junctions. *Nature*, **402**, 273–276.
39. Irie, M. (2000) Diarylethenes for memories and switches. *Chem. Rev.*, **100**, 1685–1716.
40. Feringa, B.L. (2001) *Molecular Switches*, Wiley-VCH Verlag GmbH, Weinheim.
41. Kudernac, T., Katsonis, N., Browne, W.R., and Feringa, B.L. (2009) Nano-electronic switches: light-induced switching of the conductance of molecular system. *J. Mater. Chem.*, **19**, 7168–7177.
42. Dulic, D., van der Molen, S.J., Kudernac, T., Jonkman, H.T., de Jong, J.J.D., Bowden, T.N., van Esch, J., Feringa, B.L., and van Wees, B.J. (2003) *Phys. Rev. Lett.*, **91**, 207402.
43. Van Dijk, E.H., Myles, D.J.T., Van der Veen, M.H., and Hummelen, J.C. (2006) Synthesis and properties of an anthraquinone-based redox switch for molecular electronics. *Org. Lett.*, **8**, 2333–2336.
44. Mayor, M., Weber, H.B., Reichert, J., Elbing, M., Von Hanisch, C., Beckmann, D., and Fischer, M. (2003) Electric current through a molecular rod – relevance of the position of the anchor groups. *Angew. Chem. Int. Ed. Engl.*, **42**, 5834–5838.
45. Van der Veen, M.H., Rispens, M.T., Jonkman, H.T., and Hummelen, J.C. (2004) Molecules with linear pi-conjugated pathways between all substituents: omniconjugation. *Adv. Funct. Mater.*, **14**, 215–223.
46. Wada, Y. (1999) A prospect for single molecule information processing devices. *Pure Appl. Chem.*, **71**, 2055–2066.
47. Aviram, A. (1988) Molecules for memory, logic and amplification. *J. Am. Chem. Soc.*, **110**, 5687–5692.
48. Wada, Y. (1999) Proposal of atom/molecule switching devices. *J. Vac. Sci. Technol. A*, **17**, 1399–1405.
49. Lutwyche, M.I. and Wada, Y. (1994) Estimate of the ultimate performance of the single-electron transistor. *J. Appl. Phys.*, **74**, 3654–3661.
50. Tans, S.J., Verschueren, A.R.M., and Dekker, C. (1998) Room-temperature transistor based on a single carbon nanotube. *Nature*, **393**, 49–52.
51. Bachtold, A., Hadley, P., Nakanishi, T., and Dekker, C. (2001) Logic circuits with carbon nanotube transistors. *Science*, **294**, 1317–1320.
52. Arrhenius, T.S., Blanchard-Desce, M., Dvolaitzky, M., and Lehn, J.-M. (1986) Molecular devices: caroviologens as an approach to molecular wires-synthesis and incorporation into vesicle membranes. *Proc. Natl. Acad. Sci. USA*, **83**, 5355–5359.
53. Grozema, F.C., van Duijnen, P.T., Berlin, Y.A., Ratner, M.A., and Siebbeles, L.D.A. (2002) Intramolecular charge transport along isolated chains of conjugated polymers: effect of torsional disorder and polymerization defects. *J. Phys. Chem. B*, **106**, 7791–7795.
54. Ishow, E., Gourdon, A., Launay, J.-P., Chiorboli, C., and Scandola, F. (1999) Synthesis, mass spectrometry, and spectroscopic properties of a dinuclear ruthenium complex comprising a 20 A long fully aromatic bridging ligand. *Inorg. Chem.*, **38**, 1504–1510.
55. Winters, M.U., Karnbratt, J., Eng, M., Wilson, C.J., Anderson, H.L., and Albinsson, B. (2007) Photophysics of a butadiyne-linked porphyrin dimer: influence of conformational flexibility in

de ground and first singlet excited state. *J. Phys. Chem. C*, **111**, 7192–7199.

56. Kocherzhenko, A.A., Patwardhan, S., Grozema, F.C., Anderson, H.L., and Siebbeles, L.D.A. (2009) Mechanism of charge transport along zinc porphyrin-based molecular wires. *J. Am. Chem. Soc.*, **131**, 5522–5529.
57. Kim, D. and Osuka, A. (2004) Directly linked porphyrin arrays with tunable excitonic interactions. *Acc. Chem. Res.*, **37**, 735–745.
58. Hoeben, F.J.M., Jonkheijm, P., Meijer, E.W., and Schenning, A.P.H.J. (2005) About supramolecular assemblies of pi-conjugated systems. *Chem. Rev.*, **105**, 1491–1546.
59. Sijbesma, R.P., Beijer, F.H., Brunsveld, L., Folmer, B.J.B., Hirschberg, J.H.K.K., Lange, R.F.M., Lowe, J.K.L., and Meijer, E.W. (1997) Reversible polymers formed from self-complementary monomers using quadruple hydrogen bonding. *Science*, **278**, 1601–1604.
60. El-ghayoury, A., Schenning, A.P.H.J., Van Hal, P.A., Van Duren, J.K.J., Janssen, R.A.J., and Meijer, E.W. (2001) Supramolecular hydrogen-bonded OPVS polymers. *Angew. Chem. Int. Ed. Engl.*, **40**, 3660–3663.
61. Schenning, A.P.H.J., Jonkheim, P., Peeters, E., and Meijer, E.W. (2001) Hierarchical order in supramolecular assemblies of hydrogen bonded oligo (*p*-phenylene vinylenes). *J. Am. Chem. Soc.*, **123**, 409–416.
62. Grozema, F.C., Houarner-Rassin, C., Prins, P., Siebbeles, L.D.A., and Anderson, H.L. (2007) Supramolecular control of charge transport in molecular wires. *J. Am. Chem. Soc.*, **129**, 13370–13371.
63. Frampton, M.J. and Anderson, H.L. (2007) Insulated molecular wires. *Angew. Chem. Int. Ed. Engl.*, **46**, 1028–1064.
64. Cacilli, F., Wilson, J.S., Michels, J.J., Daniel, C., Silva, C., Friend, R.H., Severin, N., Samori, P., Rabe, J.P., O'Connell, M.J., Taylor, P.N., and Anderson, H.L. (2002) Cyclodextrin-threaded conjugated polyrotaxanes as insulated molecular wires with reduced interstrand interactions. *Nat. Mater.*, **1**, 160–164.
65. James, D.K. and Tour, J.M. (2005) Molecular wires. *Top. Curr. Chem.*, **257**, 33–62.
66. Love, J.C., Estroff, L.A., Kriebel, J.K., Nuzzo, R.G., and Whitesides, G.M. (2005) Self-assembled monolayers of thiolates on metals as a form of nanotechnology. *Chem. Rev.*, **105**, 1103–1169.
67. Edwards, P.P., Gray, H.B., Lodge, M.T.J., and Williams, R.J.P. (2008) Electron transfer and electronic conduction through and intervening medium. *Angew. Chem. Int. Ed. Engl.*, **47**, 6758–6765.
68. Berlin, Y.A., Grozema, F.C., Siebbeles, L.D.A., and Ratner, M.A. (2008) Charge transfer in donor–bridge–acceptor systems: static disorder, dynamic fluctuations and complex kinetics. *J. Phys. Chem. C*, **112**, 10988–11000.
69. Landauer, R. (1957) Spatial variation of currents and fields due to localized scatterers in metallic conduction. *IBM J. Res. Dev.*, **1**, 223–231.
70. Nitzan, A. (2001) A relationship between electron-transfer rates and molecular conduction. *J. Phys. Chem. A*, **105**, 2677–2679.
71. Lafferentz, L., Ample, F., Yu, H., Hecht, S., Joachim, C., and Grill, L. (2009) Conductance of a single conjugated polymer as a continuous function of its length. *Science*, **323**, 1193–1197.
72. Giese, B., Amaudrut, J., Kohler, A.-K., Spormann, M., and Wessely, S. (2001) Direct observation of hole transfer through DNA by hopping between adenine bases and by tunneling. *Nature*, **412**, 318–320.
73. Lewis, F.D., Zhu, H., Daublain, P., Cohen, B., and Wasielewski, M.R. (2006) Hole mobility in DNA A tracts. *Angew. Chem. Int. Ed. Engl.*, **45**, 7982–7985.
74. Lewis, F.D., Zhu, H., Daublain, P., Fiebig, T., Raytchev, M., Wang, Q., and Shafirovich, V. (2006) Crossover from superexchange to hopping as the mechanism for photo-induced charge transfer in DNA hairpin conjugates. *J. Am. Chem. Soc.*, **128**, 791–800.

75. Choi, S.H., Kim, B., and Frisbie, C.D. (2008) Electrical resistance of long conjugated molecular wires. *Science*, **320**, 1482–1486.

76. Bixon, M. and Jortner, J. (2002) Long-range and very long range charge transport in DNA. *Chem. Phys.*, **281**, 393–408.

77. Berlin, Y.A., Burin, A.L., and Ratner, M.A. (2002) Elementary steps for charge transport in DNA: thermal activation vs. tunneling. *Chem. Phys.*, **275**, 61–74.

78. Giese, B. (2000) Long-distance charge transport in DNA: the hopping mechanism. *Acc. Chem. Res.*, **33**, 631–636.

79. Asaoka, S., Takeda, N., Iyoda, T., Cook, A.R., and Miller, J.R. (2008) Electron and hole transport to trap groups at the ends of conjugated polyfluorenes. *J. Am. Chem. Soc.*, **130**, 11912–11920.

80. Hoofman, R.J.O.M., de Haas, M.P., Siebbeles, L.D.A., and Warman, J.M. (1998) Highly mobile electrons and holes on isolated chains of the semiconducting polymer poly(phenylenevinylene). *Nature*, **392**, 54–56.

81. Grozema, F.C., Hoofman, R.J.O.M., Candeias, L.P., de Haas, M.P., Warman, J.M., and Siebbeles, L.D.A. (2003) The formation and recombination kinetics of positively charged poly(phenylene vinylene) chains in pulse-irradiated dilute solutions. *J. Phys. Chem. A*, **107**, 5976–5986.

82. Grozema, F.C., Siebbeles, L.D.A., Warman, J.M., Seki, S., Tagawa, S., and Scherf, U. (2002) Hole conduction along molecular wires: sigma-bonded silicon versus pi-bond-conjugated carbon. *Adv. Mater.*, **14**, 228–231.

83. Hennebicq, E., Pourtois, G., Scholes, G.D., Herz, L.M., Russell, D.M., Silva, C., Setayesh, S., Grimsdale, A.C., Mullen, K., Bredas, J.-L., and Beljonne, D. (2005) Exciton migration in rigid-rod conjugated polymers: an improved Forster model. *J. Am. Chem. Soc.*, **127**, 4744–4762.

84. Dykstra, T.E., Hennebicq, E., Beljonne, D., Gierschner, J., Claudio, G., Bittner, E.R., Knoester, J., and Scholes, G.D. (2009) Conformational disorder and ultra-fast exciton relaxation in PPV-family conjugated polymers. *J. Phys. Chem. B*, **113**, 656–667.

85. Thomas, S.W. III, Joly, G.D., and Swager, T.M. (2007) Chemical sensors based on amplifying fluorescent conjugated polymers. *Chem. Rev.*, **107**, 1339–1389.

Part I
Molecules between Electrodes

Charge and Exciton Transport through Molecular Wires. Edited by L.D.A. Siebbeles and F.C. Grozema

ISBN: 978-3-527-32501-6

2
Quantum Interference in Acyclic Molecules

Gemma C. Solomon, David Q. Andrews, and Mark A. Ratner

2.1
Introduction

The charge transport properties of molecular wires have a chameleonic character: observed characteristics depend not only on the molecular wire in question but also on the environment in which it is measured. Binding a molecular wire between two, generally metallic, electrodes affords a great range of measurements; however, this unusual environment for a molecule is not without consequence. For example, binding a molecule to electrodes will affect the structural, and consequentially vibrational, characteristics, which can be observed using inelastic electron tunneling spectroscopy [58, 112, 128]. properties of a particular class of molecular wires, acyclic cross-conjugated molecules, in a particular transport regime. The respect to voltage to obtain a vibrational spectrum of the molecule carrying the current. Theoretical models of these spectra have indicated that the electrodes act like a heterogeneous solvent, shifting the energy of vibrational modes in the functional units close to the electrode surface from those in the central part of the molecule. The environmental effect of binding to electrodes can also be seen in the electronic properties alone, and this is the regime we will consider in this chapter. Molecular transmission resonances are frequently energetically far from the Fermi energy of the metallic electrodes, which could lead to the naive expectation that charge transport in these systems would be insignificant. While the observed currents are undeniably low, a rich variety of transport phenomena are observed in these junctions.

The strength of the coupling between the molecule and electrodes, which is usually controlled chemically by allowing either physisorption or chemisorption, has a significant impact on the transport characteristics as it modulates the impact the electrodes can have on the molecular electronic structure. In the strong coupling (chemisorption) limit, molecular transmission resonances are energetically broadened by the interaction with the electrodes resulting in nonzero conductance at low bias. In the weak coupling (physisorption) limit, Coulomb-blockade limited transport is observed only, with single electron charging of the junction at energies which would not be predicted from the properties of the isolated molecule [57].

Charge and Exciton Transport through Molecular Wires. Edited by L.D.A. Siebbeles and F.C. Grozema

ISBN: 978-3-527-32501-6

In this chapter, we will focus on the strong coupling limit, molecules chemisorbed between electrodes, and examine how chemical functionalities can be used to induce dramatic variation in the off-resonant transport characteristics of molecular wires.

Single molecule electronic devices have been constructed with varied behavior [54] including switching [66], rectification [35, 72, 73], coulomb blockade [83], Kondo resonance [83], negative differential resistance (NDR) [49], and memory elements [39]. A number of measurements have established single-molecule transistor behavior in UHV conditions [29, 57, 123, 136], as well as using electrochemical gate control. [5, 24, 64, 116] For single-molecule switches, there are a number of theoretical studies on how molecular conformational change can lead to large conductance changes [120, 141], including measurements using photochromic molecules [51]. Many methods for creating molecular switches rely on, or result in, conformational change to the molecule of interest [41, 67, 77]. Recent work has highlighted how fast switching can be accomplished with hydrogen transfer in a naphthalocyanine molecule at low temperature, resulting in an on/off ratio of 2 [66].

The strong coupling limit has been of interest to researchers as chemisorption ensures that strong electronic coupling to the electrode is achieved, despite atomic scale variation in the electrode surface, ideally allowing the molecular properties to control the electronic characteristics of the junction. Measurement of the conductance of single chemisorbed molecule junctions has been obtained using repeated measurement and statistical analysis [90, 125, 131]. Despite the strong chemisorbed bond between the molecule and the electrode, geometric variation in the junction leads to variations in the conductance, both experimentally and theoretically [8]. With the extent of system-wide variation understood, to a large extent, it is then possible to consider the chemical trends that can be deduced from conductance measurements.

Molecular conductance measurements have elucidated trends that mirror those obtained from prior work on photo-induced electron transfer in donor–bridge–acceptor type systems. Relatively simple models have been employed with great success to describe observed behavior and some broad "rules of thumb" can be determined. While molecular conductance is well described by these models, there are limitations on the performance of the electronic devices that can be constructed. Necessarily, there are assumptions underlying these models and by understanding these assumptions productive avenues toward devices without these performance limitations can be elucidated. Each of these ideas will be introduced in more detail in the following subsections to provide a broad picture of the general coherent electron transport behavior of molecular wires bound to metallic electrodes.

2.1.1 Rules of Thumb

Electron transport measurements of molecules bound to electrodes have followed prior work in electron transfer and studied the properties of simple conjugated and saturated systems of various lengths. The predominant characteristic of these results is their well-behaved nature, the systems studied follow simple, unsurprising trends. From this body of knowledge three "rules of thumb" for trends in rates of electron transfer can be deduced:

1) For a particular type of molecular bridge, the rate of electron transfer will decrease with increasing bridge length.
2) The rate of electron transfer through a fully conjugated bridge will generally exceed that through a saturated bridge, frequently by a substantial amount.
3) For similar kinds of bridges the rate of electron transfer can be correlated with the energy difference between the donor and acceptor energy levels (or electrode Fermi levels) and the bridge levels, where the greater the gap the lower the rate of transfer through the system.

These "rules," as framed, are clearly not based on any general features of the fundamental physics, but rather on observed trends correlated with chemical properties of particular structures. As such, they may breakdown for any type of chemical system which is not well represented in the set of molecular wires previously studied. The class of systems that will be highlighted in this chapter are of this kind, the electron transfer properties do not follow the "rules" as the transport paths are neither simple conjugated nor saturated systems.

2.1.2
Barrier Tunneling

Despite the complex quantum nature of electron transport through molecules, the observed behavior in the off-resonant elastic tunneling regime reflects very little of this complexity. For different systems, the magnitude of the conductance varies, generally according to the "rules" above; yet, for a particular molecule the conductance mechanism is largely invariant across the measurable range. These characteristics have resulted in considerable success using a simple Simmons model [3, 4, 47, 94, 129] to describe charge transport in these junctions, where the barrier height can be controlled by the energy separation between the Fermi energy of the electrode and the closest molecular energy level. This model will be referred to as the proximate resonance barrier tunneling (PRBT) model.

The PRBT makes for an intuitive picture of charge transport in molecular wires. In the off-resonant regime, the electron is tunneling between the electrodes and the molecule is simply controlling the form of the tunnel barrier. The closer a molecular resonance to the electrode Fermi energy, the lower the barrier. There is no doubt that this model describes provides a good description of the properties of many molecular wires. The difficulty, as we will show below, is the necessary performance limitations that it puts on devices.

2.1.3
Limitations on Device Performance

So long as the PRBT model holds and the chemical nature of the molecular wire controls the form of the tunnel barrier, the chemical nature of the molecule must be manipulated to induce any sort of diversity in a molecule's transport properties, for example switching [40, 55, 81]. These processes, whilst offering their own advantages, all have distinct disadvantages with respect to the speed at which the process can occur, reversibility, repeatability and complexity of the device.

Outside the strong coupling, off-resonant tunneling regime there are obviously other ways to induce dramatic variation in the conductance characteristics, for example, charging the molecule; however, there are advantages to remain in this regime. The strong coupling regime promises well-defined junctions with molecules assembling in predetermined ways and good, reproducible, electrical contact between the molecule and the electrodes. Further, off-resonant transport may aid the longevity of the junctions as nuclear motion resulting from charging is minimized.

If dramatic variation in the conductance characteristics can be induced by electric fields alone, the performance limitations outlined above do not apply. The difficulty is that even with a third terminal or gate electrode, a molecule where the form of the tunnel barrier is controlled by the energetic location of the proximate molecular resonance will not generally exhibit dramatic conductance changes without charging [28]. Consequently, the assumptions in this model need to be interrogated so that the molecules that do not behave in this fashion can be explored.

2.1.4
Underlying Assumptions

Both the "rules of thumb" and the PRBT model, outlined above, rely on assumptions about the form of the off-resonant transmission. Specifically, the central assumption is that transmission observed near the Fermi energy will be some remnant of the tail of the energetically closest molecular resonance. Underlying this assumption is the idea that all resonances have similar energetic decay, which is not entirely correct. Examination of the transmission spectrum of almost any molecule will reveal both energetically broad and narrow resonances. The reason this assumption persists is that frequently the resonances arising from the highest occupied molecular orbital (HOMO) and lowest unoccupied molecular orbital (LUMO) do have similar decay characteristics and the energetic location and relatively slow decay of these resonances mean they dominate the off-resonant transmission.

Theoretical investigations allow the factors controlling the decay characteristics to be explored. For a particular molecule, the broadening of all transmission resonances is controlled by the strength of the interaction with the electrodes. Changing the chemical nature of the binding group or the bond lengths between the molecule and the metal surface can increase or decrease the width of all resonances as the strength of the interaction increases or decreases, respectively. The relative widths of resonances in a given molecule are also related to the strength of the interaction with the electrodes with the nature of the molecular orbital underlying each resonance controlling the magnitude. For example, an orbital localized on a side-arm of a branched structure may not interact with the electrodes to an appreciable extent and will appear as a narrow resonance in the transmission; conversely an orbital delocalized across the backbone spanning the electrodes may appear as a much broader transmission resonance. The reason that the HOMO and LUMO are frequently broadened to a similar extent is that often they are a bonding and antibonding pair of orbitals with much the same form and, consequently, the same degree of interaction with the electrodes.

Like many pervasive ideas built on empirical observation alone, there is no reason to assume that these assumptions should always hold; indeed this chapter will highlight the behavior of systems where they do not.

2.1.5 Exceptions Seen in Prior Work

The exceptions highlighted in this chapter do not constitute the only systems whose behavior falls outside the "rules of thumb" and PRBT model. One significant underlying cause of behavior that diverges from the established trends is destructive quantum interference, which has been discussed with respect to the mesoscopic systems, frequently constituting the inspiration for molecular devices, and in molecular systems.

In mesoscopic systems the quantum nature of transport is readily apparent, exemplified by the prediction [2, 115] and observation [10, 13, 130] of Aharonov–Bohm oscillations in the transport through a variety of systems. Quantum interference effects have been seen not only in cyclic structures but also in "T"-shaped semiconductor nanostructures [30, 109, 110] and predicted in arrays of quantum dots [23, 34, 98, 143].

Destructive quantum interference effects have been predicted in the transmission characteristics of Hückel models, designed to represent molecular systems [22, 25, 37, 38, 46, 56, 68, 111, 122, 126, 127]. In such models, the interference features are dramatic and in many cases dominate the spectrum; however, when density functional theory calculations are employed the results may not be nearly so stark as the model systems might suggest [38, 46]. The best known exception is the marked differences in the transmission characteristics of derivatives of benzene depending upon whether the connection to the electrodes is through the meta or para positions. These differences have been seen experimentally [70, 84] and discussed theoretically [22, 84, 95, 127, 133].

2.2 Theoretical Methods

2.2.1 Electronic Structure and Geometry

All molecular geometries were obtained by optimizing the isolated molecule in the gas phase using Q-Chem 3.0 [99] with density functional theory (DFT) using the B3LYP functional [14, 61] and 6-311G** basis. The gas phase molecules were chemisorbed (terminal hydrogens removed) to the fcc hollow site of a Au(111) surface with the Au–S bond length taken from the literature [15].

2.2.2
Transport

All transport calculations shown here were performed using implementations of the nonequillibrium Green's function formalism. In the Landauer–Imry limit, the current is given by [32, 59, 60, 132]:

$$I(V) = \frac{2e}{h} \int_{-\infty}^{\infty} dE \mathrm{Tr}[\Gamma_L(E) G^r(E) \Gamma_R(E) G^a(E)](f_L(E, V) - f_R(E, V)) \quad (2.1)$$

where $G^{r(a)}$ are the retarded (advanced) Green's functions for the molecule or extended molecule, $\Gamma_{L(R)}$ are twice the imaginary component of the self-energies for the left (right) electrodes, and $f_{L(R)}$ are the Fermi functions for the left (right) electrodes. The zero-bias transmission is given by the trace in Eq. (2.1) computed with no applied bias. There are many publications detailing the transport formalism for these types of calculations [18, 50, 85, 117, 119, 132, 138], including those which explicitly deal with the implementations we utilize [18, 85, 117, 138], and we direct the reader to these publications for a complete discussion of this aspect. One of many caveats to the applicability of these methods is that Landauer conductance calculations ignore electron correlation, and consequently may fail when resonances are approached.

Transport calculations are performed with Hückel-IV 3.0 [118, 137, 138], gDFTB [36, 85, 89] and ATK [1, 18, 102, 117]. The results from these three methods have shown to be qualitatively consistent [105] for the types of systems discussed in this chapter. We use gDFTB to analyze the symmetry components of the transport and ATK to simulate an applied gate voltage. In all cases, transport calculations have been performed with at least two of these methods.

Using three methods allows us to understand better the limitations of each and to ensure that no approximations peculiar to a particular method are leading to erroneous results. Common among the three methods are the use of one electron Hamiltonians and nonequilibrium Green's function methods for the transport. Both Hückel-IV and ATK calculate the full 3D self-consistent potential under voltage bias. Hückel-IV 3.0 uses extended Hückel electronic structure with a relatively small Au electrode pad of 3 atoms on each end of the molecule, it is also the fastest computationally. gDFTB is a tight-binding DFT program. Due to the computational benefits of parameterization, this code allows for the analysis of very large systems, including large physical electrodes. This method is also extremely fast with small electrodes; however, we use a minimum of a 4 × 4 atom unit cell with six layers in the transport direction and periodic boundary conditions. The gDFTB program has symmetry implemented, allowing a detailed analysis of the results. No gold atoms are included in the extended molecule so that the symmetry of the molecule could be used to separate the transmission into σ and π

components [108]. ATK is a commercial DFT transport package. ATK is run using a DZP (SZP for Au) basis set and the LDA functional.

The gate voltage is calculated within the ATK code by shifting the molecular part of the Hamiltonian that remains when the electrode and surface atoms are removed. This assumes an external electrostatic potential localized to the molecular region and not a physical electrode [1]. In this calculation, the gate is solely a shift in the energy levels. The actual gating effect may deviate substantially from this idealized model. Experimental methods of gating a molecule could utilize previous techniques or a third electrode in UHV [29, 57, 83, 123, 136] or with electrochemical control [5, 24, 64, 87, 116].

Throughout this chapter, we will focus on zero-bias transmission, as this is where the features of interest are most evident and the most straightforward to interpret. Naturally, such features are only of significant interest if their effects can be seen in observable quantities such as current and conductance. One characterisitic of the systems reviewed in this chapter is that the features of interest in the transmission should have considerable impact on the experimentally observable electron transport properties, essentially because the modifications on transmittance can occur near the Fermi energy.

2.2.3 Molecular Dynamics

The variation in transport behavior resulting from the changing molecular structure due to temperature-induced vibrational fluctuations is accounted for by performing transport calculations on a large number of geometries obtained by molecular dynamics simulations. The geometries are obtained using Tinker [88]. A 1 ns trajectory was run to fully equilibrate the system. For the next 100 ps, a snapshot was taken every 1 ps and the geometry was parsed.

2.2.4 Dephasing

Electronic dephasing [63, 81, 101], frequently caused by fluctuations in molecular environment or geometry, might lead to decoherence of the transmitted electron and disrupt any effects caused by interference [44, 45]. It has been shown that the traditional reactivity series in ortho, meta, and para benzene can be recast as an interference effect [44] and this effect can be erased by purely local dephasing [44, 45]. The calculation of transport dynamics is performed using the quantum Liouville equation with dephasing included by reducing the magnitude of the off-diagonal elements of the density matrix (coherence) [33, 97]. At $t = 0$, all population is placed on the donor/source site while an absorbing boundary condition on the acceptor/drain site is used to simulate irreversible electron transfer. We have calculated the decay time necessary for the total system population to decrease to 5% of the initial value.

2.3
Interference in Acyclic Cross-Conjugated Molecules

This chapter will focus on the properties and applications of one particular class of molecular wires: acyclic cross-conjugated molecules. The results highlighted have been covered in more detail in previous publications [6, 7, 104–107] and we direct readers there for further information.

2.3.1
The Chemical Nature of Cross-Conjugated Molecules

A starting point for appreciating the properties of these molecular wires is the definition of a cross-conjugated compound: "a compound possessing three unsaturated groups, two of which although conjugated to a third unsaturated center, are not conjugated to each other. The word "conjugated" is defined here in the classical sense of denoting a system of alternating single and double bonds" [86]. The synthetic literature is replete with examples of molecules containing this functionality [16, 42, 80] and the examples included here are derived from this literature.

When considering the transport properties of molecules containing a cross-conjugated functionality it is important to consider the binding orientation, as this can determine whether the electrodes are coupled by a cross-conjugated path or a linearly conjugated path. The molecule shown in Figure 2.1(b) can be bound to two electrodes in two ways depending on which two of the three thiol groups are the binding groups. If the lower two groups bind to the electrodes, there is a five carbon cross-conjugated path spanning the electrodes, with a seven carbon side chain. This path is cross-conjugated, in terms of the definition above, as the two unsaturated groups in the five carbon chain are conjugated to the unsaturated groups in the side chain but not to each other. Conversely, if the electrodes bind through one of the lower groups and the upper thiol group there is a ten carbon linearly conjugated path between the electrodes with a two carbon atom side chain.

This difference highlights that molecular transport properties are not simply molecule dependent, they are path dependent. When considering the difference between two different paths through the same molecule it is clear that the molecular orbital spectrum (Figure 2.1(a)) and the HOMO and LUMO of the system (Figures 2.1(c) and (d), respectively) show no obvious clues as to the transport differences. In the next section we will show that the nodal pattern of the molecular orbitals can be correlated with the appearance of destructive interference features in some cases; however, this is certainly not an obvious feature on first inspection. The transport properties of molecules can be correlated with their underlying chemical properties and understanding the electronic coupling and delocalization through cross-conjugated molecules gives a hint at their transport behavior.

The electronic coupling through cross-conjugated molecules can be understood from the literature on even alternant hydrocarbons [93]. Alternant hydrocarbons are

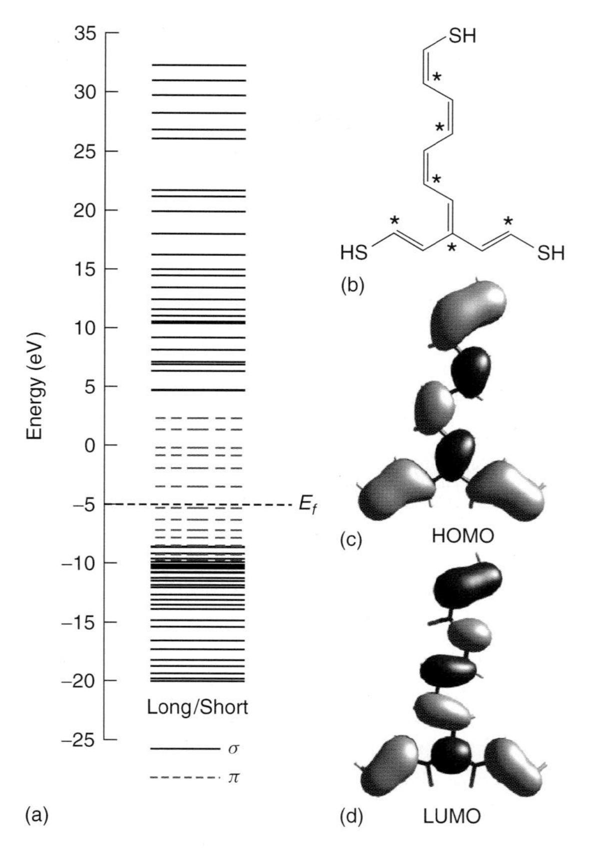

Figure 2.1 The molecular orbital energies (a) starring assignment (b) and HOMO (c) and LUMO (d) for a molecule that can bridge electrodes in two ways resulting in either a cross- or linearly conjugated path.

molecules where each neighboring conjugated carbon atom can be marked with or without a star and no two starred or unstarred atoms are bonded to each other. An example of this assignment is shown in Figure 2.1(b). The importance of this classification is that it is possible to conclude when electronic coupling will go to zero due to destructive interference based on whether the attachment points to the electrodes are either the same or differ in their starring status. Figure 2.1(b) shows that for the cross-conjugated systems path is alike coupling (the thiol attaches to two starred atoms). For the linearly conjugated path the coupling is disjoint (the thiol attaches to one starred and one unstarred atom), although this will not always be the case with multiple even numbers of subunits.

The electronic structure of an alternant hydrocarbon is characterized by orbital pairing [26, 69, 71]. That is, for every occupied orbital there is a matching virtual orbital and these form symmetric pairs above and below the mid-point of the HOMO–LUMO gap (mid-gap) and the form of these orbitals differ only in the phase on the starred atoms. The coupling through a molecule can be related to the molecular orbital coefficients [100] and it can be deduced that the mid-gap coupling is zero for alike coupling [17, 52, 92]. Cross-conjugated paths always involve alike coupling and linearly conjugated paths always involve disjoint coupling; consequently, it is anticipated that the mid-gap π coupling through cross-conjugated paths will go to zero whilst it will be nonzero for linearly conjugated paths.

Weak electronic coupling has been observed experimentally for cross-conjugated molecules [76, 124] and calculated for cyclic cross-conjugated systems [91]. This previous work has focused on the behavior of these systems close to mid-gap where it is simply concluded that cross-conjugated systems had very low coupling and linearly conjugated systems had high coupling. Here, we continue from that work and review some of the consequences of the distinctly different electronic properties of cross-conjugated and linearly conjugated systems as an applied voltage allows us to probe the behavior away from mid-gap.

The second aspect of the underlying chemical structure is electronic delocalization, which can be understood from the geometrical properties of these molecules and correlated with transport characteristics. The molecules shown in Figure 2.2 are used to illustrate the fundamental bond length differences between cross-conjugated, linearly conjugated, and saturated structures. Prior work in the area established that bond length and bond length alternation are correlated with π electron delocalization in both linear and cross-conjugated molecules [70, 127, 133]. In this analysis [6] we compare the bond lengths for 15 molecules with three different central bonding motifs; cross-conjugation, linear conjugation and a saturated carbon atom. Our analysis focuses on the delocalization of the charge between the sulfur atoms because this will be the direction of charge transport in a molecular junction. The molecules labeled **6–10** all have a double bond perpendicular to the direction of charge transport. Within this subset of molecules, **6**, **7**, and **9** meet the specific definition of cross-conjugation. Molecules **11–15** have a double bond connected in the trans orientation between the thiol terminations. Molecules **11**, **12**, and **14** are completely linearly conjugated between thiol groups. Molecules **1–5** are the same as molecules **6–10**, but with hydrogen atoms replacing the cross-conjugated double bond.

To simplify the comparison between so many molecules we focus on the bonds labeled a, b, and c in Figure 2.2. Bonds a and c are both formally single bonds and bond b is a double bond. The carbon single bonds are systematically longest in the molecules **1–5** and shortest in the molecules with linear conjugation, **11–15**. The corresponding bond lengths in the cross-conjugated molecules **6–10** are in-between the length of the saturated and unsaturated linear molecules. The changes in bond lengths agree well with previous studies completed on molecules **1**, **2**, **6**, **7**, **9** [127]. In molecules **1–5** and **11–15**, the length of bond a is not strongly correlated to the length of bond c.

To compare the carbon single-bond lengths we have calculated the percentage of double-bond character. This is done by comparing all the bond lengths with those calculated for ethene and ethane. Ethene by definition has 100% double-bond

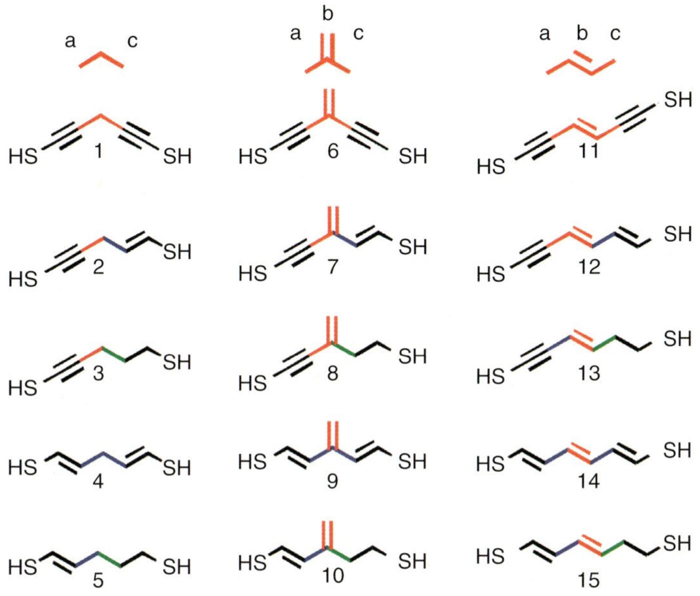

Molecule	% double bond character (ethene = %100, ethane = %0)		
	Bond (a)	Bond (b)	Bond (c)
Saturated molecules (1–5)			
1	32	x	32
2	34	x	3
3	35	x	−7
4	7	x	7
5	8	x	−2
Cross-conjugated molecules (6–10)			
6	50	89	50
7	50	89	25
8	53	93	−2
9	25	90	25
10	28	93	−2
Linearly conjugated molecules (11–15)			
11	60	87	60
12	60	87	44
13	56	93	10
14	44	87	44
15	40	93	10

Figure 2.2 A bond length analysis for a series of molecules with sections of conjugated, unconjugated and cross-conjugated carbon atoms. By definition the ethane bond length has 0% double-bond character and the ethene bond length has 100% double bond character. (Reprinted with permission from *J. Phys. Chem., C* **112**(43), 16991, 2008. Copyright 2008 American Chemical Society.)

character and ethane has 0% double-bond character. This bond length comparison shows that in a cross-conjugated molecule the π electron delocalization is less than in a linearly conjugated molecule but better than having a fully saturated carbon atom.

Using the bond length characterization to analyze this data, we come to three important conclusions. First, a cross-conjugated molecule and a linearly conjugated molecule show different behavior with respect to electron delocalization. Second, in cross-conjugated molecules the electron delocalization is reduced between cross-conjugated ends of the molecules, as is evident by the increase in both C–C bond lengths. The C–C bond length only changes due to direct coupling to an unsaturated carbon. This effect is local and the bond length changes are largely a nearest-neighbor effect. In the molecules **6–10** the C–C bond lengths (a, c) remain weakly correlated implying that the addition of the cross-conjugated bond induces very little electron delocalization across the cross-conjugated carbon atom.

2.3.2
The "Rules" Break Down

The effects of the cross-conjugated functionality on molecular transmission can be seen in Figure 2.3 by comparing 3-methylenepenta-1,4-diyne-1,5-dithiol (**1a**, cross-conjugated) with (*E*)-hexa-3-en-1,5-diyne-1,6-dithiol (**2a**) and (*Z*)-hexa-3-en-1,5-diyne-1,6-dithiol (**3a**), both linearly conjugated. Figure 2.3 shows the transmission through these three molecules separated by symmetry into the σ and π components [108]. In each case the red curve gives the σ or π component and the black curve shows the total transmission (given by the sum of the σ and π transmissions). The cross-conjugated path through **1a** results in a dramatic difference in the transmission. There is a pronounced dip in the π transport near the Fermi energy due to destructive interference and this has a number of consequences.

First, at the Fermi energy the transmission through a shorter five-carbon chain (**1a**) is only a small fraction (~6%) of the transmission through the six-carbon chains (**2a** and **3a**). This is an unexpected result as all these molecules are fully conjugated and it would be expected that the shorter ones would exhibit higher conductance. Second, the difference in the transmission occurs irrespective of the fact that the transmission resonances occur at similar energies, again contrasting with the usual trend that transmission at the Fermi energy can be correlated with the energy gap between molecular orbitals. Finally, in fully conjugated molecules it is generally assumed that at low bias conduction is dominated by transport through the π system; however in the case of **1a** it is in fact the σ system which dominates the transport close to the Fermi energy.

The PRBT model assumes that the energetically proximate molecular resonances will dominate the transmission; this is clearly incorrect when the σ transmission is the dominant component at the energies close to the electrode Fermi level. These unusual features in the zero-bias transmission lead to a large difference in the calculated current–voltage characteristics. Throughout the low-bias region the

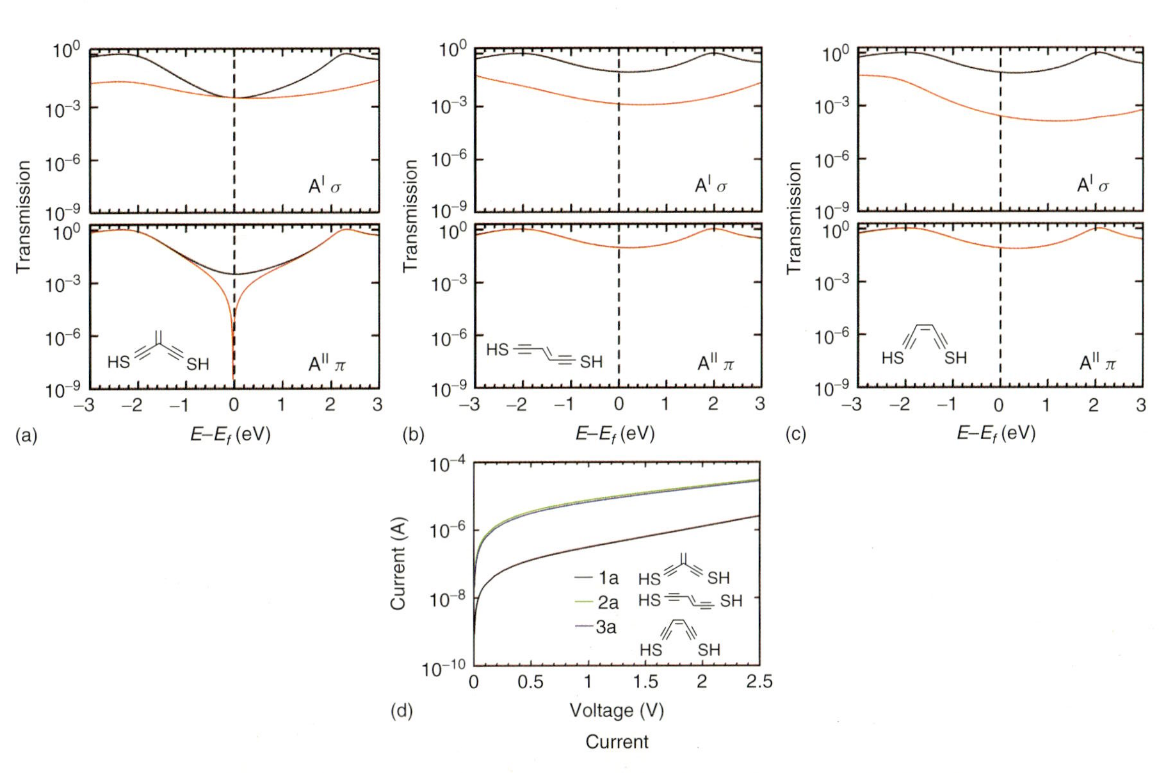

Figure 2.3 The transmission and current through a cross-conjugated molecule (**1a**) compared with its linearly conjugated counterparts (**2a** and **3a**). In each case the total transmission is shown in black and the component through either the σ or π system is shown in red. The cross-conjugated system has an antiresonance in the π transmission near the Fermi energy, providing a local minimum in the transmittance, in contrast with the high level of transmission through the linearly conjugated systems. (Reprinted with permission from *J. Am. Chem. Soc.*, **130**(51), 17301, 2008. Copyright 2008 American Chemical Society.)

current through **1a** is substantially lower than either of the two linearly conjugated systems.

With such unexpected transport characteristics resulting from seemingly benign molecules it is desirable to have a simple method for predicting when there will be interference features in molecular transmission. One condition that has been previously suggested is the presence of nonspanning nodes in the wavefunction [62]' That is, the number of nodes in the wave function will increase with increasing energy, but in 2D or 3D systems the number of nodes found along the path between the electrodes (or the donor and acceptor in chemical systems) need not increase. In the case of cross-conjugated molecules this is exactly what happens between the HOMO and the LUMO as shown in Figure 2.4. In all cases the number of nodes in the system increases from two to three; however, the new node is a nonspanning node for **1a** alone as it occurs in the side chain. Importantly, this condition will not be met in the experimentally observable range for branched linearly conjugated systems, as is evident in the transmission calculations in previous work [38]. The nonspanning nodes criterion cannot be used in a straightforward fashion to describe the interference features in phenyl rings [22, 70, 84, 95, 127, 133] or other cyclic systems, as all nodes are "spanning nodes" in a cyclic system. It may in some

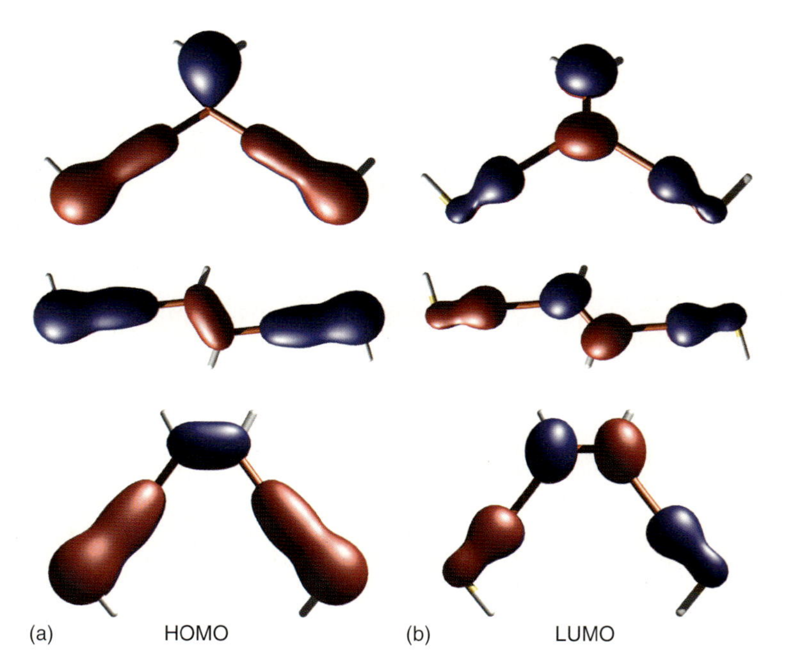

Figure 2.4 The HOMO (a) and LUMO (b) for **1a** (top), **2a** (middle), and **3a** (bottom) showing that a nonspanning node in the π system is only introduced in the case of the cross-conjugated species. (Reprinted with permission from *J. Am. Chem. Soc.*, **130**(51), 17301, 2008. Copyright 2008 American Chemical Society.)

cases be possible to derive nodal arguments by considering some part of a cyclic system.

2.4 Understanding Interference in Model Systems

With some manipulation, it is possible to rewrite the transmission (the integrand in Eq. (2.1)) as [106]:

$$
\begin{aligned}
T(E) &= \sum_{\alpha,\beta} |\sum_{i} \gamma_{\alpha,i}^{L\prime\dagger}(E) G_{i,i}^{r\prime}(E) \gamma_{i,\beta}^{R\prime}(E)|^2 \\
&= \sum_{\alpha,\beta} |\sum_{i} t_{\alpha\beta i}(E)|^2
\end{aligned}
\tag{2.2}
$$

where the index i runs over the molecular dimension and the indices α and β run over the dimensions of the electrodes (or at least the dimensions of the electrodes with nonzero coupling elements to the molecule). Here

$$
\begin{aligned}
\gamma^{L(R)}(E) &= V^{L(R)} d^{L(R)}(E) \\
\gamma^{L(R)\dagger}(E) &= d^{L(R)}(E) V^{L(R)\dagger}
\end{aligned}
\tag{2.3}
$$

where

$$
d_{\alpha(\beta)}^{L(R)}(E) = \sqrt{2\pi\,\delta\left(E - \epsilon_{\alpha(\beta)}^{L(R)}\right)}
\tag{2.4}
$$

All the quantities in Eq. (2.3) are transformed into the molecular conductance orbital basis (orbitals obtained by diagonalizing the retarded Green's function as opposed to the Hamiltonian) by the following transformations:

$$
G^{r\prime}(E) = C_r^{-1} S^{-1} G^r(E) C_r
\tag{2.5}
$$

$$
\begin{aligned}
\gamma^{L\prime}(E) &= C_r^{\dagger} S^{-1} \gamma^{L}(E) \\
\gamma^{R\prime}(E) &= C_r^{-1} \gamma^{R}(E)
\end{aligned}
\tag{2.6}
$$

As the Green's functions are now diagonal, the sum over the molecular unit involves only the single index i. Now we have the result that the total transmission is given by the sum of contributions through each of the molecular conductance orbitals ($t_{\alpha\beta i}$) from each of the eigenfunctions of G^r. Previous attempts to separate the transmission into contributions of individual molecular conductance orbitals failed to yield such a simple picture as the transmission was dominated by interference between pairs of orbitals [103].

We follow the procedure that has been outlined previously for plotting $t_{\alpha\beta i}(E)$ [106] and assume the wide-band approximation (constant density of states) for our electrodes in all plots. We plot the total transmission through this system as $\sum_{\alpha,\beta} |\sum_i t_{\alpha\beta i}(E)|^2$, although in all the examples considered here there is only a single site from each electrode coupling with the molecule.

The simplest system to illustrate the usefulness of the quantities discussed above is a two-site model system. We compare the two-site model connected to the

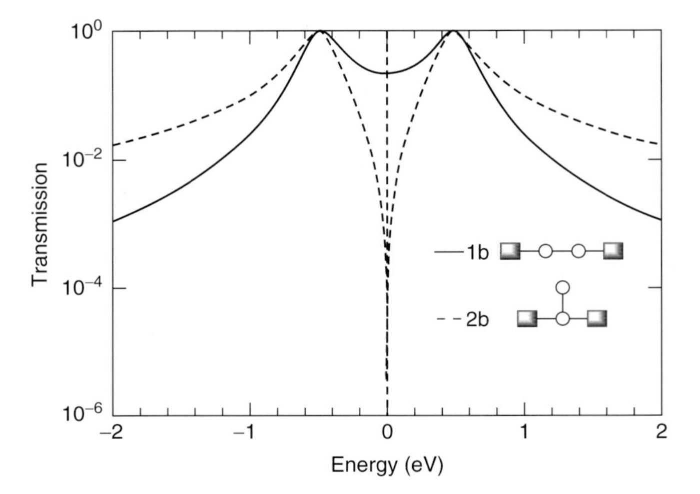

Figure 2.5 The transmission through the two two-site models considered with **1b** shown with the solid line and **2b** with the dashed line. (Reprinted with permission from *J. Chem. Phys.*, **129**(5), 054701, 2008. Copyright 2008, American Institute of Physics.)

electrodes at each end (**1b**) with the two-site model connected to both electrodes from the same site (**2b**), illustrated in the inset to Figure 2.5. These two models can be considered to be minimal models representing the cross- and linearly conjugated systems from Section 2.3.2.

The two-site model is described by a Hückel Hamiltonian

$$H_m = \begin{pmatrix} \alpha & \beta \\ \beta & \alpha \end{pmatrix} \tag{2.7}$$

with site energies α and couplings β, these are set to 0 and -0.5, respectively. The single nonzero coupling element between the site and each electrode is also set to -0.5.

Although these two models contain the same "molecule," their transmission characteristics differ considerably as shown in Figure 2.5. The two systems have the same resonances; this is to be expected as the position of these largely correlates with the position of the molecular orbitals of the isolated molecule, which are obviously identical in these two systems. But the behavior on either side of the resonances differs. Above and below the energy of the resonances, **2b** has a higher transmission than **1b**, which can be understood as **2b** has only one site in the transport direction whereas **1b** has two. Between the resonances **2b** has dramatically lower transmission than **1b** due to the large interference feature at $E = 0$. This is the model system that corresponds to the transmission behavior seen in branched structures [25, 38, 56] and cross-conjugated molecules [105]. This interference feature leads to unexpected consequences; importantly, the formalism introduced in the previous section provides the tools to understand how this occurs.

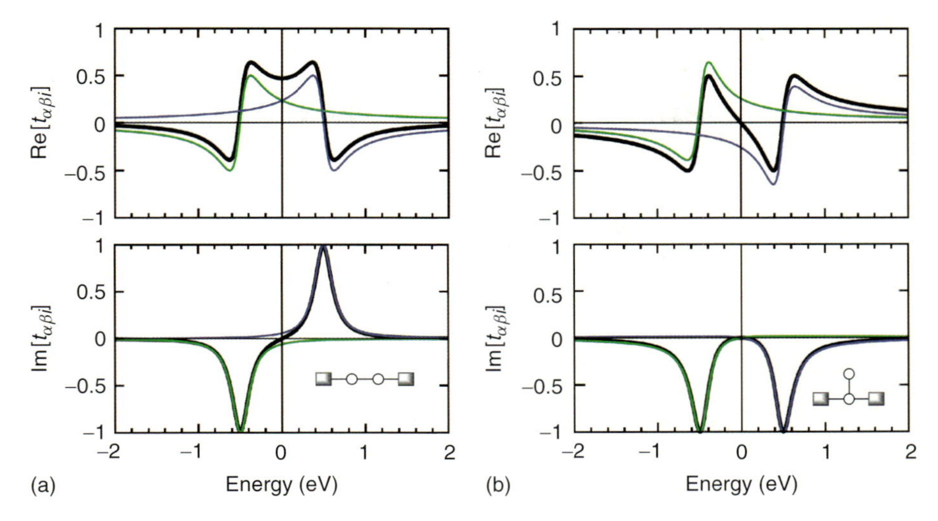

Figure 2.6 The contributions to the transmission from each molecular conductance orbital of the two two-site systems, **1b** and **2b** on (a) and (b), respectively. The contribution from the bonding orbital is shown in green, the contribution from the antibonding orbital in blue, and the sum of the two in black. In the case of **2b** both the real and imaginary components sum to zero at $E = 0$ giving rise to the antiresonance in the transmission, recall that the energy eigenvalues of the parent orbitals are ± 0.5 eV. (Reprinted with permission from *J. Chem. Phys.*, **129**(5), 054701, 2008. Copyright 2008, American Institute of Physics.)

It is now possible to separate the transmission into contributions from each of the molecular conductance orbitals. These differ from molecular orbitals in that they may be complex due to the perturbation of the electrode self-energies on the electronic structure; however, in the limit that electrode coupling goes to zero they reduce to the molecular orbitals. We will refer to the bonding and antibonding molecular conductance orbitals which arise from the underlying bonding and antibonding molecular orbitals. The transmission through each molecular conductance orbital ($t_{\alpha\beta i}$) is shown in Figure 2.6 for the two systems, they are complex quantities which are summed and squared to give the transmission as shown in Figure 2.5.

Separating the transmission into molecular conductance orbital contributions makes it clear that the only difference between **1b** and **2b** is that the antibonding orbital contribution in **2b** has a sign reversal compared with **1b**. It is remarkable that such a minimal change to a single component in the transmission produces such dramatic results and, further, that this change is correlated with a geometric change in the binding of the system.

The total transmission, as mediated by the molecule, is given by the sum of the components illustrated in Figure 2.6, squared. In the case of **1b** between the two resonances there is destructive interference in the minority, imaginary, component but constructive interference in the majority, real, component leading to nonzero transmission throughout the region. Conversely, in the case of **2b** there is destructive interference in the dominant, real, component leading to the zero transmission between the resonances as the total under the sum changes from positive to negative with increasing energy. While the sign of the sum generally has little impact on the total transmission (which goes as the sum squared) it is clear that if the sum changes sign it can have a significant impact as the transmission may go through zero and a sharp interference feature will be seen.

It is interesting to note how these Hückel model calculations on branched structures relate to real molecules. In these calculations only the π system is considered and **2b** would be considered to be a branched structure with one site in the backbone and one in a side chain. Model system calculations have not highlighted the distinction between cross-conjugated and linearly conjugated bridges but rather suggested that there was a dependence on whether the length of the side chain was odd or even [25]. This result can be recovered by observing that it is only possible to have a fully conjugated molecule with an odd length side chain when the path between the electrodes is cross-conjugated or the molecule is a radical.

In a real molecule, in contrast to a model system where only the π system is described, the total transmission will not go to zero at an antiresonance unless the antiresonance exists in all symmetry components of the transmission. In the case of cross-conjugated molecules, the antiresonance close to the Fermi energy is in the π system, which would otherwise be the majority component of the total transmission. Consequently, the minority component of the transmission close to the Fermi energy, the σ component, then dominates and defines the "floor" of the dip in the total transmission. Generally, however, model system calculations will capture the essential elements of molecular transport and will describe the π component of the transmission well [104]. The antiresonance arises from the unusual transport characteristics coming from the central ethylene unit creating the cross-conjugated path; other unsaturated groups appended from this unit simply act as conjugated spacers. Significantly, this means the ethylene unit can be connected to any number of spacer groups and still maintain its function: it is a local effect.

2.5 Using Interference for Devices

2.5.1 Schematic Designs

As a starting point for designing molecular electronic devices based on cross-conjugated functionalities, we examine schematic transmission characteristics and the device properties that result. A perfect single-channel conductor has a conductance of $1G_0 = 1/12.9\,\text{k}\Omega$. This limit of $1G_0$ is a direct result from quantum theory [21, 31], has been measured in chains of Au atoms [53, 82, 134] at high bias of $\sim 2\,\text{V}$ [135, 142], and has been verified in calculations [114, 140]. On the zero bias transmission plots shown here, a transmission equal to 1 at the Fermi energy corresponds to a zero bias conductance of $1G_0$. With a definitive upper limit for single channel conductance, it is clear that the perfect single channel switch would have a conductance of $1G_0$ in the on state and be a perfect insulator (a conductance of $0G_0$) in the off state.

Most of the molecular systems studied to date operate as effective single channel conductors. This can be deduced from the transmission maximum rarely going much above one at resonance. The variation in molecular transmission with electron injection energy means that they can be considered a switch or transistor (because a change in bias or gate voltage leads to a change in conductance), albeit not necessarily a useful one. We have outlined criteria for an ideal switch [7]: (1) an infinite ratio of the on/off current, (2) a subthreshold swing [139] of 0 mV/decade indicating a switch that abruptly changes from the off state to the on state at a defined threshold voltage, (3) fast switching times that do not rely on nuclear motion but only changes in the relative electron density, (4) reproducibility and stability [41] by minimizing charging and geometric organization, and (5) low-bias operation, energetically separated from a molecular resonance. While many other factors including leakage current, threshold voltage, and cost are extremely important, we focus our discussion on the five criteria listed above, specifically the on/off ratio and the subthreshold swing.

We start by considering an idealized transmission plot for a system with an interference feature at the Fermi energy, but is otherwise identical to an ideal single channel conductor, both of which are shown in Figure 2.7(a). Integrating over a window defined by the changing chemical potentials of the electrode with applied bias (Figure 2.7(b)) gives the observed current (Figure 2.7(c)). The conductance through these two systems is shown in Figure 2.7(d). Here the system with the interference feature switches from a low conductance "off" state to a high conductance "on" state with increasing voltage. For a system with some sort of notched transmission such as this, manipulating this interference feature can clearly result in switching behavior.

There are a number of scenarios which can be envisaged for how a system with an interference feature can be switched, and they can be distinguished by the way they impact on the transmission. In all cases the device is in the "off" state in the absence of the applied perturbation, as indicated by the original transmission shown in Figure 2.8(a). Transistor function can be achieved by an applied gate voltage which can shift the position of the interference feature away from the Fermi energy turning the device "on," as shown in Figure 2.8(b). Figures 2.8(c) and (d) represent two different mechanisms for achieving sensor or memory function. In Figure 2.8(c), a chemical reaction, photoisomerization or possibly a change in the number of electrons on the molecule, causes a very large change in the transmission function. Where there previously was a large interference feature resulting in very low transmission, there is now simply a modest reduction in the off resonant transmission. This change amounts to a change in the chemical structure, for example, a cross-conjugated path being chemically modified resulting in a linearly conjugated path, without destructive interference effects, between the electrodes. A large change in the molecule of interest would make reproducible

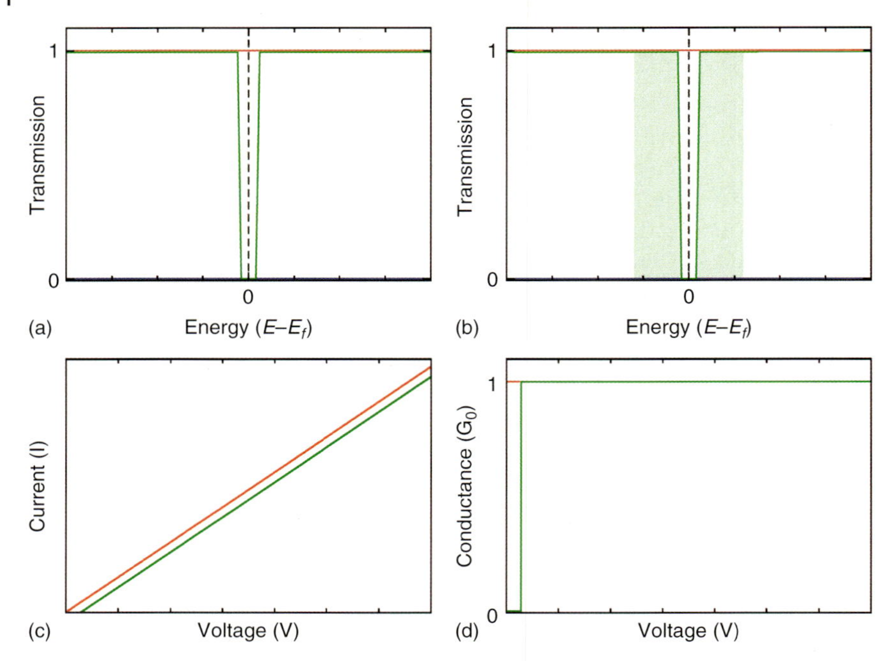

Figure 2.7 The transmission and current/voltage behavior of a voltage switch is shown schematically, where the conductance goes from 0 to 1G_0 as a function of voltage. The red line represents an ideal single channel conductor and the green line represents the system of interest. In (a), the system's transmission goes to zero near the Fermi energy and 1G_0 at all other energies. (b) To calculate the current and conductance of (a), the transmission plot (assuming invariance to applied voltage) is integrated between the chemical potentials of the leads, shown as the shaded region. (c) The current/voltage behavior realized by integrating the transmission plot in (b). The conductance as a function of voltage is shown in (d). (Reprinted with permission from *J. Am. Chem. Soc.*, **130**(51), 17309, 2008. Copyright 2008 American Chemical Society.)

switching more difficult to control in a device. While not ideal for fast repetitive switching, this method of charging a device may provide a route to creating a functional memory device with distinct on/off states representing bit storage or a single-use sensor. Figure 2.8(d) represents a shift of the molecular orbital energies upon chemical or physical binding of a molecular group, switching the molecule from an off-conductance state to an on conductance state. It is also conceivable that smaller shifts in the transmission spectrum would allow sequential detection of multiple molecules. Calculations showing both the effects of gating (b) as well as the tuning of a transmission feature (d) will be shown in the following sections.

2.5.2 Tuning Interference

In order to achieve the sort of functionality shown schematically in the previous subsection, we need to use relationships between chemical structure and transport

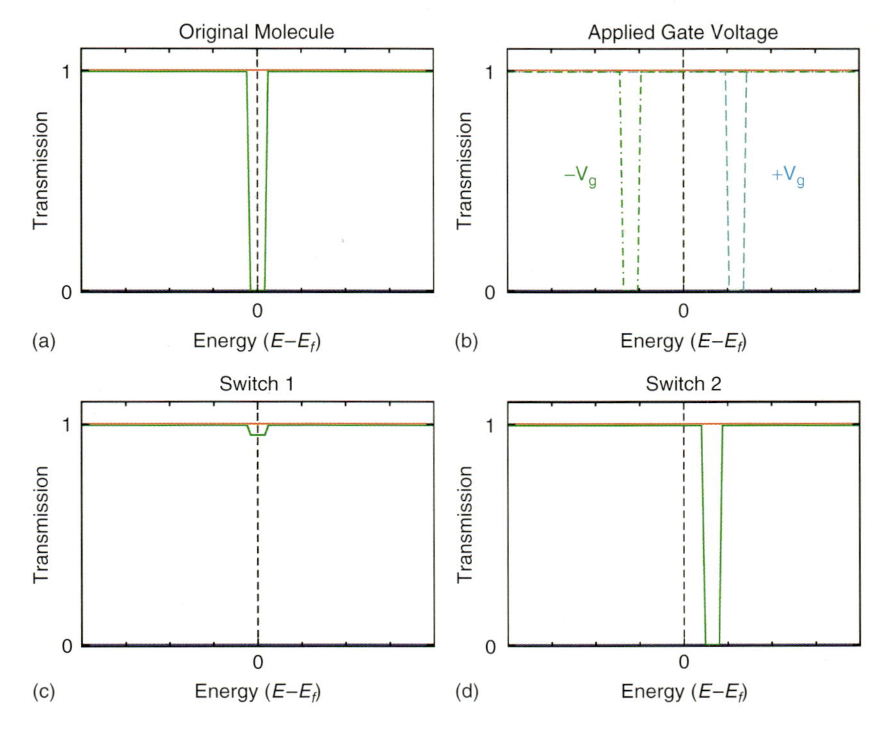

Figure 2.8 The transmission of our hypothetical system with distinct on and off states is shown again as the green line in (a). In plots (b), (c), and (d) the molecule has been switched in three different ways from a low conductance state to a high conductance state at low-bias voltage. In (b) we show a situation where an applied gate voltage might shift the dip in the transmission spectrum away from the Fermi energy shown as the green and blue dashed lines. In (c) a large change in the molecule occurs causing the transmission to drastically change. (d) A molecular interaction event or change in electron density in the molecule could shift the transmission spectrum. (Reprinted with permission from *J. Am. Chem. Soc.*, **130**(51), 17309, 2008. Copyright 2008 American Chemical Society.)

properties to design molecular devices. All molecules have transmission resonances which will control the energies at which the transmission maxima are reached and the limits of the "on" state. The energetic location of these resonances can be controlled by many well-known means, for example, extending the conjugation length to decrease the HOMO–LUMO gap or adding electron donating or withdrawing groups to shift the energetic position of the resonances. A less well-known aspect is tuning the characteristics of the interference feature to control the "off" state. There are two essential elements to the interference feature: its depth and its energetic position.

The depth of the interference feature is controlled by the residual σ system transport after the π system transport is removed. It is well known that σ transport decreases with increasing molecular length; however not all means of extension are equal. In Figure 2.9 we show two series of molecules with increasing

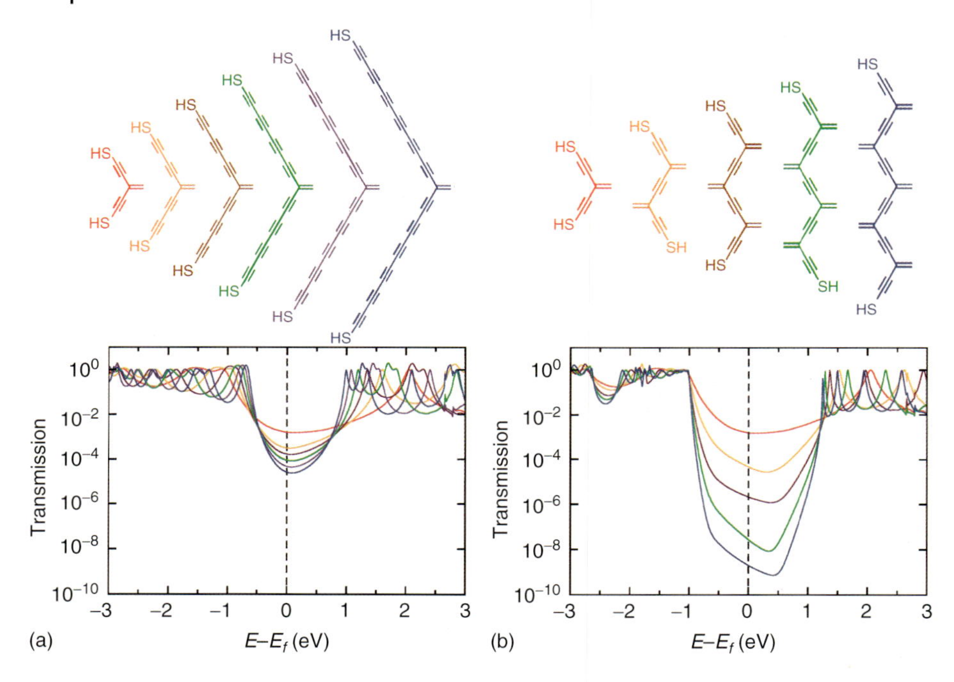

Figure 2.9 (a) and (b) Behavior of the HOMO and LUMO energies and the interference minimum with increasing molecular length. In the series of molecules shown in (a), one triple bond is added to both ends of the molecule. This increased conjugation length leads to a smaller HOMO–LUMO gap and a small decrease in the transmission minimum. In (b), a cross-conjugated unit and a triple bond are added leading to little change in the HOMO–LUMO gap but a large decrease in the transmission minimum. (Reprinted with permission from *J. Am. Chem. Soc.*, **130**(51), 17309, 2008. Copyright 2008 American Chemical Society.)

carbon backbone length. In (a) we add a pair of triple bonds symmetrically to a cross-conjugated molecule. Shown in (b) is a series of molecules where the cross-conjugated backbone has been synthesized [42, 121], made with an increasing number of cross-conjugated units separated by triple bonds. While the increasing length in both these systems does reduce the transmission in the "off" state, it is clearly not simply the number of carbon atoms in the chain that is significant. In (a) the decrease in transmission is ~0.3 orders of magnitude for each four atom repeat unit added, compared with ~1.5 orders of magnitude for each three carbon atom repeat unit added in (b).

The comparison of these two systems in Figure 2.9 highlights the interesting possibilities with cross-conjugated systems that have not been explored with other types of molecular wires. The relative stability of the HOMO–LUMO gap size with increasing length in (b) is indicative of the cross-conjugated unit breaking electron delocalization [19, 20, 43, 76]. Possibly the most tantalizing opportunity is the insight that cross-conjugated molecules can give into the properties of the underlying σ system in conjugated molecules.

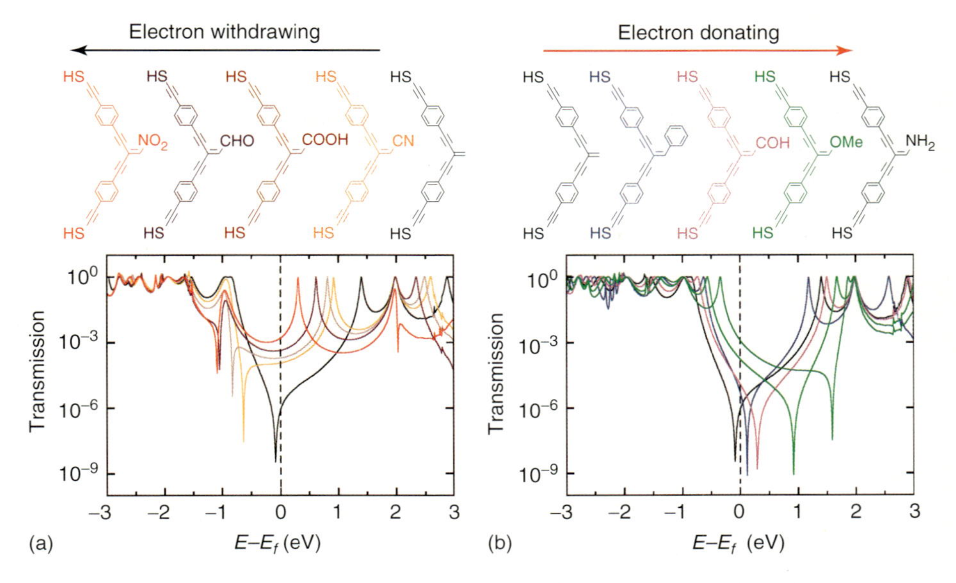

Figure 2.10 Electron donating and withdrawing groups attached to the cross-conjugated unit. The interference feature is shown to be tuned ±1.5 eV from the Fermi level by changing the electron density on the cross-conjugated bond. (Reprinted with permission from *J. Am. Chem. Soc.*, **130**(51), 17309, 2008. Copyright 2008 American Chemical Society.)

The energetic location of the interference feature can be controlled by changes in electron density across the system. Electron donating and withdrawing groups, substituted off the central ethylene unit can systematically shift the position of the interference feature across the measurable range, as shown in Figure 2.10 where the calculations are performed using Hückel-IV. The sensitivity of the system to the effects of substituents means that even commonly used protecting groups can shift the position of the interference feature, thereby altering the system's function [7]. The sensitivity of the system to substituent effects is also reflected in a lack of good agreement between theoretical methods as to what the magnitude of the shift is in each case [7], although the trends are common among the methods. This lack of precise agreement actually presents an interesting opportunity for systems which might provide discriminating benchmarks for testing theory against experiment.

2.5.3 Transistor

The sensitivity of the interference feature to changes in electron density on the molecule means that the system is ideally suited for transistor function. There have been a number of experimental measurements on single-molecule transistors [27, 65, 144]; however, in the strong coupling limit molecules will not always show much change in their elastic transport properties with the application of a gate voltage [28]. In Figure 2.11 we show the response to a gate voltage (c)–(e) calculated using ATK for a series of molecular wires shown in (a). The transmission properties for these molecules are shown in (b) using both ATK (solid line) and Hückel-IV (dotted line) to show how well these methods can compare across the many orders of magnitude that the transmission properties span. While all of the systems exhibit a response to the applied gate voltage, Figure 2.11(f) shows that it is only the cross-conjugated molecule that results in a large change in conductance with

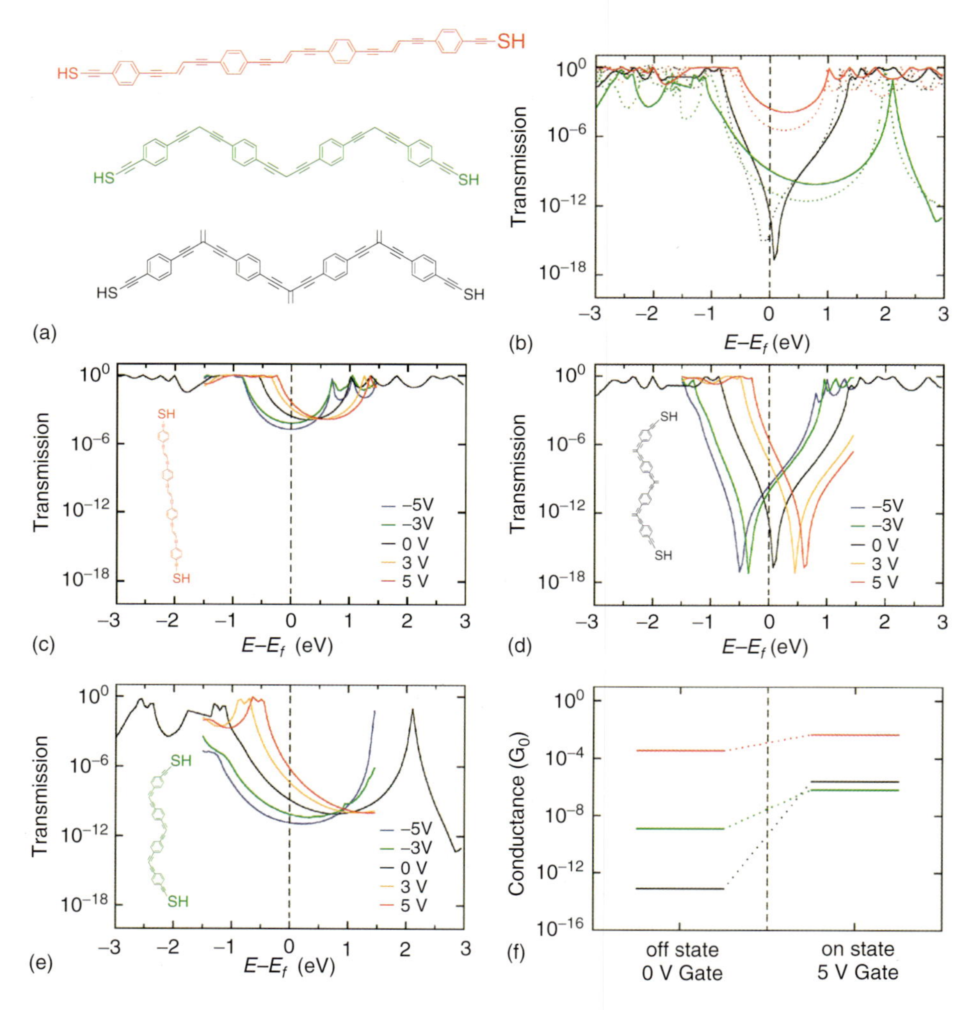

Figure 2.11 Calculations of conductance change with applied gate voltage in three test molecules. In (b), the transmission spectrum calculated for the molecules shown in (a) includes a promising potential molecular transistor, the cross-conjugated molecule in black. The solid lines are calculated using ATK and the dashed lines calculated using Hückel-IV. (c), (d), and (e) show the effect of gate voltage on the transmission. (f) shows the calculated change in conductance between on and off state of ~8 orders of magnitude in the cross-conjugated molecule. (Reprinted with permission from *J. Am. Chem. Soc.*, **130**(51), 17309, 2008. Copyright 2008 American Chemical Society.)

gate bias, here the "off" state is defined as no gate perturbation to the system and the "on" state by a 5-V gate bias. The relatively featureless mid-gap transmission of the molecules with linear conjugation (c) and broken conjugation (e), result in little change when the transmission is shifted by the application of the gate voltage. This is highlighted in (f). In the system with broken conjugation, shown in green, the conductance changes 3 orders of magnitude, and the fully conjugated molecule, shown in red, changes 1.5 orders of magnitude. Conversely, the sharp dip of the interference feature in the transmission of the cross-conjugated molecule makes this system very sensitive to shift induced by the gate electrode. The conductance through the cross-conjugated molecule changes by 8 orders of magnitude with an applied gate voltage of 5 V.

These results are shown for a representative system but could be further optimized by engineering the interference minimum to occur directly at the Fermi level. The on state could possibly be increased 5 orders of magnitude in the case of the cross-conjugated molecule by increasing gate voltage shifting the HOMO orbital closer to the Fermi level; however, approaching resonance increases the probability of electron charging and molecular rearrangement. The cross-conjugated molecule has a subthreshold swing of ~625 mV/decade (calculated by using the transmission difference between 0 and 5 V gate voltage), which is ~2.5 and ~5.25 times lower than in the conjugated and saturated molecules. The comparison between these rates is important because the conversion between the gate voltage used in the calculations and the gate electrode in the experiment is nontrivial [1] and will have a large effect on the absolute value of the subthreshold swing. The large dynamic range, sensitivity to incident electron energy, and switching based on changes in electron density make cross-conjugated molecules promising candidates for molecular transistors.

2.5.4 Rectifier

The field of single-molecule electronics is rooted in the proposal of a single-molecule rectifier [12]. In the intervening years, many experimental attempts have been made [74] to measure a rectification ratio in single-molecule transport, with marginal success in comparison with solid-state devices. The essential element of a single-molecule rectifier is some sort of chemical functionality that leads to an asymmetric response to applied bias. This can be as a result of asymmetry in the molecule, or more successfully in the binding to the two electrodes [75, 78, 79, 90, 96]. While these approaches have had some indisputable success, experimental and theoretical investigations show molecular rectifiers with rectification ratios typically $\ll 100$ [74, 90, 96, 113], these results are a far cry from typical solid-state rectification ratios that can be $>10^5$. It has been suggested that as long as something similar to the PRBT model holds, the rectification ratio for single molecules will never be greater than 100 [9]. As we have already highlighted in this chapter, the PRBT model does not hold for cross-conjugated molecules and consequently by engineering asymmetry into these systems, interesting rectification properties can be exploited.

The systems that have been highlighted in the previous sections of this chapter have dealt with the case of a single (sometimes degenerate) interference peak in symmetric molecules. Rather than following the previous approaches of using asymmetry in binding to the electrodes, we design molecules with asymmetric interference features which should yield much greater asymmetry in measured current–voltage characteristics. The understanding derived from the investigations highlighted in Section 2.5.2 are critical for this molecular design as both the zero-bias transmission and the voltage response have to be controlled.

We illustrate our design principle with model calculations on candidate molecules. As was seen in Section 2.5.2 these calculations are very sensitive to the theoretical method and the Fermi level placement (band lineup), which means that small changes in the electronic structure can have large consequences in the measured response.

The asymmetry in our proposed class of rectifiers is engineered by using multiple functional groups that generate interference features, as shown in Figure 2.12. This molecule consists of two cross-conjugated units (or other groups that produce interference features, e.g., meta-substituted benzene) with split interference features, separated by a conjugated spacer. The split interference features are controlled by differing chemical substitution on the two cross-conjugated units to give an interference feature on either side of the Fermi energy. With an applied bias across the molecule, the interference positions are expected to move toward each other or away from each other as shown in Figure 2.12(a). This is a result of the interference response in cross-conjugated molecules to electron donation and withdrawal. In one case the applied bias acts in the same direction as the substituent effect, shifting the interference features away from the Fermi energy and in the opposite bias direction the applied voltage opposes the substituent effect, shifting the interference features toward the Fermi energy.

We calculate this response with Hückel-IV as shown in Figure 2.12. This molecule has a cross-conjugated unit with a methyl ether and a cross-conjugated unit with carboxyl termination. In this example, the carboxyl-terminated cross-conjugated unit is the more electron-withdrawing group and the ether-terminated unit is the more electron-donating group. In Figure 2.12(c), the transmission through this molecule is shown at three different voltage points $-1, 0, +1$. The interference dips move toward the Fermi energy at negative bias and shift away from the Fermi energy at positive bias. In Hückel-IV, we calculate a rectification ratio of 249 at 1.2 V, whereas in the other methods the maximum rectification ratios are calculated to be 18.6 at 1.0 V in gDFTB and 17.6 at 0.6 V in ATK.

The rectification ratio is controlled by the relationship between how low the interference features make the transmission as they move toward the Fermi energy, compared with how high the transmission is as the interference features move away from the Fermi energy. In all our transport calculations, increasing the conjugated spacer length in the center of the molecule leads to a corresponding increase in the rectification ratio: increasing length decreases the minimum transmission as a result of the interference feature. The magnitude of the transmission between the split interference features can be increased dramatically by having

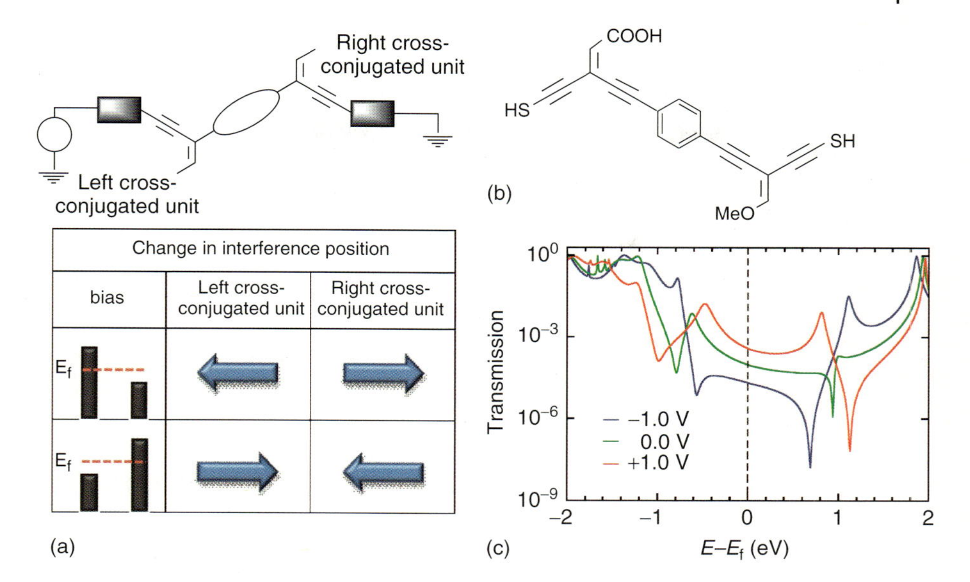

Figure 2.12 In molecules with multiple cross-conjugated units, an applied bias voltage will split the interference features. This splitting occurs because an applied bias has an electron donating or withdrawing effect that moves the interference position as shown in Figure 2.10. Shown in (a), the more positive an electrode is, the more it moves the interference feature of the closest cross-conjugated group to lower energy and conversely the more negative an electrode, the more it moves the interference feature to higher energy. In (b), we show an asymmetric molecule with the corresponding transmission plots shown in (c). The asymmetry causes two antiresonance features at different energy. At negative bias these antiresonance features move together and at positive applied bias they move apart.

a molecular resonance near the Fermi level with interference features separated equally energetically above and below this resonance.

One strategy for maximizing the rectification ratio in our model rectifiers has been to use functionalities that contain oxygen as, in our calculations, these groups have resonances near the Fermi level. A group that has been investigated both experimentally [11] and theoretically [91] is the anthraquinone functional group, a cyclic cross-conjugated group. This group has the characteristics of interest, an interference feature below the Fermi energy and a localized resonance just above the Fermi energy [11]. To create a molecule with a resonance split by two antiresonance peaks, we asymmetrically add a cross-conjugated unit. As shown in Figure 2.13, we have taken the anthraquinone functional group, added a large conjugated spacer and two methyl-terminated cross-conjugated groups (the second cross-conjugated unit orients the sulfur termination toward the Au electrodes).

In Figure 2.13, we show the current voltage behavior and the rectification ratio for our proposed rectifier. As the interference features come together with negative bias the current decreases from 0.2 to 0.8 V, while in the positive bias, the current increases as the interference dips move apart. The rectification ratio increases steadily from 0 to 0.8 V where it quickly falls off. At 0.8 V the rectification ratio

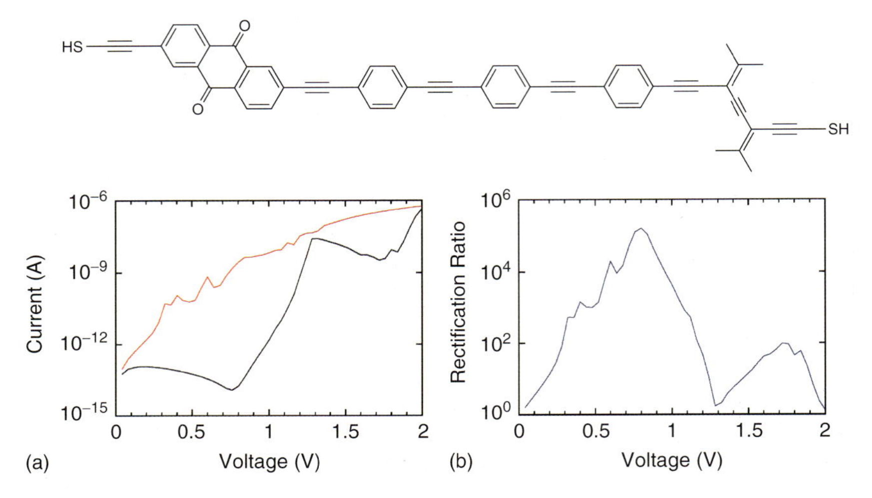

Figure 2.13 Rectifier behavior maximized in the Hückel IV transport code. In (a), the positive bias current/voltage behavior is shown in red and the negative bias current/voltage behavior is shown in black. In (b), the rectification ratio as a function of voltage shows a peak of >150,000 at 0.8 V applied voltage. (Reprinted with permission from *J. Am. Chem. Soc.*, **130**(51), 17309, 2008. Copyright 2008 American Chemical Society.)

of $>150,000$ is orders of magnitude higher than other published single-molecule rectifier calculations or experiments, without relying on asymmetric binding to the electrodes.

The design of this system was influenced by an objective of maximizing the rectification ratio as calculated in Hückel-IV. In ATK and gDFTB the maximum rectification ratio is not nearly so large, but the differences can be simply understood from the underlying differences in the predicted transmission for this molecule [7]. The strength of our approach to rectifier design is the extreme sensitivity of the system to changes in electric field; however, this is also a challenge for theory as the subtle differences between the methods are highlighted in significant differences in predicted observables. The differences between the three theoretical methods are sufficiently large that realizing a useful device by these principles may require experimental optimization.

2.5.5
Negative Differential Resistance

In many molecules with asymmetrically placed interference features some region of NDR will be seen in the current–voltage characteristics, as the interference feature or features move toward the Fermi energy with increasing bias. This behavior is seen in Figure 2.13(a) where the current in the negative voltage direction decreases 1 order of magnitude between 0.2 and 0.7 V. NDR is a behavior that has been sought in molecular electronic devices and this approach provides a new avenue toward its realization.

Interference features in the electronic transmission provide a different approach for tuning conductance characteristics in molecules by providing a way to switch off transport. In previous work on NDR in molecules on silicon surfaces, the conductance dip is attributed to the conduction band edge passing a molecular resonance [48, 49]. The effect of an interference feature is analogous; however, it is a component of the molecular transmission that is moving into the measurement window and suppressing transmission rather than an effect from the electrodes density of states. The advantage of using an interference feature is that there is no indication that this will lead to charging of the molecule, and consequently geometric relaxation, whereas shifting a band edge may result in these structural changes and consequently long-term instability of the device.

2.6 Probing the Limits of Calculations: Important Real-World Phenomena

The value of this theoretical work rests in part on whether it can be translated into real devices. For this to be achieved, none of the assumptions in the theoretical work must fail to such an extent that they undermine the device performance. In this section we consider what we anticipate are the main assumptions that might undermine our predictions, namely, peculiarities of the theoretical method, the thermal effects of conformational freedom and dephasing, molecular short circuits and transport through inelastic and excited state channels.

2.6.1 Theoretical Method

The peculiarities of the theoretical method present somewhat of a thorny issue for our analysis. In this chapter we have already shown instances where there is good agreement among the three codes we use, for example, Figure 2.11 and this issue was also discussed in detail previously [7, 105]. Conversely, we have also discussed situations where there are substantial discrepancies between the predictions of the three computational approaches [7]. The important point to note here is that all the computational methods predict interference features for cross-conjugated molecules, the disparity arises as the methods predict different energetic locations for the resonances and different behavior with voltage. As such, we would suggest that these differences are really quantitative rather than qualitative. Clearly, until these methods can be more precisely benchmarked against experiment, care is required in the interpretation of their predictions.

2.6.2 Dephasing

Previous work has shown that pure local dephasing is capable of destroying interference [44, 45, 101]. A dephasing process can be fluctuations in the system,

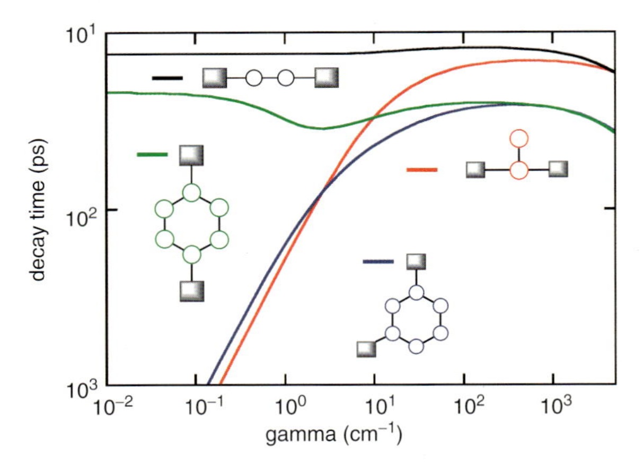

Figure 2.14 Electronic dephasing effects calculated for meta and para benzene compared with linear and cross-conjugated site representations. The decay time is defined as the time necessary for the total population to decrease 95%. Gamma is a parameter representing the dephasing strength in the model. (Reprinted with permission from *J. Phys. Chem. C*, **112**(43), 16991, 2008. Copyright 2008 American Chemical Society.)

on the timescale of the electron transfer, which destroy the coherence between paths through the molecule. We model dephasing processes in cross-conjugated molecules by considering Hückel models of cross- and linearly conjugated molecules, as was previously done for meta- and para-substituted phenyl systems [44, 45].

Figure 2.14 shows that increasing the dephasing in the system will increase transport through systems dominated by destructive quantum interference (the meta benzene and cross-conjugated systems) and decrease transport through the more highly conductive systems (para-substituted benzene and the linearly conjugated system). Interestingly, the strength of the dephasing interaction has to be stronger for the decay time to coalesce for the cross- and linearly conjugated systems than for the phenyl systems. This provides a hint as to why dephasing effects might not be catastrophic for our predictions about the transport properties of cross-conjugated molecules: the effects of dephasing are dependent on the length scale of the system. The cross-conjugated unit that is responsible for the interference effects in these systems is so small, and the effect is so local that if it is possible to see strong interference effects in any system, this is a likely candidate. The strongest indications that dephasing will not be a problem in cross-conjugated molecules are the room temperature experimental measurements of the low level of electronic communication through meta-substituted benzene, a comparably sized system whose transmission is dominated by a large destructive interference feature [70, 84].

2.6.3
Conformational Freedom

The mechanism by which conformational flexibility can affect the conductance characteristics is that a real measurement will not be made on the perfectly

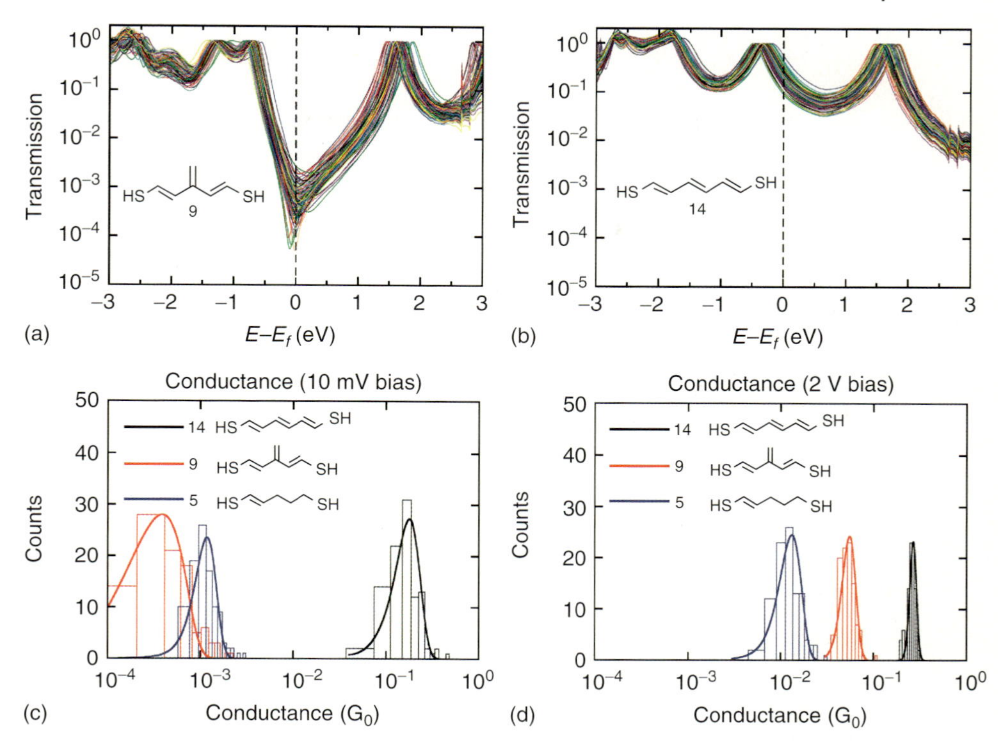

Figure 2.15 Molecular dynamics results for molecules **5**, **9**, **14**. Plots (a) and (b) show 100 transmission traces calculated for molecules **9** and **14**. Plots (c) and (d) show Gaussian fits to the distribution of conductance values calculated at 10 mV and 2 V. Reprinted with permission from *J. Phys. Chem. C*, **112**(43), 16991, 2008. Copyright 2008 American Chemical Society.

symmetric optimized geometries. We use molecular dynamics simulations to investigate the effects of the geometric fluctuations associated with ambient temperature by generating an ensemble of geometries and analyzing the conductance characteristics. Here we consider a situation where the molecular geometry is fluctuating, but we still assume that it is static on the timescale of the electron transfer. That is, the molecular fluctuations are slow with respect to the electron transfer time. Figures 2.15(a) and (b) show the variation in transmission that is anticipated for 100 different geometries of cross- and linearly conjugated molecules respectively (molecules **9** and **14**). These results are translated into predicted conductance for these systems as well as a partially saturated molecule (molecule **5**) and histograms of these results are shown for low and high bias, Figures 2.15(c) and (d). Conformational flexibility modifies the conductance of all three molecules. The most important finding of this study is that conformational variation does not destroy the interference feature in the cross-conjugated system and, consequently, the low- and high-conductance states of this system can be accessed with increasing bias.

2.6.4 Short Circuits

In the strong coupling limit the strong chemical bond between the molecule of interest and the electrode defines the path for electron transport through the system. Electron transport "through bonds" is generally so much greater than transport "through space" that it is easy to assume that the largest electronic coupling element between the two electrodes will have to follow the chemical bonds of the system. When structural or functional groups are present which induce destructive interference, however, this assumption may not hold.

Figure 2.16 shows the transport through three molecules. The first two, para(S) and meta(S), have thiol groups which chemisorb to the electrode surface (with terminal hydrogens removed) resulting in strong coupling to the electrode. The meta(S) system has two meta-substituted phenyl rings, which result in an interference feature near the Fermi energy and lead to a computed zero-bias transmission

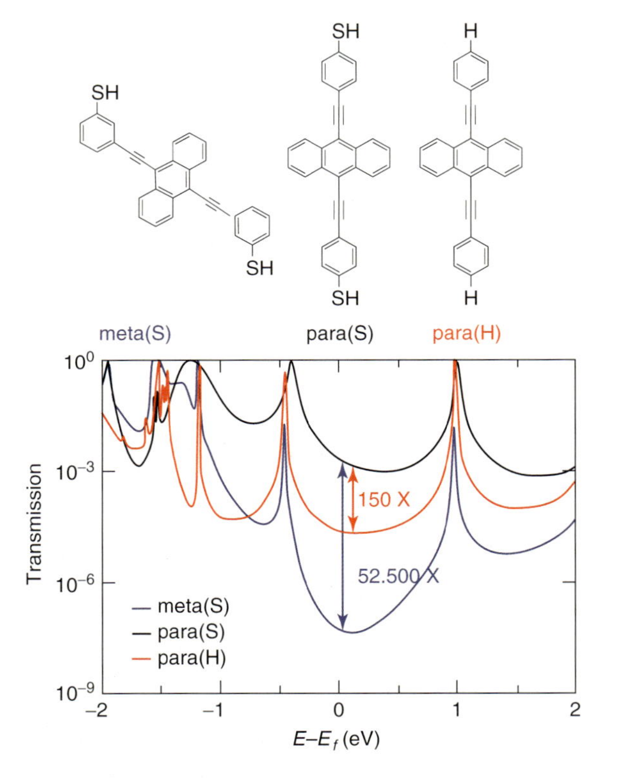

Figure 2.16 In the para(S) and meta(S) molecules, we calculate a zero-bias conductance ratio of >50,000, in stark contrast to the para (S)/para(H) ratio of 150. The H–Au bond length is set to 2.35 Å. The low-bias conductance of para(H) can be orders of magnitude higher than meta(S).

$\sim$52, 000 times lower than para(S). The third system, para(H), has no thiol groups and is simply placed between the electrodes with a similar orientation to the para(S) system as though, perhaps, it were held there by some other functionality (for example meta-substituted thiol groups). The point we wish to illustrate here is that, depending on the electrode orientation, a meta-substituted system such as meta(S) may be short circuited by transport through the unsubstituted hydrogen atoms at the para positions. Care has to be taken to effectively insulate highly conductive paths through a molecule if a path dominated by interference effects is to be measured.

2.6.5
Transport Through Inelastic and Excited State Channels

It is possible that transport through inelastic or excited state channels may not be low in cross-conjugated molecules. If this is in fact the case, it may be that these pathways (which are generally of lesser importance) may dominate the transport properties, in an analogous fashion to pathways with "through space" coupling dominating the electronic transport as illustrated above. The precise magnitude of these effects remains to be seen, but the possibility warrants further investigation and should be considered for the design and interpretation of experimental work.

2.7
Conclusions

The unusual electron transport properties of acyclic cross-conjugated molecules set them apart from the majority of systems studied to date. Their transport behavior does not follow the usual "rules of thumb," nor can they be readily described by a naive application of the PRBT model. They are members of a limited class of molecules whose off-resonant transport is dominated by destructive quantum interference effects.

When bound to electrodes, molecular wires offer a range of possibilities for single-molecule electronic devices. Optimizing chemical functionalities to achieve desirable performance characteristics has, however, proved to be nontrivial. Acyclic cross-conjugated molecules offer an interesting prospect for future investigation with these theoretical studies suggesting that a diverse range of conductance characteristics can be obtained from relatively minor perturbations. In particular, acyclic cross-conjugated molecules may yield favorable device performance with changes in electric field alone.

As with any theoretical work, the realization of such devices, or lack thereof, will reveal whether the assumptions underlying this work constitute a reasonable base for modeling molecular electron transport dominated by interference effects. In any event, these systems should provide a clear, discriminating test system against which theoretical methods can be benchmarked. The best case scenario for these predictions is the realization of molecular electronic devices with competitive performance characteristics, the worst case is dramatic disparities between theory and experiment highlighting the inadequacies of the theoretical methods. The beauty is that either way interference effects and acyclic cross-conjugated molecules offer exciting future avenues for research into molecular wires.

Acknowledgments

We thank the MURI program of the DoD, the NCN, and MRSEC programs of the NSF and the ONR and NSF Chemistry divisions for support.

References

1. Atomistix toolkit version 2.0.4. Atomistix A/S (*www.atomistix.com*).
2. Aharonov, Y. and Bohm, D. (1959) Significance of electromagnetic potentials in the quantum theory. *Phys. Rev.*, **115**(3), 485.
3. Akkerman, H. B., Blom, P. W. M., de Leeuw, D. M. and Boer, B. (2006) Towards molecular electronics with large-area molecular junctions. *Nature*, **441**(7089), 69.
4. Akkerman, H. B., Naber, R. C. G., Jongbloed, B., van Hal, P. A., Blom, P. W. M., de Leeuw, D. M. and Boer, B. (2007) Electron tunneling through alkanedithiol self-assembled monolayers in large-area molecular junctions. *PNAS*, **104**(27), 11161.
5. Albrecht, T., Guckian, A., Ulstrup, J. and Vos, J. G. (2005) Transistor effects and in situ stm of redox molecules at room temperature. *Nanotechnol. IEEE Transact.*, **4**(4), 430.
6. Andrews, D. Q., Solomon, G. C., Goldsmith, R. H., Hansen, T., Wasielewski, M. R., Van Duyne, R. P. and Ratner, M. A. (2008) Quantum interference: the structural dependence of electron transmission through model systems and cross-conjugated molecules. *J. Phys. Chem. C*, **112**(43), 16991.
7. Andrews, D. Q., Solomon, G. C., Van Duyne, R. P. and Ratner, M. A. (2008) Single molecule electronics: increasing dynamic range and switching speed using cross-conjugated species. *J. Am. Chem. Soc.*, **130**(51), 17309.
8. Andrews, D. Q., Van Duyne, R. P. and Ratner, M. A. (2008) Stochastic modulation in molecular electronic transport junctions: molecular dynamics coupled with charge transport calculations. *Nano Lett.*, **8**(4), 1120.
9. Armstrong, N., Hoft, R. C., McDonagh, A., Cortie, M. B. and Ford, M. J. (2007) Exploring the performance of molecular rectifiers: limitations and factors affecting molecular rectification. *Nano Lett.*, **7**(10), 3018.
10. Aronov, A. G. and Sharvin, Yu. V. (1987) Magnetic flux effects in disordered conductors. *Rev. Mod. Phys.*, **59**(3), 755.
11. Ashwell, G. J., Urasinska, B., Wang, C., Bryce, M. R., Grace, I. and Lambert, C. J. (2006) Single-molecule electrical studies on a 7 nm long molecular wire. *Chem. Commun.*, (45), 4706.
12. Aviram, A. and Ratner, M. A. (1974) Molecular rectifiers. *Chem. Phys. Lett.*, **29**(2), 277.
13. Bachtold, A., Strunk, C., Salvetat, J.-P., Bonard, J.-M., Forro, L., Nussbaumer, T. and Schonenberger, C. (1999) Aharonov–Bohm oscillations in carbon nanotubes. *Nature*, **397**(6721), 673.
14. Becke, A. D. (1993) Density-functional thermochemistry. iii. the role of exact exchange. *J. Chem. Phys.*, **98**, 5648.
15. Bilic, A., Reimers, J. R. and Hush, N. S. (2005) The structure, energetics, and nature of the chemical bonding of phenylthiol adsorbed on the Au(111) surface: implications

for density-functional calculations of molecular-electronic conduction. *J. Chem. Phys.*, **122**(9), 094708.
16. Boldi, A. M., Anthony, J., Gramlich, V., Knobler, C. B., Boudon, C., Gisselbrecht, J. P., Gross, M. and Diederich, F. (1995) Acyclic tetraethynylethene molecular scaffolding – multinanometer-sized linearly conjugated rods with the poly(triacetylene) backbone and cross-conjugated expanded dendralenes. *Helv. Chim. Acta*, **78**(4), 779.
17. Borden, W. T. and Davidson, E. R. (1977) Effects of electron repulsion in conjugated hydrocarbon diradicals. *J. Am. Chem. Soc.*, **99**(14), 4587.
18. Brandbyge, M., Mozos, J.-L., Ordejón, P., Taylor, J. and Stokbro, K. (2002) Density-functional method for nonequilibrium electron transport. *Phys. Rev. B*, **65**(16), 165401.
19. Bruschi, M., Giuffreda, M. G. and Luthi, H. P. (2002) trans versus geminal electron delocalization in tetra- and diethynylethenes: a new method of analysis. *Chem. Eur. J.*, **8**(18), 4216.
20. Bruschi, M., Giuffreda, M. G. and Luthi, H. P. (2005) Through versus cross electron delocalization in polytriacetylene oligomers: a computational analysis. *ChemPhysChem.*, **6**(3), 511.
21. Bruus, H. and Flensberg, K. (2004) *Many-Body Quantum Theory in Condensed Matter Physics: An Introduction*, Oxford University Press, USA.
22. Cardamone, D. M., Stafford, C. A. and Mazumdar, S. (2006) Controlling quantum transport through a single molecule. *Nano Lett.*, **6**(11), 2422.
23. Chakrabarti, A. (2006) Electronic transmission in a model quantum wire with side-coupled quasiperiodic chains: Fano resonance and related issues. *Phys. Rev. B*, **74**(20), 205315.
24. Chen, F., He, J., Nuckolls, C., Roberts, T., Klare, J. E. and Lindsay, S. (2005) A molecular switch based on potential-induced changes of oxidation state. *Nano Lett.*, **5**(3), 503.
25. Collepardo-Guevara, R., Walter, D., Neuhauser, D. and Baer, R. (2004) A Hückel study of the effect of a molecular resonance cavity on the quantum conductance of an alkene wire. *Chem. Phys. Lett.*, **393**(4–6), 367.
26. Coulson, C. A. and Rushbrooke, G. S. (1940) *Proc. Camb. Phil. Soc.*, **36**, 1931.
27. Dadosh, T., Gordin, Y., Krahne, R., Khivrich, I., Mahalu, D., Frydman, V., Sperling, J., Yacoby, A. and Bar-Joseph, I. (2005) Measurement of the conductance of single conjugated molecules. *Nature*, **436**(7051), 677.
28. Danilov, A., Kubatkin, S., Kafanov, S., Hedegard, P., Stuhr-Hansen, N., Moth-Poulsen, K. and Bjornholm, T. (2008) Electronic transport in single molecule junctions: control of the molecule-electrode coupling through intramolecular tunneling barriers. *Nano Lett.*, **8**(1), 1.
29. Danilov, A. V., Kubatkin, S. E., Kafanov, S. G. and Bjornholm, T. (2006) Strong electronic coupling between single C60 molecules and gold electrodes prepared by quench condensation at 4 K. A single molecule three terminal device study. *Faraday Discuss.*, **131**, 337.
30. Datta, S. (1989) Quantum devices. *Superlattices Microstruct.*, **6**(1), 83.
31. Datta, S. (2005) *Quantum Transport: Atom to Transistor*, Cambridge University Press, Cambridge.
32. Datta, S. (1997) *Electronic Transport in Mesoscopic Systems*, Cambridge University Press, Cambridge.
33. Davis, W. B., Wasielewski, M. R., Ratner, M. A., Mujica, V. and Nitzan, A. (1997) Electron transfer rates in bridged molecular systems: a phenomenological approach to relaxation. *J. Phys. Chem. A*, **101**(35), 6158.
34. Ladron de Guevara, M. L. and Orellana, P. A. (2006) Electronic transport through a parallel-coupled triple quantum dot molecule: Fano resonances and bound states in the continuum. *Phys. Rev. B*, **73**(20), 205303.
35. Elbing, M., Ochs, R., Koentopp, M., Fischer, M., Hanisch, C., Weigend, F., Evers, F., Weber, H. B. and Mayor, M. (2005) Molecular electronics special feature: a single-molecule diode. *PNAS*, **102**(25), 8815.
36. Elstner, M., Porezag, D., Jugnickel, G., Elsner, J., Haugk, M., Frauenheim,

T., Suhai, S. and Seifert, G. (1998) Self-consistent-charge density-functional tight-binding method for simulations of complex materials properties. *Phys. Rev. B*, **58**, 7260.

37. Ernzerhof, M., Bahmann, H., Goyer, F., Zhuang, M. and Rocheleau, P. (2006) Electron transmission through aromatic molecules. *J. Chem. Theory Comput.*, **2**(5), 1291.

38. Ernzerhof, M., Zhuang, M. and Rocheleau, P. (2005) Side-chain effects in molecular electronic devices. *J. Chem. Phys.*, **123**(13), 134704.

39. Flood, A. H., Stoddart, J. F., Steuerman, D. W. and Heath, J. R. (2004) Chemistry: enhanced: whence molecular electronics?. *Science*, **306**(5704), 2055.

40. Galperin, M., Ratner, M. A. and Nitzan, A. (2007) Molecular transport junctions: vibrational effects. *J. Phys. Condens. Matter*, **19**(10), 103201.

41. Galperin, M., Ratner, M. A., Nitzan, A. and Troisi, A. (2008) Nuclear coupling and polarization in molecular transport junctions: beyond tunneling to function. *Science*, **319**(5866), 1056.

42. Gholami, M. and Tykwinski, R. R. (2006) Oligomeric and polymeric systems with a cross-conjugated framework. *Chem. Rev.*, **106**(12), 4997.

43. Giuffreda, M. G., Bruschi, M. and Lüthi, H. P. (2004) Electron delocalization in linearly conjugated systems: a concept for quantitative analysis. *Chem. Eur. J.*, **10**(22), 5671.

44. Goldsmith, R. H., Wasielewski, M. R. and Ratner, M. A. (2006) Electron transfer in multiply bridged donor-acceptor molecules: dephasing and quantum coherence. *J. Phys. Chem. B*, **110**(41), 20258.

45. Goldsmith, R. H., Wasielewski, M. R. and Ratner, M. A. (2007) Scaling laws for charge transfer in multiply bridged donor/acceptor molecules in a dissipative environment. *J. Am. Chem. Soc.*, **129**(43), 13066.

46. Goyer, F., Ernzerhof, M. and Zhuang, M. (2007) Source and sink potentials for the description of open systems with a stationary current passing through. *J. Chem. Phys.*, **126**(14), 144104.

47. Grave, C., Risko, C., Shaporenko, A., Wang, Y., Nuckolls, C., Ratner, M. A., Rampi, M. A. and Zharnikov, M. (2007) Charge transport through oligoarylene self-assembled monolayers: interplay of molecular organization, metal-molecule interactions, and electronic structure. *Adv. Funct. Mater.*, **17**(18), 3816.

48. Guisinger, N. P., Basu, R., Greene, M. E., Baluch, A. S. and Hersam, M. C. (2004) Observed suppression of room temperature negative differential resistance in organic monolayers on si(100). *Nanotechnology*, **15**(7), S452.

49. Guisinger, N. P., Greene, M. E., Basu, R., Baluch, A. S. and Hersam, M. C. (2004) Room temperature negative differential resistance through individual organic molecules on silicon surfaces. *Nano Lett.*, **4**(1), 55.

50. Hall, L. E., Reimers, J. R., Hush, N. S. and Silverbrook, K. (2000) Formalism, analytical model, and a priori Green's-function-based calculations of the current-voltage characteristics of molecular wires. *J. Chem. Phys.*, **112**(3), 1510.

51. He, J., Chen, F., Liddell, P. A., Andrasson, J., Straight, S. D., Gust, D., Moore, T. A., Moore, A. L., Li, J., Sankey, O. F. and Lindsay, S. M. (2005) Switching of a photochromic molecule on gold electrodes: single-molecule measurements. *Nanotechnology*, **16**(6), 695.

52. Hush, N. S. *http://www.physics.uq.edu.au/cmp-workshop/forms/hush.ppt*, 2006.

53. Itakura, K., Yuki, K., Kurokawa, S., Yasuda, H. and Sakai, A. (1999) Bias dependence of the conductance of Au nanocontacts. *Phys. Rev. B*, **60**(15), 11163.

54. Joachim, C., Gimzewski, J. K. and Aviram, A. (2000) Electronics using hybrid-molecular and mono-molecular devices. *Nature*, **408**(6812), 541.

55. Joachim, C. and Ratner, M. A. (2005) Molecular electronics special feature: molecular electronics: some views on transport junctions and beyond. *PNAS*, **102**(25), 8801.

56. Kalyanaraman, C. and Evans, D. G. (2002) Molecular conductance of dendritic wires. *Nano Lett.*, **2**(5), 437.

57. Kubatkin, S., Danilov, A., Hjort, M., Cornil, J., Bredas, J.-L., Stuhr-Hansen, N., Hedegard, P. and Bjornholm, T. (2003) Single-electron transistor of a single organic molecule with access to several redox states. *Nature*, **425**(6959), 698.

58. Kushmerick, J. G., Lazorcik, J., Patterson, C. H., Shashidhar, R., Seferos, D. S. and Bazan, G. C. (2004) Vibronic contributions to charge transport across molecular junctions. *Nano Lett.*, **4**(4), 639.

59. Landauer, R. (1957) *IBM J. Res. Dev.*, **1**, 223.

60. Landauer, R. (1970) Electrical resistance of disordered one-dimensional lattices. *Phil. Mag.*, **21**(172), 863.

61. Lee, C., Yang, W. and Parr, R. G. (1988) Development of the Colle-Salvetti correlation-energy formula into a functional of the electron density. *Phys. Rev. B*, **37**, 785.

62. Lee, H. W. (1999) Generic transmission zeros and in-phase resonances in time-reversal symmetric single channel transport. *Phys. Rev. Lett.*, **82**(11), 2358.

63. Leggett, A. J., Chakravarty, S., Dorsey, A. T., Fisher, M. P. A., Garg, A. and Zwerger, W. (1987) Dynamics of the dissipative two-state system. *Rev. Mod. Phys.*, **59**(1), 1.

64. Li, X., Xu, B., Xiao, X., Yang, X., Zang, L. and Tao, N. (2006) Controlling charge transport in single molecules using electrochemical gate. *Faraday Discuss.*, **131**, 111.

65. Liang, W., Shores, M. P., Bockrath, M., Long, J. R. and Park, H. (2002) Kondo resonance in a single-molecule transistor. *Nature*, **417**(6890), 725.

66. Liljeroth, P., Repp, J. and Meyer, G. (2007) Current-induced hydrogen tautomerization and conductance switching of naphthalocyanine molecules. *Science*, **317**(5842), 1203.

67. Loppacher, Ch, Guggisberg, M., Pfeiffer, O., Meyer, E., Bammerlin, M., LÃ … thi, R., Schlittler, R., Gimzewski, J. K., Tang, H. and Joachim, C. (2003) Direct determination of the energy required to operate a single molecule switch. *Phys. Rev. Lett.*, **90**(6), 066107.

68. Maiti, S. K. (2007) Quantum transport through polycyclic hydrocarbon molecules. *Phys. Lett. A*, **366**(1-2), 114.

69. Mallion, R. B. and Rouvray, D. H. (1990) The golden jubilee of the Coulson-Rushbrooke pairing theorem. *J. Math. Chem.*, **5**(1), 1.

70. Mayor, M., Weber, H. B., Reichert, J., Elbing, M., von Hänisch, C., Beckmann, D. and Fischer, M. (2003) Electric current through a molecular rod – relevance of the position of the anchor groups. *Angew. Chem. Int. Ed.*, **42**(47), 5834.

71. McWeeny, R. (1979) *Coulson's Valence*, 3rd edition, Oxford University Press, Oxford.

72. Metzger, R. M. (2003) Unimolecular electrical rectifiers. *Chem. Rev.*, **103**(9), 3803.

73. Metzger, R. M., Chen, B., Hopfner, U., Lakshmikantham, M. V., Vuillaume, D., Kawai, T., Wu, X., Tachibana, H., Hughes, T. V., Sakurai, H., Baldwin, J. W., Hosch, C., Cava, M. P., Brehmer, L. and Ashwell, G. J. (1997) Unimolecular electrical rectification in hexadecylquinolinium tricyanoquinodimethanide. *J. Am. Chem. Soc.*, **119**(43), 10455.

74. Metzger, R. M. (2006) Unimolecular rectifiers: present status. *Chem. Phys.*, **326**(1), 176.

75. Miller, O. D., Muralidharan, B., Kapur, N. and Ghosh, A. W. (2008) Rectification by charging: contact-induced current asymmetry in molecular conductors. *Phys. Rev. B*, **77**(12), 125427.

76. Moonen, N. N. P., Pomerantz, W. C., Gist, R., Boudon, C., Gisselbrecht, J. P., Kawai, T., Kishioka, A., Gross, M., Irie, M. and Diederich, F. (2005) Donor-substituted cyanoethynylethenes: pi-conjugation and band-gap tuning in strong charge-transfer chromophores. *Chem. Eur. J.*, **11**(11), 3325.

77. Moresco, F., Meyer, G., Rieder, K.-H., Tang, H., Gourdon, A. and Joachim, C. (2001) Conformational

changes of single molecules induced by scanning tunneling microscopy manipulation: a route to molecular switching. *Phys. Rev. Lett.*, **86**(4), 672.

78. Mujica, V., Kemp, M., Roitberg, A. and Ratner, M. (1996) Current-voltage characteristics of molecular wires: eigenvalue staircase, coulomb blockade, and rectification. *J. Chem. Phys.*, **104**(18), 7296.

79. Mujica, V., Ratner, M. A. and Nitzan, A. (2002) Molecular rectification: why is it so rare?. *Chem. Phys.*, **281**(2-3), 147.

80. Nielsen, M. B., Schreiber, M., Baek, Y. G., Seiler, P., Lecomte, S., Boudon, C., Tykwinski, R. R., Gisselbrecht, J. P., Gramlich, V., Skinner, P. J., Bosshard, C., Gunter, P., Gross, M. and Diederich, F. (2001) Highly functionalized dimeric tetraethynylethenes and expanded radialenes: strong evidence for macrocyclic cross-conjugation. *Chem. Eur. J.*, **7**(15), 3263.

81. Nitzan, A. (2001) Electron transmission through molecules and molecular interfaces. *Annu. Rev. Phys. Chem.*, **52**(1), 681.

82. Ohnishi, H. and Takayanagi, K. (1998) Quantized conductance through individual rows of suspended gold atoms. *Nature (London)*, **395**(6704), 780.

83. Park, J., Pasupathy, A. N., Goldsmith, J. I., Chang, C., Yaish, Y., Petta, J. R., Rinkoski, M., Sethna, J. P., Abruna, H. D., McEuen, P. L. and Ralph, D. C. (2002) Coulomb blockade and the Kondo effect in single-atom transistors. *Nature*, **417**(6890), 722.

84. Patoux, C., Coudret, C., Launay, J. P., Joachim, C. and Gourdon, A. (1997) Topological effects on intramolecular electron transfer via quantum interference. *Inorg. Chem.*, **36**(22), 5037.

85. Pecchia, A. and Carlo, A. Di. (2004) Atomistic theory of transport in organic and inorganic nanostructures. *Rep. Prog. Phys.*, **67**(8), 1497.

86. Phelan, N. F. and Orchin, M. (1968) Cross conjugation. *J. Chem. Educ.*, **45**(10), 633.

87. Pobelov, I. V., Li, Z. and Wandlowski, T. (2008) Electrolyte gating in redox-active tunneling junctions: an electrochemical stm approach. *J. Am. Chem. Soc.*, **130**(47), 16045.

88. Ponder, J. W. (2004) *Tinker - Software Tools for Molecular Design*, 4.2 edition, Washington University School of Medicine, St. Louis, MO.

89. Porezag, D., Frauenheim, T., Kohler, T., Seifert, G. and Kaschner, R. (1995) Construction of tight-binding-like potentials on the basis of density-functional theory: application to carbon. *Phys. Rev. B*, **51**, 12947.

90. Reichert, J., Ochs, R., Beckmann, D., Weber, H. B., Mayor, M. and Löhneysen, H. v (2002) Driving current through single organic molecules. *Phys. Rev. Lett.*, **88**(17), 176804.

91. Reimers, J. R., Hall, L. E., Crossley, M. J. and Hush, N. S. (1999) Rigid fused oligoporphyrins as potential versatile molecular wires. 2. b3lyp and scf calculated geometric and electronic properties of 98 oligoporphyrin and related molecules. *J. Phys. Chem. A*, **103**(22), 4385.

92. Richardson, D. E. and Taube, H. (1983) Electronic interactions in mixed-valence molecules as mediated by organic bridging groups. *J. Am. Chem. Soc.*, **105**(1), 40.

93. Salem, L. (1966) *The molecular orbital theory of conjugated systems.*, W. A. Benjamin, New York.

94. Salomon, A., Boecking, T., Chan, C. K., Amy, F., Girshevitz, O., Cahen, D. and Kahn, A. (2005) How do electronic carriers cross Si-bound alkyl monolayers?. *Phys. Rev. Lett.*, **95**(26), 266807.

95. Sautet, P. and Joachim, C. (1988) Electronic interference produced by a benzene embedded in a polyacetylene chain. *Chem. Phys. Lett.*, **153**(6), 511.

96. Scott, G. D., Chichak, K. S., Peters, A. J., Cantrill, S. J., Stoddart, J. F. and Jiang, H. W. (2006) Mechanism of enhanced rectification in unimolecular Borromean ring devices. *Phys. Rev. B*, **74**(11), 113404.

97. Segal, D. and Nitzan, A. (2001) Steady-state quantum mechanics of thermally relaxing systems. *Chem. Phys.*, **268**(1–3), 315.

98. Shangguan, W. Z., Au Yeung, T. C., Yu, Y. B. and Kam, C. H. (2001) Quantum transport in a one-dimensional quantum dot array. *Phys. Rev. B*, **63**(23), 235323.

99. Shao, Y., Molnar, L. F., Jung, Y., Kussmann, J., Ochsenfeld, C., Brown, S. T., Gilbert, A. T. B., Slipchenko, L. V., Levchenko, S. V., O'Neill, D. P., DiStasio Jr, R. A., Lochan, R. C., Wang, T., Beran, G. J. O., Besley, N. A., Herbert, J. M., Lin, C. Y., Voorhis, T. V., Chien, S. H., Ryan, A. S., Steele, P., Rassolov, V. A., Maslen, P. E., Korambath, P. P., Adamson, R. D., Austin, B., Baker, J., Byrd, E. F. C., Dachsel, H., Doerksen, R. J., Dreuw, A., Dunietz, B. D., Dutoi, A. D., Furlani, T. R., Gwaltney, S. R., Heyden, A., Hirata, So, Hsu, C.-P., Kedziora, G., Khalliulin, R. Z., Klunzinger, P., Lee, A. M., Lee, M. S., Liang, W. Z., Lotan, I., Nair, N., Peters, B., Proynov, E. I., Pieniazek, P. A., Rhee, Y. M., Ritchie, J., Rosta, E., Sherrill, C. D., Simmonett, A. C., Subotnik, J. E., Woodcock Iii, H. L., Zhang, W., Bell, A. T. and Chakraborty, A. K. (2006) Advances in methods and algorithms in a modern quantum chemistry program package. *Phys. Chem. Chem. Phys.*, **8**(27), 3172.

100. Siddarth, P. and Marcus, R. A. (1992) Calculation of electron-transfer matrix elements of bridged systems using a molecular fragment approach. *J. Phys. Chem.*, **96**(8), 3213.

101. Skinner, J. L. and Hsu, D. (1986) Pure dephasing of a two-level system. *J. Phys. Chem.*, **90**(21), 4931.

102. Soler, J. M., Artacho, E., Gale, J. D., Garcia, A., Junquera, J., Ordejón, P. and Sanchez-Portal, D. (2002) The siesta method for ab initio order-n materials simulation. *J. Phys.: Condens. Matter*, **14**(11), 2745.

103. Solomon, G. C., Gagliardi, A., Pecchia, A., Frauenheim, T., DiCarlo, A., Reimers, J. R. and Hush, N. S. (2006) Molecular origins of conduction channels observed in shot-noise measurements. *Nano Lett.*, **6**(11), 2431.

104. Solomon, G. C., Andrews, D. Q., Van Duyne, R. P. and Ratner, M. A. (2009) Electron transport through conjugated molecules: when the pi system only tells part of the story. *ChemPhysChem*, **10**(1), 257.

105. Solomon, G. C., Andrews, D. Q., Goldsmith, R. H., Hansen, T., Wasielewski, M. R., Van Duyne, R. P. and Ratner, M. A. (2008) Quantum interference in acyclic systems: conductance of cross-conjugated molecules. *J. Am. Chem. Soc.*, **130**(51), 17301.

106. Solomon, G. C., Andrews, D. Q., Hansen, T., Goldsmith, R. H., Wasielewski, M. R., Van Duyne, R. P. and Ratner, M. A. (2008) Understanding quantum interference in coherent molecular conduction. *J. Chem. Phys.*, **129**(5), 054701.

107. Solomon, G. C., Andrews, D. Q., Van Duyne, R. P. and Ratner, M. A. (2008) When things are not as they seem: quantum interference turns molecular electron transfer "rules" upside down. *J. Am. Chem. Soc.*, **130**(25), 7788.

108. Solomon, G. C., Gagliardi, A., Pecchia, A., Frauenheim, T., Carlo, A. D., Reimers, J. R. and Hush, N. S. (2006) The symmetry of single-molecule conduction. *J. Chem. Phys.*, **125**(18), 184702.

109. Sols, F., Macucci, M., Ravaioli, U. and Hess, K. (1989) On the possibility of transistor action based on quantum interference phenomena. *Appl. Phys. Lett.*, **54**(4), 350.

110. Sordan, R. and Nikolic, K. (1996) The nonlinear transport regime of a T-shaped quantum interference transistor. *Appl. Phys. Lett.*, **68**(25), 3599.

111. Stafford, C. A., Cardamone, D. M. and Mazumdar, S. (2007) The quantum interference effect transistor. *Nanotechnology*, **18**(42), 424014.

112. Stipe, B. C., Rezaei, M. A. and Ho, W. (1998) Single-molecule vibrational spectroscopy and microscopy. *Science*, **280**(5370), 1732.

113. Stokbro, K., Taylor, J. and Brandbyge, M. (2003) Do Aviram-Ratner diodes rectify?. *J. Am. Chem. Soc.*, **125**(13), 3674.

114. Stokbro, K., Taylor, J., Brandbyge, M. and Ordejón, P. (2003) Transiesta: a spice for molecular electronics. *Ann. N.Y. Acad. Sci.*, **1006**, 212. Molecular Electronics III):

115. Douglas Stone, A. (1985) Magnetoresistance fluctuations in mesoscopic wires and rings. *Phys. Rev. Lett.*, **54**(25), 2692.

116. Tao, N. (2005) Measurement and control of single molecule conductance. *J. Mater. Chem.*, **15**(32), 3260.

117. Taylor, J., Guo, H. and Wang, J. (2001) Ab initio modeling of quantum transport properties of molecular electronic devices. *Phys. Rev. B*, **63**(24), 245407.

118. Tian, W., Datta, S., Hong, S., Reifenberger, R., Henderson, J. I. and Kubiak, C. P. (1998) Conductance spectra of molecular wires. *J. Chem. Phys.*, **109**(7), 2874.

119. Todorov, T. N. (2002) Tight-binding simulation of current-carrying nanostructures. *J. Phys.: Condens. Matter*, **14**(11), 3049.

120. Troisi, A. and Ratner, M. A. (2004) Conformational molecular rectifiers. *Nano Lett.*, **4**(4), 591.

121. Tykwinski, R. R. and Zhao, Y. (2002) Cross-conjugated oligo(enynes). *Synlett*, **2002**(12), 1939.

122. van der Veen, M. H., Rispens, M. T., Jonkman, H. T. and Hummelen, J. C. (2004) Molecules with linear π-conjugated pathways between all substituents: omniconjugation. *Adv. Funct. Mater.*, **14**(3), 215.

123. van der Zant, H. S. J., Kervennic, Y.-V., Poot, M., O'Neill, K., Groot, Z., Thijssen, J. M., Heersche, H. B., Stuhr-Hansen, N., Bjornholm, T., Vanmaekelbergh, D., van Walree, C. A. and Jenneskens, L. W. (2006) Molecular three-terminal devices: fabrication and measurements. *Faraday Discuss.*, **131**, 347.

124. van Walree, C. A., Kaats-Richters, V. E. M., Veen, S. J., Wieczorek, B., van der Wiel, J. H. and van der Wiel, B. C. (2004) Charge-transfer interactions in 4-donor 4′-acceptor substituted 1,1-diphenylethenes. *Eur. J. Org. Chem.*, **2004**(14), 3046.

125. Venkataraman, L., Klare, J. E., Nuckolls, C., Hybertsen, M. S. and Steigerwald, M. L. (2006) Dependence of single-molecule junction conductance on molecular conformation. *Nature*, **442**(7105), 904.

126. Walczak, K. (2004) The role of quantum interference in determining transport properties of molecular bridges. *Central Eur. J. Chem.*, **2**(3), 524.

127. Walter, D., Neuhauser, D. and Baer, R. (2004) Quantum interference in polycyclic hydrocarbon molecular wires. *Chem. Phys.*, **299**(1), 139.

128. Wang, W., Lee, T., Kretzschmar, I. and Reed, M. A. (2004) Inelastic electron tunneling spectroscopy of an alkanedithiol self-assembled monolayer. *Nano Lett.*, **4**(4), 643.

129. Wang, W., Lee, T. and Reed, M. A. (2003) Mechanism of electron conduction in self-assembled alkanethiol monolayer devices. *Phys. Rev. B*, **68**(3), 035416.

130. Webb, R. A., Washburn, S., Umbach, C. P. and Laibowitz, R. B. (1985) Observation of he Aharonov–Bohm oscillations in normal-metal rings. *Phys. Rev. Lett.*, **54**(25), 2696.

131. Xu, B. and Tao, N. J. (2003) Measurement of single-molecule resistance by repeated formation of molecular junctions. *Science*, **301**(5637), 1221.

132. Xue, Y., Datta, S. and Ratner, M. A. (2002) First-principles based matrix Green's function approach to molecular electronic devices: general formalism. *Chem. Phys.*, **281**(2-3), 151.

133. Yaliraki, S. N. and Ratner, M. A. (2002) Interplay of topology and chemical stability on the electronic transport of molecular junctions. *Ann. N.Y. Acad. Sci.*, **960**(1), 153.

134. Yanson, A. I., Rubio Bollinger, G., van den Brom, H. E., Agrait, N. and van Ruitenbeek, J. M. (1998) Formation and manipulation of a metallic wire of single gold atoms. *Nature*, **395**(6704), 783.

135. Yasuda, H. and Sakai, A. (1997) Conductance of atomic-scale gold contacts under high-bias voltages. *Phys. Rev. B*, **56**(3), 1069.

136. Yu, L. H. and Natelson, D. (2004) Transport in single-molecule transistors: Kondo physics and negative differential resistance. *Nanotechnology*, **15**(10), S517.

137. Zahid, F., Paulsson, M. and Datta, S. In Morkoc, H., editor, *Advanced Semiconductor and Organic Nano-Techniques*, volume 3, page 1. Academic Press, New York, 2003.

138. Zahid, F., Paulsson, M., Polizzi, E., Ghosh, A. W., Siddiqui, L. and Datta, S. (2005) A self-consistent transport model for molecular conduction based on extended Hückel theory with full three-dimensional electrostatics. *J. Chem. Phys.*, **123**(6), 064707.

139. Zahid, F., Paulsson, M., Polizzi, E., Ghosh, A. W., Siddiqui, L. and Datta, S. (2005) A self-consistent transport model for molecular conduction based on extended Hückel theory with full three-dimensional electrostatics. *J. Chem. Phys.*, **123**(6), 064707.

140. Zahid, F., Paulsson, M. and Datta, S. (2003) Electrical conduction through molecules in *Advanced Semiconductors and Organic Nano-Techniques Part III: Physics and Technology of Molecular and Biotech Systems*, H., Morkoc (ed) Elsevier Academic Press, pp. 1. In editor, volume III, pageCAN 140:431824 76-0 Electric Phenomena Purdue University, West Lafayette, IN, USA. Conference; General Review written in English.

141. Zhang, C., He, Y., Cheng, H.-P., Xue, Y., Ratner, M. A., Zhang, X. G. and Krstic, P. (2006) Current-voltage characteristics through a single light-sensitive molecule. *Phys. Rev. B*, **73**(12), 125445.

142. Zheng, T., Jia, H., Wallace, R. M. and Gnade, B. E. (2006) Characterization of conductance under finite bias for a self-assembled monolayer coated au quantized point contact. *Appl. Surf. Sci.*, **253**(3), 1265.

143. Zheng, Y., Lu, T., Zhang, C. and Su, W. (2004) Antiresonance of electron tunneling through a quantum dot array. *Physica E: Low-dimensional Systems and Nanostructures*, **24**(3–4), 290.

144. Zhitenev, N. B., Meng, H. and Bao, Z. (2002) Conductance of small molecular junctions. *Phys. Rev. Lett.*, **88**(22), 226801.

3
Hopping Transport in Long Conjugated Molecular Wires Connected to Metals

Seong Ho Choi and C. Daniel Frisbie

3.1
Introduction

Long, π-conjugated molecules are often described as "molecular wires," a term that implies these molecules can transport charge efficiently over long distances [1–5]. Here "long" corresponds to perhaps ~5–100 nm, greater than typical tunneling lengths, but tiny in comparison to length scales in the everyday world. Interest in molecular wires has at least two principal motivations. Some workers have postulated that molecular conductors may be useful in future nanoelectronic devices where the inherent small size of molecules and their tunable electronic properties, sensitivity to specific analytes, and propensity for self-organization seem potentially advantageous [4–8]. On the other hand, fundamental studies of conduction in molecular wires offer opportunities to elucidate microscopic conduction mechanisms operative in films of conjugated polymers. This in turn may positively impact efforts to optimize these materials for applications in solar energy conversion [9, 10], electroluminescent displays [11–13], and printed electronics [14, 15], for example.

The past decade has witnessed remarkable progress [16–22] in our ability to characterize electrical conduction in molecules using a simple testbed called a *molecular junction*. In a molecular junction, single molecules or an ensemble of molecules are connected between closely spaced metal electrodes, Figure 3.1. In these nanostructures, the conductivity of the molecular wires is determined directly by recording the current–voltage (I–V) characteristics. There are now perhaps a half dozen different methods for forming reproducible molecular junctions that are summarized in this book and in several recent reviews [18–22]. The ability to contact molecules easily and reproducibly enables systematic measurements of conduction as a function of molecular architecture [17, 23–25], molecular length [25–27], applied voltage (electric field) [28], temperature [29–37], contact metallurgy [27], and surface linker chemistry [26, 32, 38–42]. By combining organic synthesis with such systematic measurements, exciting opportunities are created for probing structure–transport relationships in molecular wires in a manner

Charge and Exciton Transport through Molecular Wires. Edited by L.D.A. Siebbeles and F.C. Grozema

ISBN: 978-3-527-32501-6

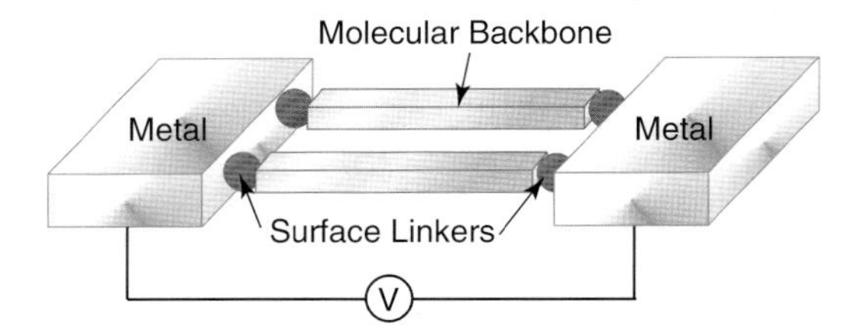

Figure 3.1 Schematic representation of a metal–molecule–metal junction.

analogous to classical physical organic chemistry approaches for understanding reaction mechanisms [43].

In this context, it is important to recognize that understanding of direct current (DC) conduction through molecules builds on several decades of prior work in the chemical community on electron-transfer (ET) processes within molecules dissolved in solution. Rate constants have been measured for intramolecular ET reactions on a wide range of systematically varied donor–bridge–acceptor (D–B–A) molecules [2, 44–48]. It is well established that ET rates depend on both the redox levels of the D and A components [44, 49], the bonding architecture, and length of the bridge [50–54], as well as dynamic motions (vibrations) within the molecule [48, 55]. Furthermore, it has been demonstrated that different ET mechanisms (e.g., tunneling versus hopping) can be distinguished by varying the temperature [48]. A great body of literature on structure–property correlations for intramolecular ET reactions has accumulated [2, 44–55], and predictions about steady-state DC conduction in molecular junctions can be made based on this prior work [56, 57].

However, the presence of metallic contacts in molecular junctions also introduces significant electronic perturbations that make these solid-state experiments fundamentally different from solution ET studies. Injection of charge into molecules from metals naturally depends on image forces and dipoles that can exist at metal–molecule interfaces, as well as on metal–molecule coupling and the energetic position of the Fermi level with respect to the molecular orbitals [57–59]. Additionally, the population of electrons that are available for transport across a molecular junction is tuned by the applied bias, which is clearly different from the situation in isolated D-B-A molecules in which single electrons (or holes) are made available for transfer via an initial excitation step. Finally, steady-state DC conduction in molecules may lead to charge correlation phenomena (e.g., Coulomb blockading [60–62]) that are not germane to typical solution ET experiments. As described in other chapters of this book, solution ET studies allow the use of ultrafast spectroscopy to observe ET dynamics that cannot be accessed in molecular junctions. Thus, solution ET and solid-state DC conduction experiments are best viewed as complementary approaches. Collectively, they produce a more complete picture of the problem of electron transport in molecular wires.

The focus of this chapter is on DC electrical conduction in molecular junctions, especially junctions incorporating long conjugated molecular wires in which charges are injected into molecular orbitals and driven along the molecular backbone by an applied electric field – the so-called hopping transport regime. Elegant experiments involving singe electron tunneling processes in molecular junctions [60–62], for example, will not be covered as our emphasis here is on longer molecules in which multiple hopping events occur. Significant portions of the chapter will cover changes in transport mechanisms that are driven by applied bias, temperature, or molecular length. We begin by reviewing basic charge transport mechanisms as comprehension of these is required to understand the broad spectrum of current–voltage behavior exhibited in molecular junctions. We conclude with an outlook on future hopping transport experiments in molecular wires.

3.2
Charge Transport Mechanisms

The charge transport mechanism of a wire can be revealed by how its electrical resistance varies with length and temperature. For example, the resistance of a conventional metallic wire is directly proportional to its length and increases with temperature generally following a polynomial [63]. This is a direct result of the diffusive (incoherent) carrier transport mechanism in the metal. However, this particular scaling need not hold for conduction through molecular wires connected to metal contacts. It is well understood that for sufficiently short molecules, electrons can tunnel between the two contacts and in this case the junction resistance increases exponentially with molecular length and is only weakly temperature dependent [24–27, 29, 64–69]. The tunneling mechanism is often "nonresonant" in that the tunneling electron energies are not precisely matched with the molecular orbital energies; however, the frontier molecular orbitals still assist the tunneling process (i.e., they lower the junction resistance) by lowering the effective tunneling barrier [41].

For longer wires at moderate temperatures, the rate of tunneling is strongly suppressed and instead charge can be injected into frontier orbitals of the wire molecules and transported by an incoherent hopping mechanism [68, 69]. In this case, the transport is thermally activated and the length dependence of resistance is predicted to be linear [68–72]. This is the regime of central interest for molecular wire studies. However, it is important to note that the charge transport mechanism in a junction depends on many factors including applied bias [28, 33]. Large biases may result in field emission, a tunneling mechanism that occurs when the tunneling barrier is strongly perturbed by the applied electric field [73–75]. In the following sections, the predictions of simple tunneling, field emission, and hopping models are reviewed in more detail as a forerunner to the discussion of recent results on DC electrical transport in molecular wires.

3.2.1
Tunneling (Direct Tunneling and Field Emission)

In coherent tunneling, electron waves traverse a potential barrier without any change in phase, and propagate again on the other side of the barrier. Figure 3.2

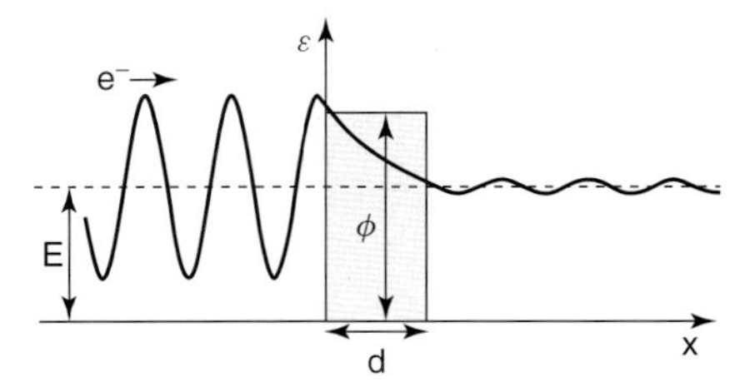

Figure 3.2 Transmission of electron wavefunction through a potential barrier. The probability density to the right of the barrier is decreased due to the attenuation of the wave through the barrier. A classical particle would have zero probability of penetrating the barrier; however, the quantum behavior of the electron permits transmission.

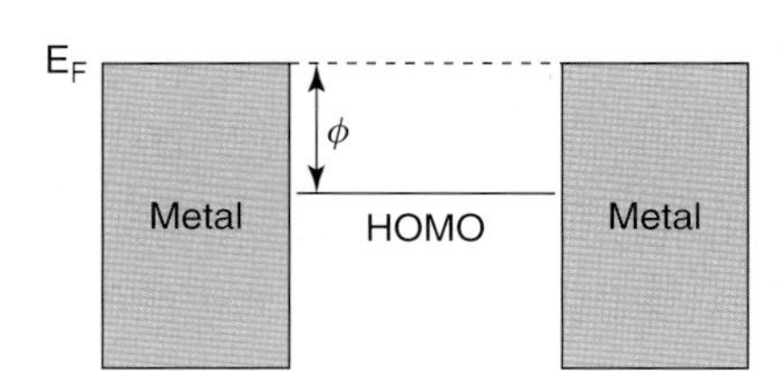

Figure 3.3 Simplified energy level diagram for a molecular electronic junction. For most junctions studied to this point, the junction Fermi level lies closer to the molecular HOMO, defining the barrier height, ϕ, as the energy offset between the HOMO and the Fermi level. In reality, the barrier height is not constant across the molecule, but varies with the energy of each site in the molecular wire.

illustrates electron transmission through a rectangular barrier. A rectangular barrier is clearly an extremely crude approximation for a molecular junction, but the simplicity of this model and its qualitative predictions have made it widely used for initial analysis of molecular conduction in the tunneling regime [65].

In a common approximation for molecular junctions, the tunneling barrier has a height equal to the offset between the Fermi level and the closest molecular energy level (see Figure 3.3), and a width equal to the length of the molecule. For systems where the Fermi level is far (>1 eV) from a molecular level (nonresonant tunneling), this approximation is perhaps reasonable. As the average barrier height decreases, the precise structure of electronic states in the junction becomes increasingly important, and the description of the junction as a single barrier clearly breaks down [76, 77].

According to the Wentzel–Kramers–Brillouin (WKB) approximation [64], the amount of wave attenuation shown in Figure 3.2 is affected by the height (ϕ) and length (d) of the barrier. In the 1960s, Simmons derived an equation for current passing through an arbitrary tunneling barrier using the WKB approximation, and it is reduced to the following expression in the very low-bias range [65]:

$$J = \frac{e^2 V}{4\pi^2 \hbar^2 d}(2m\phi)^{1/2} \exp\left[\frac{-2d}{\hbar}(2m\phi)^{1/2}\right] \tag{3.1}$$

where J is the current density, ϕ is the mean barrier height, e is the elementary charge, d is the barrier length, and V is the applied bias.

Equation (3.1) can be rewritten in terms of resistance (R) and simplified to a formulation useful for experiments, where resistance exhibits exponential dependence on the barrier length [24–27, 78]:

$$R = R_0 \exp(\beta d) \tag{3.2}$$

Here R_0 is the effective contact resistance, and $\beta = 2(2m\phi)^{1/2}/\hbar$ is a structure-dependent attenuation factor. (The factor β also appears in the theory of variable range hopping in disordered media, where it is interpreted as the inverse of the charge localization length [79, 80].) According to Eq. (3.2), the fit parameters R_0 and β are extracted from a semilog plot of resistance versus molecular length; β is the slope of the best-fit line to the data points and R_0 is the y-intercept [27].

Length-dependent measurements are thus extremely useful to determine the efficiency of electron tunneling through molecular bonds. More efficient tunneling is characterized by lower β-values. Conjugated molecules with delocalized π electrons often exhibit β-values in the range of 2.0–5.0 nm^{-1}, while saturated chains have β between 7.0 and 10.0 nm^{-1} [17, 25]. The contact resistance R_0 reflects the type of metal and the surface linker chemistry used to bind the molecules in the junction [27, 78, 81]. The importance of contact resistance in a molecular junction can be assessed directly by length-dependent measurements. Junction measurements on molecules of only one length leave open substantial questions regarding what fraction of the measured resistance (at a given bias) is due to contact effects versus the transport along the molecular backbone. The important role of length-dependent transport measurements will be emphasized again in Section 3.3 of this chapter.

Equation (3.2) works well for coherent nonresonant tunneling in the low-bias regime, and this equation has also been derived from a combination of the Landauer formula and the Green's function formulation [27], which provides a clear connection between the transmission probability and the conductance [58, 66, 68, 82–84]. The importance of the Green's function formulation stems in part from the ability to factorize the total transmission probability into the molecule–metal and the molecular backbone components [84]. This formalism also takes into account the mixing of molecular orbitals with the metal electron states at the contacts. The coupling of molecular levels to metal leads to broadening of the molecular states, an important effect that modifies the overall junction transmission. In junctions where tunneling is the dominant charge transport mechanism, current is predicted to have very weak dependence on temperature (Table 3.1) [68, 69].

Table 3.1 Charge transport mechanisms in a molecular junction.

Conduction mechanism	Characteristic behavior	Temperature dependence	Length dependence	Bias dependence
Nonresonant tunneling	$R \propto \exp\left(\frac{2d\sqrt{2m\phi}}{\hbar}\right)^{a}$	Weak	$\ln R \propto d$	$J \propto V$
Field emission	$J \propto V^2 \exp\left(-\frac{4d\sqrt{2m\phi^3}}{3q\hbar V}\right)$	Weak	$J \propto \exp(-d)$ At a given bias	$\ln\left(\frac{J}{V^2}\right) \propto \frac{1}{V}$
Hopping	$R \propto d \exp\left(\frac{\phi}{kT}\right)$	$\ln R \propto \frac{1}{T}$	$R \propto d$	$J \propto V$

[a]The characteristic behavior described here is for the low-bias regime. J is the current density.

Equation (3.1) describes current through a rectangular barrier in the low-bias limit, and can be simplified to

$$I \propto V \exp\left(-\frac{2d\sqrt{2m_e\phi}}{\hbar}\right) \tag{3.3}$$

which is a useful form, as will be demonstrated below. At the opposite limit, when the applied bias exceeds the barrier height, the barrier shape switches from rectangular to triangular, and the current–voltage dependence is described as follows:

$$\ln\left(\frac{I}{V^2}\right) \propto -\frac{4d\sqrt{2m_e\phi^3}}{3\hbar q V}\left(\frac{1}{V}\right) \tag{3.4}$$

Tunneling through a triangular barrier is termed "field emission" or Fowler–Nordheim tunneling [85]. In field emission, a plot of $\ln(I/V^2)$ against $1/V$ is expected to yield a line with a negative slope, the absolute value of which will depend on the barrier height based on Eq. (3.4).

A transition from direct tunneling to field emission is evident in a variety of molecular junctions [74, 75]. To examine experimentally the transition from direct tunneling to field emission requires recasting Eq. (3.3) in terms of the variables $\ln(I/V^2)$ and $1/V$ so it can be directly compared to Eq. (3.4). The resulting equation is as follows:

$$\ln\left(\frac{I}{V^2}\right) \propto \ln\left(\frac{1}{V}\right) - \frac{2d\sqrt{2m_e\phi}}{\hbar} \tag{3.5}$$

From Eq. (3.5), a plot of $\ln(I/V^2)$ against $1/V$ exhibits a logarithmic growth in the low-bias regime. Therefore, a transition from logarithmic growth to linear decay in a plot of $\ln(I/V^2)$ against $1/V$ is a signature of a transition from direct tunneling to field emission.

Figure 3.4 shows a Fowler–Nordheim plot for an Au–anthracenethiol–Au tunnel junction [74]. The inset displays the same data plotted on standard current–voltage axes. The dashed line in Figure 3.4 denotes the voltage required for transition

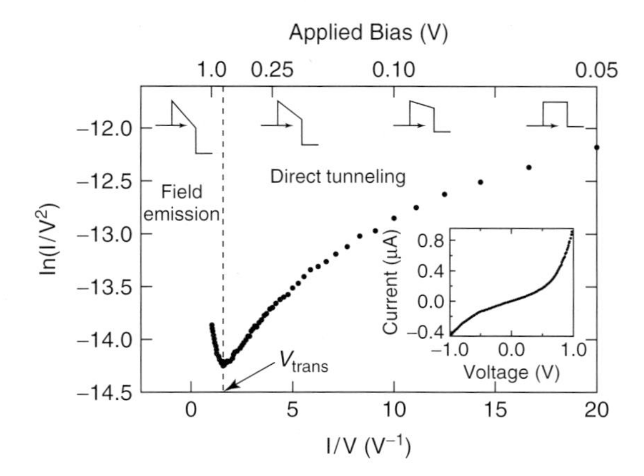

Figure 3.4 Solid circles represent the average of 100 $I-V$ curves for a Au–anthracenethiol–Au junction. The dashed line corresponds to the voltage at which the tunneling barrier transitions from trapezoidal to triangular (V_{trans}). Also shown are representations of the barrier shape at various values of applied bias. The inset shows current–voltage data on standard axes. (Adapted from Ref. [74] with permission; Copyright 2006 by the American Physical Society.)

from direct tunneling to field emission for anthracenethiol, which is referred as the transition voltage (V_{trans}). The shape of the curve in the two bias regions matches the shape predicted by Eqs. (3.4) and (3.5) (linear decrease at high bias and logarithmic growth at low bias). Furthermore, the specific value of V_{trans} provides a means of experimentally estimating the height of the original rectangular barrier of a given molecule. Note that V_{trans} remains an estimate, and not an exact measure of the barrier height because the original tunneling equation as outlined in Eq. (3.2) does not explicitly account for voltage drops at the contacts or the image potential of the tunneling electron. Nevertheless, the point of Figure 3.4 is that proper analysis of $I-V$ characteristics can reveal the onset of field emission in molecular junctions. This will be important in the analysis of molecular wire transport in Section 3.3.

3.2.2
Hopping

For conjugated wire molecules greater than a few nanometers in length, the coherent tunneling model discussed above breaks down, particularly near room temperature. Instead of direct metal-to-metal tunneling, the charge conduction process may be viewed as a series of discrete steps involving first injection of charge into the molecular orbitals, field-induced drift of the charge carrier down the length of the molecule, and finally extraction of the charge into the receiving contact. In this incoherent limit, the charge has a relatively long residence time on the molecule and consequently there is substantial vibronic coupling between molecular motions and the charge, that is, polarons are formed. One can roughly estimate the charge residence time with knowledge of the molecule–metal coupling energy. If this

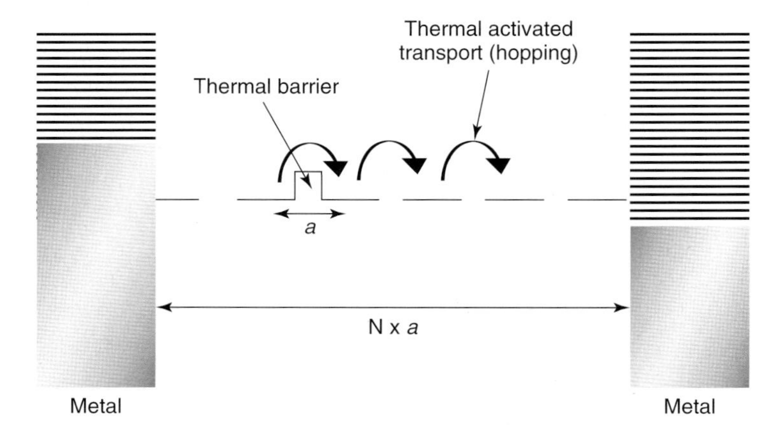

Figure 3.5 Schematic energy diagram for incoherent hopping through N sites separated by distance *a* on a molecular wire.

coupling is on the order of 0.1 eV [86], the *uncertainty principle* suggests that the residence time will be >40 fs. This time is comparable to the timescales of typical vibrations (10 fs to 1 ps) [1], meaning vibronic coupling and polaron formation will likely occur. A better way to estimate the carrier residence time is to use actual current density data for molecular junctions, as will be shown later.

The polaronic charge carrier moves along the molecular backbone by hopping between localized states (orbitals), Figure 3.5. This hopping process is thermally activated around room temperature. Indeed many researchers working on molecular wires implicitly define hopping conduction as thermally activated, although in the solid-state physics literature hopping is generically defined as transport between localized states regardless of whether the hop involves thermal activation or tunneling [70–72, 87]. In the hopping regime around room temperature, one expects that the junction resistance might follow classical Arrhenius behavior [46, 71],

$$R = R_0 \exp\left(\frac{E_a}{kT}\right) \tag{3.6}$$

where E_a is an activation energy and k is the Boltzmann constant. This Arrhenius-type thermal activation is readily explicable in the Marcus picture of electron transfer [46, 51, 88]. Thermal motion of nuclei within the molecular wire (e.g., a bond rotation) results in a favorable geometry that facilitates electronic coupling and migration of charge. In this picture, the activation energy (E_a) corresponds to the energy required to reach the transition state. The precise temperature dependence of transport will likely depend on the specific molecular wires under study (e.g., the density of states distribution) and certainly this currently remains an open question [79, 80, 89].

Hopping transport is not predicted to exhibit exponential length dependence characteristic of coherent tunneling, but instead scales linearly with the length in the low-bias regime (Table 3.1). The weaker length dependence facilitates the transport of charge over greater distances, and thus it is the hopping regime that might be considered most "wire-like." Linear length dependence is also a signature of ohmic conduction where the charge transport is field driven. However, hopping conduction at higher applied biases can deviate from ohmic behavior in principle; for example, it is interesting to consider the role of space charge limited conduction, which is commonly observed in semiconductors, and for which current normally scales as the applied bias squared [90, 91].

3.2.3
Recent Results on Molecular Wires Connected to Metals

In light of the foregoing discussion on transport mechanisms, it is clear that the predominance of any mechanism – direct tunneling, field emission, or hopping – will depend on numerous factors including the electronic structure of the molecular wires and their length, the applied bias, and temperature. The interplay of these factors has yet to be sorted out experimentally and theoretically. However, there have been a few reports on observations of mechanistic transitions in molecular wires connected to metals. Thermal effects on DC conduction through nitro-substituted oligo-phenylene-ethynylene (OPE) molecules have been reported in pioneering work by Selzer *et al.*, Figure 3.6 [30, 36, 37]. Red curves in Figure 3.6(b) shows Arrhenius plots of current versus temperature for a single molecule OPE junction in which there is a characteristic transition from temperature-independent behavior at low temperatures, where conduction is dominated by tunneling, to temperature-dependent hopping behavior at high temperatures. The activation energy in the hopping regime corresponds very well with theoretical calculations of barriers for rotations of the rings in the nitro-substituted OPE [30]. Thus, Selzer and colleagues argued that above the transition temperature of $\sim$100 K, the onset of torsional fluctuation of the phenyl rings leads to vibronic coupling that suppresses tunneling and facilitates a hopping process. They also suggested that local heating of the molecules, resulting from IV power dissipation in the junction, may play a role in the thermally activated hopping [37].

Further, they have investigated the effect of local environment by comparing the temperature-dependent conduction in two testbeds with the same OPE molecule – one based on an isolated OPE and the other on a self-assembled monolayer (SAM) between two gold nanowire segments ("in-wire" junctions, Figure 3.6(a)) [36]. Figure 3.6(b) shows total current of the isolated molecule junction (gray curves) compared to the normalized current-per-molecule for an in-wire junction (black curves) as a function of temperature (10–300 K). While currents in the in-wire junction display temperature independence over entire temperature range, thermal activated behavior is exhibited only above 100 K in the single-molecule junction. The failure to observe a transition to hopping in the in-wire junction was attributed to a higher barrier for the rotation of the phenylene rings, due to a restricted volume for torsional modes in a close-packed SAM matrix. The issue of intermolecular interactions in molecular junction measurements remains an important question.

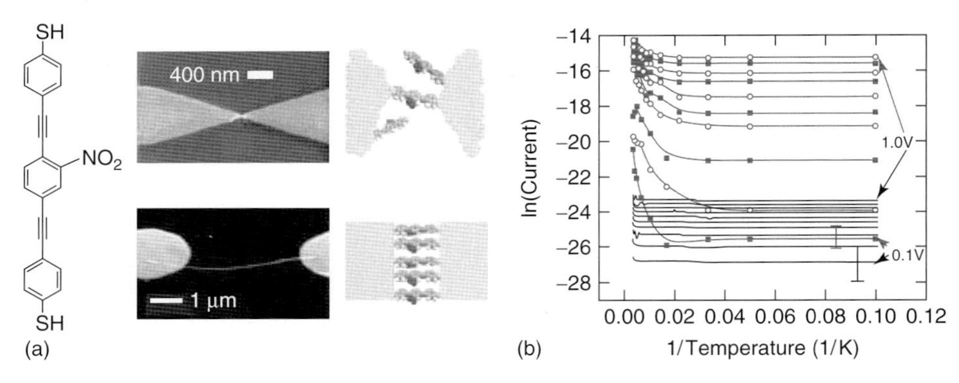

Figure 3.6 (a) Schematic of a nitro-substituted OPE (left), illustration of an individual molecular junction and its corresponding SEM image (center up), and of an in-wire molecular junction and SEM image of a junction aligned between two large-area electrodes (center low). (b) Comparison of current versus temperature behavior in in-wire SAM junctions and isolated molecular junctions as a function of bias. The bias increment between curves is 0.1 V, and the bias of the lowest curve is 0.1 V. (Adapted from Ref. [36] with permission; Copyright 2006 by the American Chemical Society.)

Tao and colleagues [32] have observed thermally activated hopping in redox active perylene tetracarboxylic dimides (PTCDIs) molecules connected between an STM tip and a gold substrate, Figure 3.7. In these experiments, charge was electrochemically induced on the PTCDI molecules and conduction through the charged molecules was thermally activated, consistent with a hopping mechanism (Figure 3.7(b)). However, when the same molecules were probed without inducing charge, the temperature dependence of conduction was much weaker, indicative of a tunneling mechanism (Figure 3.7(c)).

Very recently, Rampi and colleagues have reported the preparation and characterization of extremely long molecular wires, up to 40 nm in length. These wires were synthesized using metal ion coordination chemistry (Figure 3.8(a)) [92]. Although the authors have not yet reported the temperature dependence of conduction, a multistep charge hopping process between the metal centers in the backbone is implicated by the linear length dependence they observed for the conduction (Figures 3.8(b) and (c)).

Even longer conjugated molecular wires with lengths greater than 100 nm were recently prepared by *in situ* polymerization of dibromoterfluorene monomers on an Au (111) surface [94]. The conductance of a single polyfluorene chain was measured as a function of distance between an STM tip attached to one end of the molecule and the Au surface (Figures 3.9(a) and (b)). However, the conductance at small bias voltages was only measurable over an adjustable distance of 0–4 nm because the current levels for longer lengths of this wire molecule were below the apparatus detection limit. Over the short range of lengths that was measured, there was an exponential attenuation of current with length, indicative of tunneling (Figure 3.9(c)).

In the following section, we describe the preparation and characterization of a different conjugated molecular wire system that clearly exhibits a mechanistic

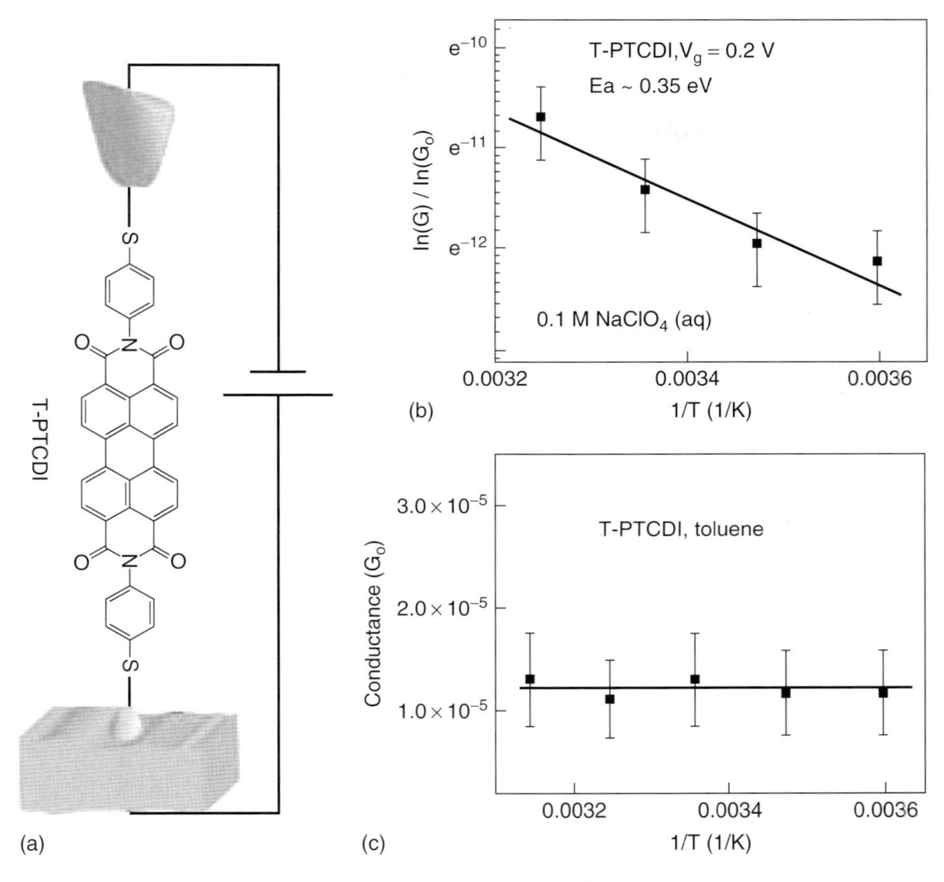

Figure 3.7 (a) Schematics of T-PTCDI in an STM break junction. Solvent (not shown) surrounds the molecule. (b) Arrhenius plot of conductance versus inverse temperature for T-PTCDI that has been electrochemically reduced in electrolytes. (c) A semilog plot of conductance versus inverse temperature for a T-PTCDI molecule in nonpolar solvent. (Adapted from Ref. [32] with permission; Copyright 2007 by the American Chemical Society.)

transition from tunneling to hopping as a function of wire length [33]. The synthetic flexibility of these wires opens considerable opportunities for systematic investigation of molecular wire conduction as a function of controlled molecular architecture.

3.3 Oligophenylene Imine Molecular Wires: A Flexible System for Examining the Physical Organic Chemistry of Hopping Conduction in Molecules

A versatile technique for the controlled fabrication of π-conjugated phenylene imine (azomethine) oligomers bound to a substrate was recently reported by

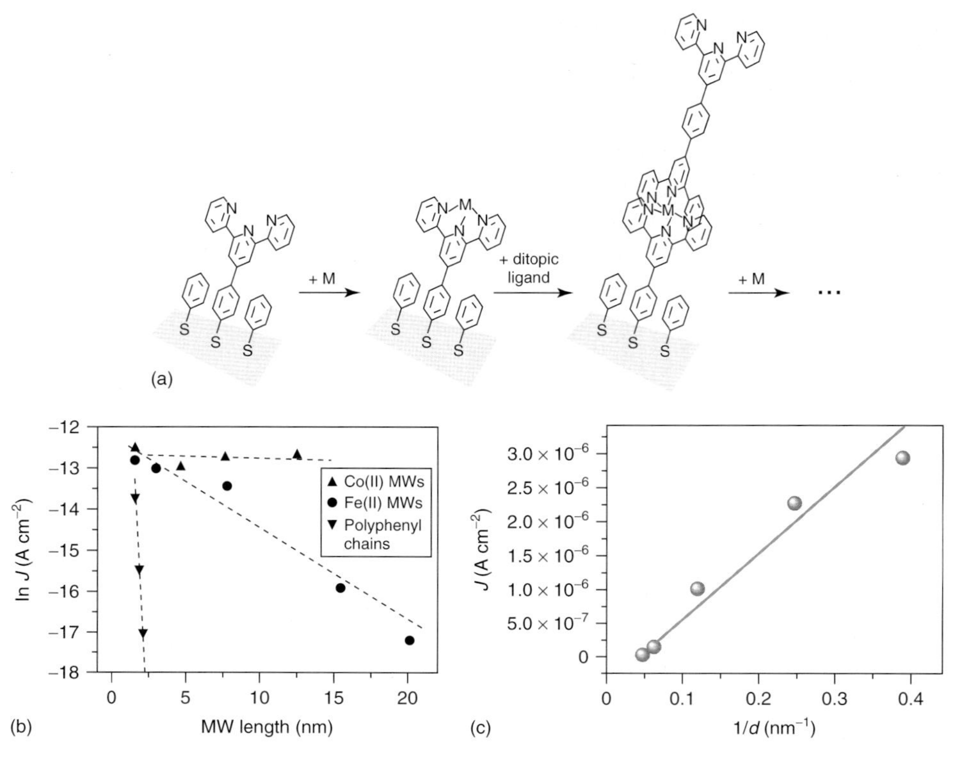

Figure 3.8 (a) Schematic representation of the stepwise growth of metal (M) coordinated molecular wires on a metal surface. (b) A semilog plot of current versus molecular length for the Fe(II) and Co(II)-based molecular wires. Filled down triangles: Data reported from previously published work for junctions incorporating polyphenyl-based molecular wires [93]. (c) A linear plot of current versus inverse molecular length for the Fe(II)-based molecular wires. (Adapted from Ref. [92] with permission; Copyright 2009 by the Nature Publishing Group.)

the current authors who followed a procedure originally described by Rosink *et al.* (Figure 3.10(a)) [33, 95]. The growth procedure begins by adsorption of a monolayer of 4-aminothiophenol on gold. Benzenedicarboxaldehyde is then added in solution, allowing formation of a second layer by reaction of one aldehyde group with the amine bound to the surface. The resulting imine extends conjugation over both the individual layers. By reaction with phenylenediamine, a third layer can be formed that further extends the degree of conjugation. Thus, by a sequence of alternating addition of dialdehyde and diamine, the length of the oligomer is continually extended. In this stepwise fashion, long conjugated oligophenyleneimine (OPI) molecules are readily built, controlling the orientation. The surface density of the wire molecules remains essentially constant with increasing length as long as the reaction nearly completes at each step. Each OPI-p wire terminated with $-NH_2$ or $-CHO$ groups can be end-capped with benzaldehyde or aniline, respectively [33].

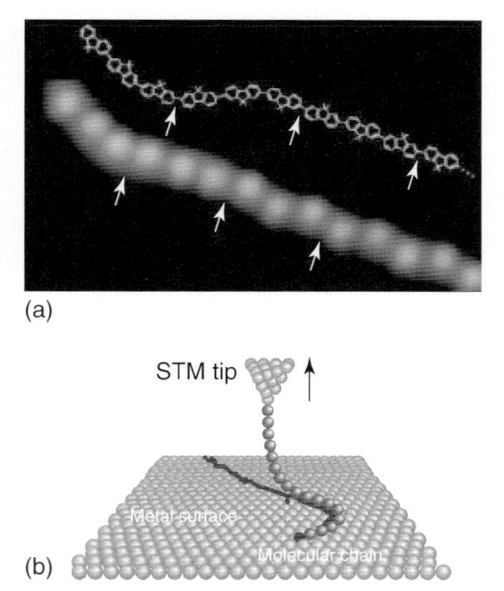

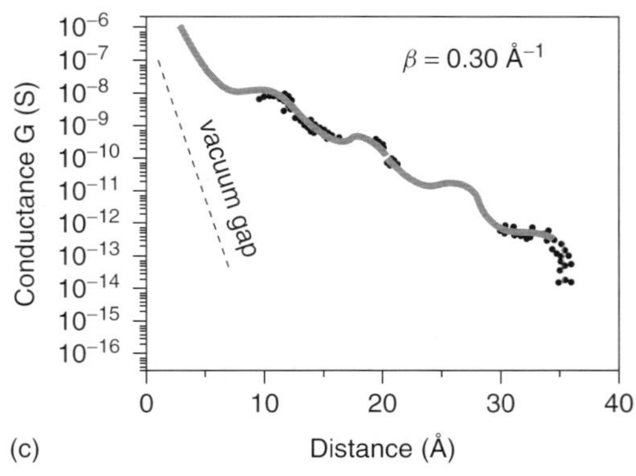

Figure 3.9 (a) STM image (5.9 by 3.6 nm) of a single polyfluorene chain end with its chemical structure superimposed (using a different scaling). (b) Scheme of the chain pulling procedure: after contacting a molecular chain to the STM tip, the chain can be lifted from the surface in a ropelike manner upon tip retraction because of its flexibility and weak interaction with the substrate. (c) A semilog plot of conductance versus distance between the tip attached to the chain and the metal substrate, which exhibits an exponential decay with a β value of 3.0 nm^{-1}. The characteristic oscillations with a period of about 10 Å reflect mechanical deformation of the molecular wire, as it slides over the metal surface during the pulling process. (Adapted from Ref. [94] with permission; Copyright 2009 by the AAAS.)

The end-capping provides a consistent terminal group throughout the OPI series that facilitates reproducible electrical characterization.

The growth of OPI wires is conveniently monitored using reflection–absorption Fourier transform infrared spectroscopy (RAIRS). RAIRS data, shown in Figure 3.10(b), reveal the alternate appearance and disappearance of carbonyl stretches (1710 cm^{-1}) and symmetric amine stretches (3350 cm^{-1}) in the OPI-p wires, which verifies the imination mechanism and indicates near quantitative reaction of all exposed reactive endgroups. The intensities of imine stretching (1620 cm^{-1}) and the benzene ring vibrational mode (1500 cm^{-1}) increase with the number of repeat units, as expected. Complete end-capping to form finished OPI wires is confirmed by the disappearance of the terminal group vibrational modes (not shown) [33].

The main advantage of this stepwise imine chemistry lies in its compatibility with a variety of molecular building blocks. Wires may be built with precisely controlled chain architecture, and at the end of this chapter we suggest some different building blocks designed to examine the role of steric interactions and redox properties on wire transport. Recently, we have taken the next step to demonstrate the versatility of the method by building wires from larger, more conjugated building blocks. Specifically, alternate addition of naphthalene and

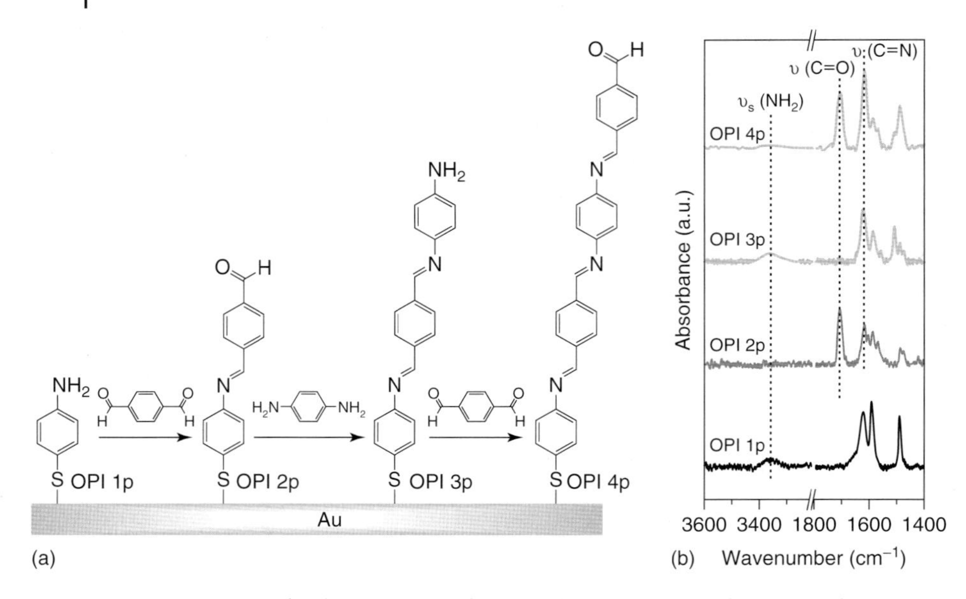

Figure 3.10 (a) Molecular structure and synthetic route to OPI-p monolayers on gold substrates (shown up to fourth layers). (b) The corresponding RAIRS spectra. Vertical dashed lines indicate positions of symmetric amine stretches (NH_2, 3350 cm^{-1}), carbonyl stretches (C=O, 1710 cm^{-1}), and imine stretches (C=N, 1620 cm^{-1}). Peaks for NH_2 and C=O appear in the uncapped OPI-p wires with odd and even numbers of repeat units, respectively. Subsequent capping of OPI-p molecules with either aniline or benzaldehyde produce the finished OPI wires that were subsequently measured. (Modified from Ref. [33] with permission; Copyright 2008 by the AAAS.)

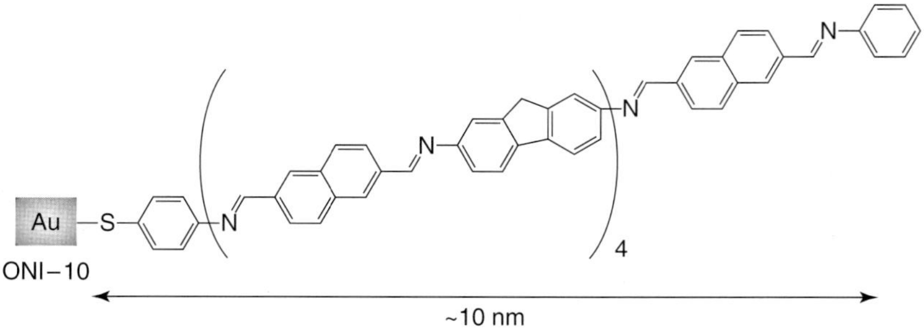

Figure 3.11 Molecular structure of ONI-10, the longest ONI wire that was prepared by stepwise additions of naphthalene-2,6-dicarboxaldehyde and fluorene-2,7-diamine.

fluorene blocks has been employed to build oligonaphthalenefluoreneimine (ONI) wires up to 10 nm in length (Figure 3.11). Preliminary results confirm that the addition chemistry also works cleanly for growth of these ONI wires and this work will be reported elsewhere.

OPI wires ranging in length from 1.5 to 7.3 nm have been reported and their transport properties were characterized in detail as a function of molecular length, temperature, and applied bias [33]. Here we cover the experimental details of the characterization of OPI wires, including the conducting probe atomic force microscopy (CP-AFM) approach employed for electrically contacting them. We describe the $I-V$ behavior in detail, focusing, in particular, on transitions in the conduction mechanism from tunneling to hopping to field emission.

3.3.1
Forming Junctions using a Soft, Spring-Loaded Contact

A variety of strategies have been devised for forming nanoscale molecular junctions, in which molecules are delicately engaged between two metal contacts [18–22]. Each method has inherent advantages and disadvantages, generally relating to issues of yield, reproducibility, ease of formation, and the number of molecules incorporated. Over the last 10 years, we and others have demonstrated that CP-AFM is a particularly convenient and reproducible approach [24–27, 74, 78, 81, 96–99].

In CP-AFM, a metal-coated AFM tip is brought into contact with a SAM of molecular wires under controlled load (see Figure 3.12(a)) [100, 101]. The metal-coated tips are prepared by vapor deposition of a 5-nm chromium adhesion layer followed by a 100-nm metal layer. The probe position is adjusted through an optical feedback system using a laser that reflects off the back surface of the AFM tip and in this way the load on the junction can be precisely controlled; the load is typically several nano-Newton and the tip–SAM contact areas are on the order of $10\,nm^2$ for a tip with a 50-nm radius of curvature. $I-V$ characteristics of the molecular wire monolayer are measured by sweeping the tip voltage.

One can view the force microscope operated in this manner as essentially a molecular-probe station analogous to device-probe stations used in semiconductor

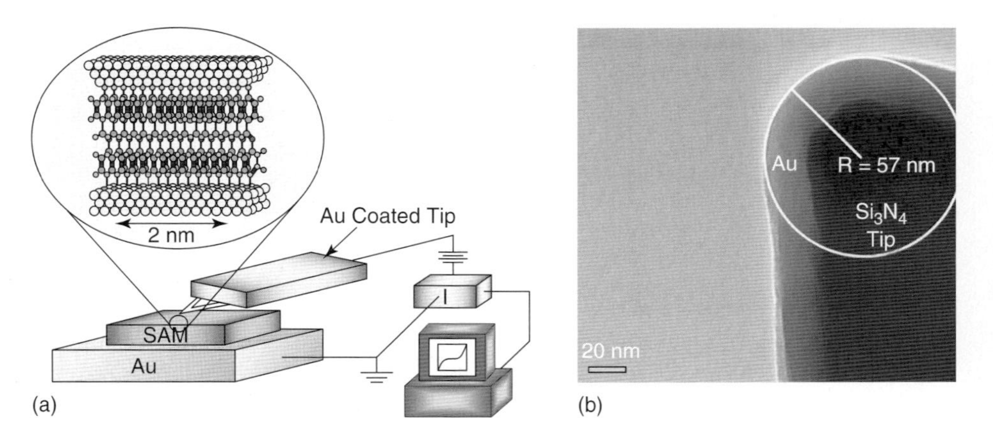

Figure 3.12 (a) Measurement of molecular wire resistance using conducting probe atomic force microscopy (CP-AFM). A gold-coated tip is brought into contact with a monolayer on a gold substrate. (b) Transmission electron microscopy (TEM) image of a gold-coated AFM tip.

electronics. This CP-AFM or molecular-probe station approach has several attractive features. First, it is inherently a soft contact, nondestructive method. By virtue of the flexible cantilever attached to the probe tip, the top contact is spring loaded and the force is easily adjusted so that permanent (plastic) deformation of the SAM layer is avoided. Second, the CP-AFM method is simple and fast. No micro- or nanofabrication processes are necessary, meaning that junctions can be made and tested in the time span of a few hours or less. Third, the dimensions of the junctions are inherently small because of the small radius of curvature of the AFM tip (see Figure 3.12(b)). This reduces the probability of electrical shorts and means that conduction can be probed through a small number of molecules ($\sim$100) [27]. Fourth, CP-AFM offers significant experimental flexibility in that the tip and substrate can be coated with different metals, which allows the role of metal contacts on junction $I-V$ behavior to be assessed [27]. Finally, the technique is also compatible with variable temperature measurements, which is important for verifying transport mechanisms [33]. These collective attributes make the CP-AFM testbed a very powerful and versatile method for molecular-wire experiments.

3.3.2 Electrical Transport in OPI Wires

3.3.2.1 Length Dependence of Resistance

Figure 3.13(a) shows a semilog plot of resistance (R) versus molecular length (L) for OPI wires, where each data point represents the average of 10 measurements. The resistance was determined using Au-coated CP-AFM tips in contact with the OPI wires grown on Au. A current–voltage sweep generally yields a sigmoidally shaped $I-V$ curve as will be shown later, but over a small voltage range (± 0.3 V) the response is linear. The resistances shown in Figure 3.13 are low voltage resistances determined over this small range. Significantly, a clear transition of the length dependence of low voltage resistance is observed near 4 nm (OPI 5), indicating that the conduction mechanism is different in short (OPI 1–5) and long OPI wires (OPI 6–10). In short wires, the linear fit in Figure 3.13(a) indicates that the data are well described by Eq. (3.2) for nonresonant tunneling. The β-value is found to be 3 nm^{-1}, which is within the range of β-values of typical conjugated molecules [17, 25, 49]. For long OPI wires, a much flatter resistance versus molecular length relation ($\beta \sim 0.9\ \mathrm{nm}^{-1}$) is exhibited. The extremely small β suggests that the principal transport mechanism is hopping [2, 102–104]. In addition, a plot of R versus L for long wires is linear (see Figure 3.13(b)), which is expected for hopping transport (Table 3.1) and indicates that Eq. (3.2) does not apply for long wires.

3.3.2.2 Temperature Dependence of Resistance

Although a change in transport mechanism is apparent in the R versus L plot, the temperature dependence is the key to verify different transport mechanisms. In the hopping regime, transport is activated. Of course, it is natural to ask what specific nuclear motions control the hopping rate.

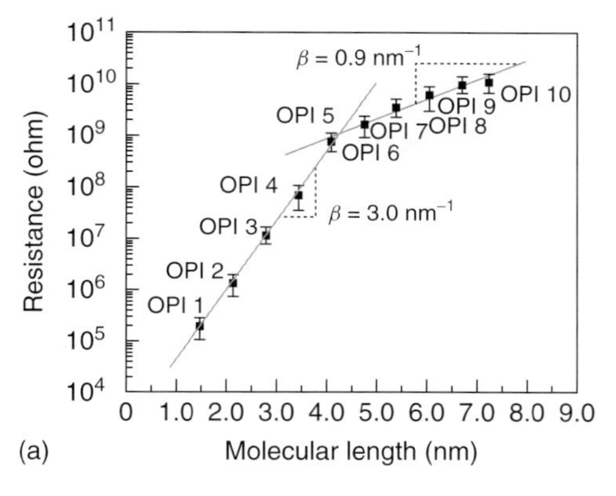

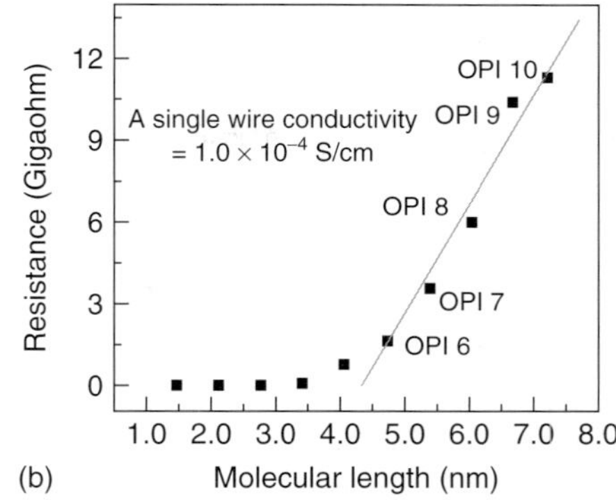

Figure 3.13 (a) Semilog plot of R versus L for the gold/OPI wire/gold junctions. Each data point is the average differential resistance obtained from 10 I–V traces in the range −0.3 to +0.3 V. Error bars represent one standard deviation. The straight line is a linear fit to the data according to Eq. (3.2). (b) A linear plot of R versus L for the gold/wire/gold junctions. Straight lines are linear fits to the data. (Adapted from Ref. [33] with permission; Copyright 2008 by the AAAS.)

Temperature-dependent CP-AFM measurements on OPI-4, 5, 6, and 10 were performed in an environmentally controlled (low humidity) AFM equipped with a variable temperature stage. Figure 3.14 shows an Arrhenius plot of resistance versus $1/T$ for each of the wires. Clearly, the resistance for OPI-4 and 5 is independent of temperature from 246 to 333 K, as expected for tunneling. However, both OPI-6 and 10 display strongly thermally activated transport characteristic of hopping. The activation energies determined from the slopes of the data are identical at 0.28 eV (6.5 kcal/mol) for both OPI 6 and 10, which implies that the same molecular motion contributes to the intramolecular hopping process in both wires. Contact effects are not responsible for the activated transport, as straightforward calculation shows that the injection efficiency is >99% [105]. Collectively, the data in Figures 3.13 and 3.14 provide unambiguous evidence for a mechanistic transition from tunneling to hopping near 4 nm in OPI wire length.

A key question concerns the nature of the hopping sites in the long wires and the origin of the 0.28-eV activation energy. An important point in this regard is that the ultraviolet (UV)-visible absorption data on OPI wire SAMs show that the conjugation does not extend over the entire wire. The optical gap (E_g) reduces with molecular length up to OPI-3, and then remains constant at 2.6 eV for longer OPI wires [33]. This result indicates that the π-conjugation extends over three repeating units via the imine linkage and that longer wires contain weakly linked conjugated subunits. From prior work, it is known that electronic delocalization is limited in aromatic oligoimines because of the nonzero dihedral angle between the benzene ring and the imine bonds, that is, the wire molecules are not flat and the π-conjugation is broken [106]. Indeed, the activation energy for twisting of the phenyl rings has been measured to be 0.3 eV [107, 108], in remarkable agreement with the activation energy measured in the junction transport experiments.

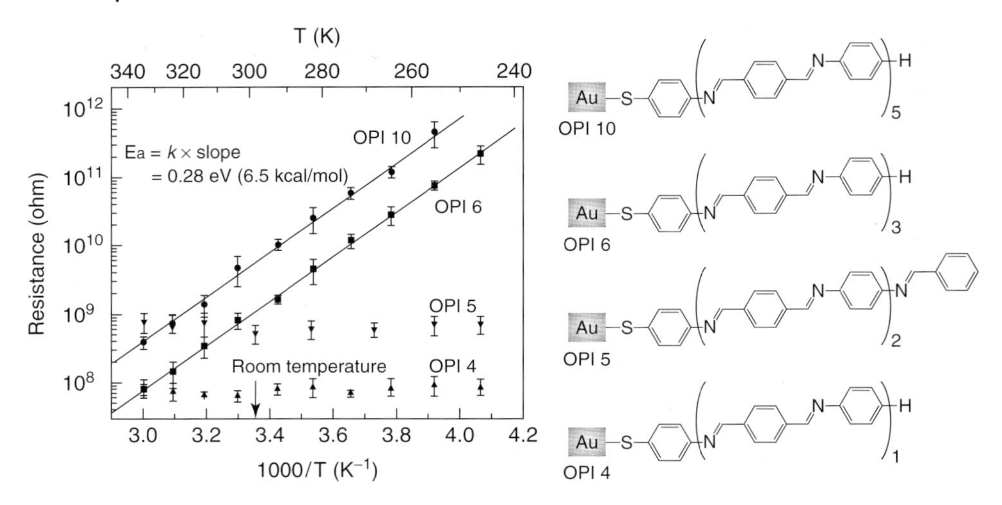

Figure 3.14 Arrhenius plots for resistance versus temperature data for OPI-4, 5, 6, and 10. Each data point is the average differential resistance obtained at six different locations on the samples in the range −0.2 to +0.2 V. Error bars represent one standard deviation. Straight lines are linear fits to the data according to Eq. (3.6). (Adapted from Ref. [33] with permission; Copyright 2008 by the AAAS.)

A cartoon illustrating the proposed charge hopping process in OPI wires is shown in Figure 3.15. The picture emerges that charge (likely holes) injected from the positive contact into the long OPI wires is localized to a subunit consisting of approximately three repeat units. Field-induced drift of the charge down the backbone requires that the dihedral angle between conjugated subunits decreases to near 0°, a process that transiently extends the conjugation along the molecule and allows the charge to hop. This nuclear motion is thermally activated and concerted, involving rotation of the phenyl rings coupled with alteration of the C=N bond length [108]. This physical model of conduction gated by torsional motion of conjugated subunits is very much consistent with prior studies of electron transfer in D–B–A compounds [48], and it underscores the hugely important role of nuclear motion in charge conduction in molecular wires.

One can further check the consistency of this model by calculating the average residence time for charge on OPI wire and comparing that time to the timescale associated with the torsional fluctuation. At +0.2 V bias, the current through OPI-6 is 2×10^{-10} A. For this calculation, we take the number of molecules in the junction to be ~100 [27] so that the current per molecule is 2×10^{-12} A, which is approximately 10^7 electrons/second per molecule. This implies that average charge residence times are on the order of 0.1 μs, far slower than the period of the torsional mode (~50 fs) [109], as expected.

3.3.2.3 Electric Field Dependence of Charge Transport

To have a complete understanding of transport in a molecular wire, it is also necessary to characterize the voltage and electric field dependence of the I–V

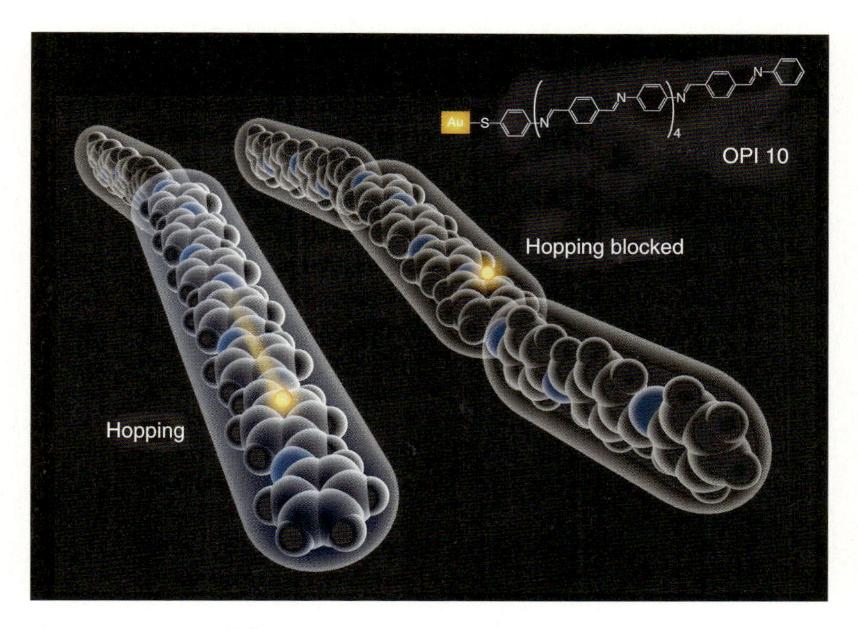

Figure 3.15 Proposed illustration for OPI-10. A charge (yellow ball) is localized at the conjugation subunit where three repeats are strongly coupled (right). Once molecular motions couples transiently the conjugation subunits, a charge can be driven down the wire (left).

characteristics [28]. Applying a bias across a junction may perturb the electronic structure of the wire as the electric fields may be very large, of order 10^5-10^7 V/cm; certainly there can be field-driven changes in conduction mechanisms. In the case of OPI wires, the situation is admittedly somewhat complex, as will be seen, though the field, temperature, and length dependence are all self-consistent.

Figure 3.16 displays a compilation of $I-V$ data for OPI wires plotted on different axes. Figures 3.16(a) and (b) show the $I-V$ curves of OPI-4 and 10, representative of short and long OPI wires. Both curves are sigmoidally shaped, but there is also clearly a qualitative difference in the transition from low to high voltages; the OPI-10 wire appears to show a more abrupt increase in current beyond 1.0 V. Figures 3.16(c) and (d) show semilog plots of current versus voltage and electric field, respectively, for all 10 OPI wires. For short wires (OPI 1–5) in Figure 3.16(c), there are large decreases in current with length, whereas these changes become much smaller for the longer OPI wires (6–10). The electric field dependence in Figure 3.16(d) reveals that for the long OPI wires the curves nearly collapse on top of one another. This is expected for hopping transport because it is inherently field driven; the tunneling process in short wires, on the other hand, is voltage driven, and the log I versus electric field plots for these wires remain distinctly separated.

Surprisingly, the Fowler–Nordheim plots for OPI-4 and 10 in Figures 3.16(e) and (f) reveal that there is a striking change in conduction behavior with applied bias. This change in behavior is not readily apparent in either the log $I-V$ or the

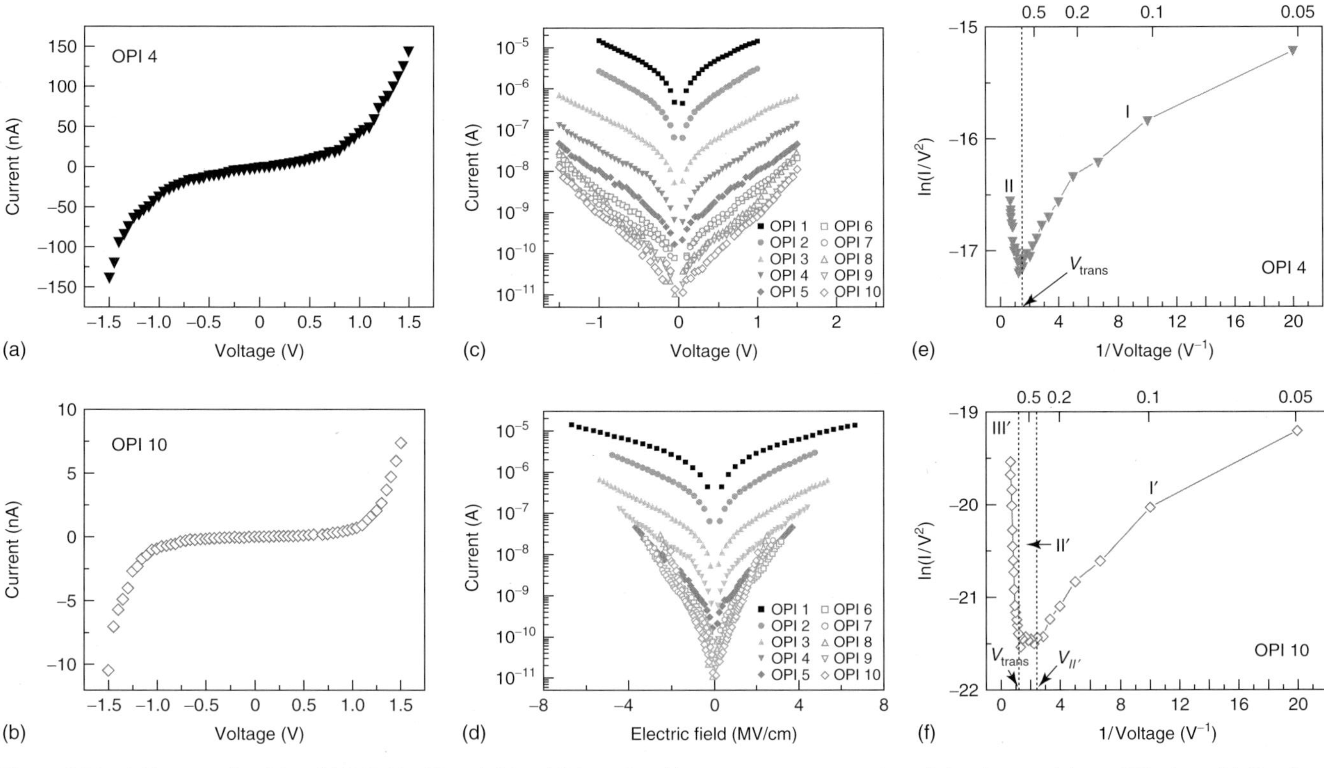

Figure 3.16 *I–V* curves for (a) gold/OPI-4/gold and (b) gold/OPI-10/gold junctions, representative of the short and long OPI wires. (c) Semilog plot of *I* versus voltage (*V*) for OPI 1–10. (d) Semilog plot of *I* versus electric field (*E*) for OPI 1–10. (e) Fowler–Nordheim plot for the OPI-4. Two distinct regimes (I and II) are clearly observable with a minimum at V_{trans}, indicating the switch from tunneling to field emission. (f) Fowler–Nordheim plot for the OPI-10. Three distinct regimes (I′, II′, and III′) are evident. (Adapted from Ref. [33] with permission; Copyright 2008 by the AAAS.)

Table 3.2 The analysis of I–V characteristics for selected OPI wires.

Monolayer	n in I, I′ $I \propto V^n$	n in II′ $I \propto V^n$	V_{trans} (V) (E_{trans} (MV/cm))	$V_{II'}$ (V) ($E_{II'}$ (MV/cm))	ϕ_{FE}(eV)
OPI-1	1.1	–	0.95 [6.3]	–	–
OPI-2	1.1	–	0.85 [4.0]	–	–
OPI-4	1.3	–	0.75 [2.2]	–	–
OPI-6	1.2	2.6	1.00 [2.1]	0.5 [1.0]	0.3–0.5
OPI-8	0.9	2.3	0.90 [1.5]	0.40 [0.7]	0.3–0.6
OPI-10	0.9	2.6	0.95 [1.3]	0.40 [0.6]	0.3–0.5

Modified from Ref. [33] with permission; Copyright 2008 by the AAAS.

I–V characteristics. Considering first OPI-4 (Figure 3.16(e)), the current scales logarithmically with $1/V$ for low voltages (regime I) as expected from Eq. (3.5), indicative of direct tunneling. Above the transition voltage V_{trans} (regime II), the current scales linearly with $1/V$ with a negative slope. This is a clear evidence for the onset of field emission (Eq. (3.4)). As noted earlier, the transition point, V_{trans} is an estimate of the effective barrier height at $V = 0$, and the V_{trans} values for OPI wires are listed in Table 3.2. The decrease in V_{trans} with length in short wires indicates that the estimated barrier height decreases as expected due to the increase in conjugation length.

The Fowler–Nordheim plot for OPI-10 in Figure 3.16(f) is similar but reveals *three* distinct transport regimes (I′, II′, and III′). In the low-bias regime I′, the current scales logarithmically with $1/V$, also characteristic of ohmic hopping conduction and consistent with the length- and temperature-dependence data. The negative slope in the high-voltage regime III′ suggests that field emission also occurs in OPI-10 (similar results were obtained for OPI-6–9). From the slope in regime III′, the emission barrier height (ϕ_{FE}) is estimated to be in the range of 0.3–0.5 eV assuming carrier effective mass ratios in the range 0.1–1.0, which are typical for molecular junctions (see Table 3.2). The intermediate transport regime II′ may correspond to a space charge limited transport regime, though this point will have to be clarified with further experiments.

Interpretation of I–V data for OPI-4 and 10 can be further aided by yet one more type of analysis, in this case log I–log V plots shown in Figure 3.17. These plots also clearly reveal the same two regimes of transport for OPI-4 and the same three regimes for OPI-10. Direct comparison can be made between the Fowler–Nordheim plots in Figure 3.16 and the log–log plots in Figure 3.17. The two types of figures are entirely consistent showing clear transitions in I–V behavior at the same voltages. However, it is worthwhile noting that analyzing the data in these different formats is crucial to elucidating mechanisms. For example, it would have been difficult to identify field emission as a key mechanism from Figure 3.17 alone; the Fowler–Nordheim plots were essential to that.

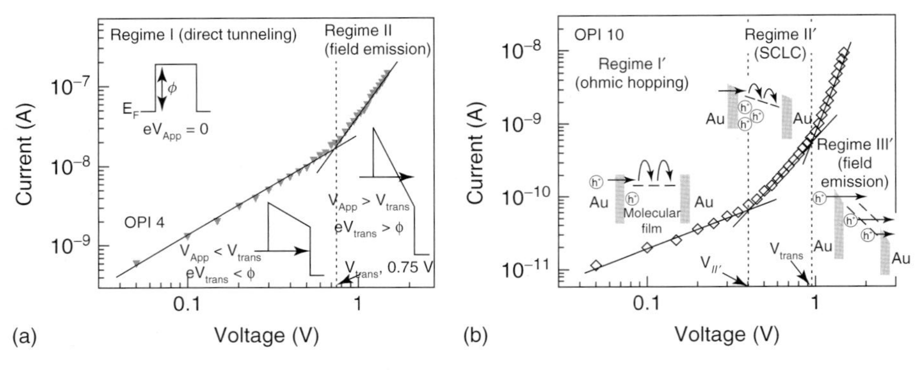

Figure 3.17 (a) Log–log plot of the average of 10 I–V traces for the gold/OPI-4/gold. Insets depict the changing tunneling barriers with the applied bias. At zero bias, the tunneling barrier height (ϕ) is set by the Fermi level-molecular orbital energy offset. For biases less than the barrier height, the shape of the barrier is trapezoidal. At an applied bias greater than V_{trans}, the barrier shape becomes triangular, and the transport mechanism changes to field emission. Note that theses cartoons are drawn for the case of electron transport mediated by an unoccupied orbital; hole transport mediated by an occupied orbital is equivalent. (b) Log–log plot of the average of 10 I–V traces for the gold/OPI-10/gold. Three different regimes are exhibited. The insets represent the possible conduction behavior in each regime. Note that theses cartoons are drawn for the case of hole carriers present in the wires (h^+ represents hole carriers). (Adapted from Ref. [33] with permission; Copyright 2008 by the AAAS.)

For short wires such as OPI-4, the change in mechanism from tunneling to field emission has already been recognized in a variety of molecular systems. In simple terms, it arises from a change of the tunneling barrier shape (see inset illustrations in Figure 3.17(a)) [74, 75]. Linear I–V behavior is expected for tunneling in the low-bias regime where to first approximation the metal–wire–metal junction can be modeled as a simple trapezoidal tunneling barrier (Eq. (3.1)). At high bias (regime II), the electric field changes the shape of the tunneling barrier from trapezoidal to triangular. In this case, the I–V behavior can be described by the Fowler–Nordheim relation, Eq. (3.4).

For long OPI wires, not all aspects of the bias dependence are clear. The linear dependence on V in the low-voltage regime (I′) is consistent with ohmic hopping. The principal issue in this regime is the steady-state concentration of carriers (likely holes). Are carriers present in the wires at equilibrium (no bias) when the junction is formed? Such carriers would likely arise by equilibration/charge transfer with the contacts [110], but understanding the carrier generation/injection mechanism in molecular wires in the hopping regime is an important unresolved issue. As the applied bias increases and reaches $V_{\mathrm{II}'}$, it is possible that the contacts inject additional charges into the wires such that the mechanism switches to the so-called space charge limited current (SCLC) regime (II′); in this regime, the dependence of current on voltage would be expected to be superlinear [90, 91, 111]. At high bias (regime III′), the transport mechanism is pretty clearly dominated by field emission. Observation of weak temperature dependence of conduction in this regime could confirm this.

3.3.3 Summary of OPI Wire Experiments to Date

To summarize, DC electrical conduction has been measured in a set of conjugated OPI wires having systematically controlled lengths between 1.5 and 7.3 nm. At low-bias voltages, the predicted transition from tunneling to hopping transport is exhibited at a molecular length of 4 nm, as supported by striking and mutually consistent changes of the length, temperature, and electric field dependence of the transport characteristics. At high biases, field emission occurs in both short and long wires. Preliminary results indicate that a set of ONI wires (Figure 3.11) having systematically controlled lengths between 2.1 and 10.3 nm also exhibit similar transport characteristics. The reproducibility of the measurements and the flexibility of the imine wire synthesis chemistry together create significant opportunities to improve fundamental understanding of electrical transport in conjugated molecules. We conclude by considering some future research directions based on both the imine wire architecture [33] and the recently reported metal complex approach [92].

3.4 Outlook: Probing the Physical Organic Chemistry of Hopping Conduction

Recent successes summarized in this chapter in building long molecular wires and measuring their DC conduction properties suggest that a wide variety of experiments can be undertaken to probe the influence of molecular structure on hopping transport. There is now the very real and exciting possibility that we can understand electrical transport in molecular systems with unprecedented detail. One might describe this new research field as the physical organic chemistry of hopping conduction, in which classical methods of systematic structure variation are linked to measurable changes in rates of hopping transport. As explained in the introduction, the importance of exploring the relationship between molecular structure and hopping transport lies in future opportunities to exploit molecules in nanoelectronics and in aiding efforts to optimize conjugated polymer materials for the applications in hybrid, flexible, or printed electronics. In this final part of this chapter, we consider several new wire designs aimed at understanding the roles of (i) intramolecular steric interactions, (ii) electronic structures, and (iii) redox properties on hopping conduction.

Figure 3.18 shows candidate molecular wire structures based on oligoimines and metal complexes. Our prior measurements on OPI wires indicate that the interruption of conjugation in molecular wires is an important factor in hopping rates. Indeed, it is known that polyarylazomethines generally exhibit a twisted conformation mostly due to steric hinderance between hydrogen atoms on adjacent

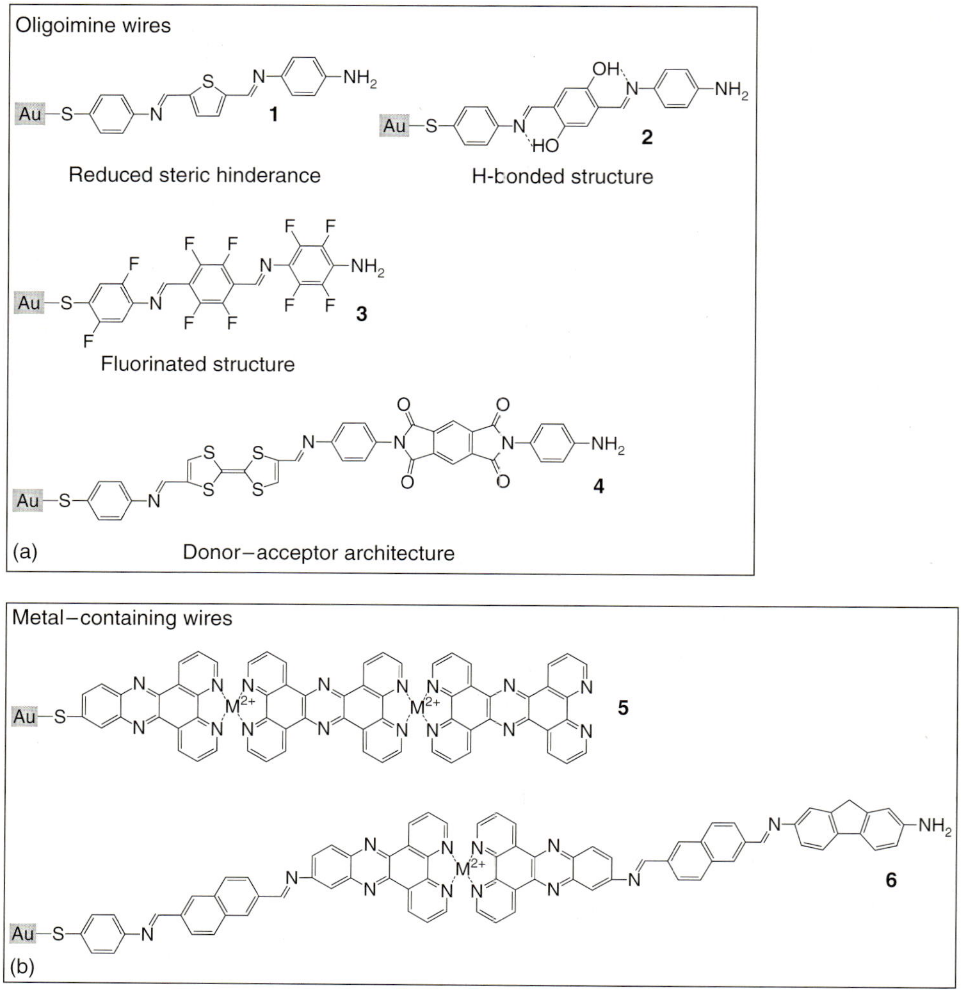

Oligoimine wires

Reduced steric hinderance

H-bonded structure

Fluorinated structure

(a) Donor–acceptor architecture

Metal–containing wires

(b)

Figure 3.18 Alternative molecular wire structures. (a) Oligoimine systems: wire 1 is designed to release steric strain in the wire. Wire 2 is designed to planarize the dihedral angle with hydrogen bonds. In wire 3, the fluorine substituents will lower the LUMO level perhaps allowing electron (as opposed to hole) conduction. Wire 4 is a donor–acceptor architecture that may lead to a low band gap wire with higher conductivity. (b) Metal-containing systems: wire 5 can be prepared by metal coordination chemistry and may be highly conductive by virtue of the multiple metal centers. Wire 6 contains a single redox center that might lead to charge correlation effects (Coulomb blockades) and interesting *I*–*V* behavior.

aromatic rings [112, 113]. UV-vis data of OPI wires previously indicated that electronic delocalization extends only over three repeating units because of the nonplanar conformation [33]. Replacing a phenylene with the smaller thiophene ring, as in wire structure **1**, will release the strain between adjacent rings allowing a more coplanar geometry and increasing the overall conjugation length. In fact, it has already been reported that incorporating thiophene rings significantly reduced the dihedral angle around the imine linkage in conjugated azomethine polymers [114, 115]. Several repeat units of the molecule 1 structure could be connected to create long wires.

Another strategy for manipulating the dihedral angle of the imine linkage is to use intramolecular hydrogen bonding between a hydroxyl group and the nitrogen in the imine bond as in wire structure **2**. The energy gain from intramolecular hydrogen bonding, forming quasi six-membered rings, may be large enough to compensate the steric strain energy associated with a more coplanar configuration. It has already been demonstrated that intramolecular hydrogen bonding can enable increased π-conjugation in extended systems [116].

To investigate the influence of electronic structure on hopping transport, one can imagine developing conjugated oligoimine systems substituted with electron-donating or withdrawing substituents (structure **3**) or wires consisting of alternating donor and acceptor blocks (structure **4**). Fluorine substitution is chosen for structure **3** because it has a strong electronegative inductive effect on π-systems and is not sterically bulky [117–120]. It can be anticipated that fluorine substitution will stabilize the frontier molecular orbitals (i.e., create a negative shift in the HOMO and LUMO energies) with respect to the Fermi level of the metal contacts. This in turn may result in electron (as opposed to hole) injection into the wire orbitals, allowing examination of electron hopping. It has been recently reported that fluoroarene–thiophene oligomers exhibit n-type mobility in field effect transistors [119], and the LUMO of perfluorinated distyrylbenzenes (DSBs) is lowered by 0.5 eV compared to DSB, while the optical band gap nearly remains the same [120].

Another controlling factor for hopping transport is carrier density. OPI wires exhibited relatively low conductivity ($\sim 1.0 \times 10^{-4}$ S/cm) partially due to the low carrier density. Utilizing donor–acceptor architectures within the molecular backbone (as in structure **4**) may increase the carrier concentration present in the wires (Figure 3.18(a)). In the context of polymer electronics, donor–acceptor systems cause partial intramolecular charge transfer that enables manipulation of the electronic structure (HOMO/LUMO levels), leading to low band gap semiconducting polymers (<2.0 eV) [10, 121, 122]. In low band gap systems, the oxidation and reduction are achieved at lower potentials, facilitating both electron and hole injection into the system [122]. This strategy may yield higher single chain conductivity by exhibiting self-doping effects on molecular wires.

Finally, the metal complex approach to building molecular wires (structures **5** and **6**) offers opportunities to make wires with multiple well-defined redox states (Figure 3.18(b)). Already it has been shown that long metal complex wires are highly conductive [92], which is attributed to the presence of d-electrons that can be delocalized over the ligands. Such structures offer opportunities to examine wire conduction as a function of metal type (d-electron count), ligand design, and the position of redox sites. Because multiple charges can be accommodated by the redox sites, correlative effects (charge–charge repulsion) may lead to interesting $I-V$ behavior. Locating a single redox center in the middle of a wire molecule may allow Coulomb blockading effects to be observed at room temperature.

Many other wire architectures having a range of intentionally designed electrical transport behaviors, such as rectification, can be imagined. The ability to synthesize and measure conduction in molecular wires easily and reproducibly has opened up an exciting new research area for chemists and physicists.

References

1. Nitzan, A. and Ratner, M.A. (2003) Electron transport in molecular wire junctions. *Science*, **300**, 1384–1389.
2. Davis, W.B., Svec, W.A., Ratner, M.A. and Wasielewski, M.R. (1998) Molecular-wire behaviour in *p*-phenylenevinylene oligomers. *Nature*, **396**, 60–63.
3. Schwab, P.F.H., Levin, M.D. and Michl, J. (1999) Molecular rods. 1. Simple axial rods. *Chem. Rev.*, **99**, 1863–1934.
4. Joachim, C., Gimzewski, J.K. and Aviram, A. (2000) Electronics using hybrid-molecular and mono-molecular devices. *Nature*, **408**, 541–548.
5. Carroll, R.L. and Gorman, C.B. (2002) The genesis of molecular electronics. *Angew. Chem. Int. Ed.*, **41**, 4378–4400.
6. Heath, J.R. and Ratner, M.A. (2003) Molecular electronics. *Phys. Today*, **56**, 43–49.
7. Kwok, K.S. and Ellenbogen, J.C. (2002) Moletronics: future electronics. *Mater. Today*, **5**, 28–37.
8. Chen, Y. *et al.* (2003) Nanoscale molecular-switch crossbar circuits. *Nanotechnology*, **14**, 462–468.
9. Li, G. *et al.* (2005) High-efficiency solution processable polymer photovoltaic cells by self-organization of polymer blends. *Nat. Mater.*, **4**, 864–868.
10. Peet, J. *et al.* (2007) Efficiency enhancement in low-bandgap polymer solar cells by processing with alkane dithiols. *Nat. Mater.*, **6**, 497–500.
11. Sirringhaus, H., Tessler, N. and Friend, R.H. (1998) Integrated optoelectronic devices based on conjugated polymers. *Science*, **280**, 1741–1744.
12. Shen, Z. *et al.* (1997) Three-color, tunable, organic light-emitting devices. *Science*, **276**, 2009–2011.
13. Greenham, N.C. *et al.* (1993) Efficient light-emitting diodes based on polymers with high electron affinities. *Nature*, **365**, 628–630.
14. Yan, H. *et al.* (2009) A high-mobility electron-transporting polymer for printed transistors. *Nature*, **457**, 679–686.
15. Cho, J.H. *et al.* (2008) Printable ion-gel gate dielectrics for low-voltage polymer thin-film transistors on plastic. *Nat. Mater.*, **7**, 900–906.
16. Ruitenbeek, J. *et al.* (2005) in *Introducing Molecular Electronics*, Part II Experiment (eds G. Cuniberti, G. Fagas, and K. Richter), Springer, Heidelberg, Germany.
17. Salomon, A. *et al.* (2003) Comparison of electronic transport measurements on organic molecules. *Adv. Mater.*, **15**, 1881–1890.
18. Mantooth, B.A. and Weiss, P.S. (2003) Fabrication, assembly, and characterization of molecular electronics components. *Proc. IEEE*, **91**, 1785–1802.
19. Wang, W.Y., Lee, T.H. and Reed, M.A. (2005) Electronic transport in molecular self-assembled monolayer devices. *Proc. IEEE*, **93**, 1815–1824.

20. Chen, F. *et al.* (2007) Measurement of single-molecule conductance. *Annu. Rev. Phys. Chem.*, **58**, 535–564.
21. Selzer, Y. and Allara, D.L. (2006) Single-molecule electrical junctions. *Annu. Rev. Phys. Chem.*, **57**, 593–623.
22. McCreery, R.L. (2004) Molecular electronic junctions. *Chem. Mater.*, **16**, 4477–4496.
23. Venkataraman, L. *et al.* (2006) Dependence of single-molecule junction conductance on molecular conformation. *Nature*, **442**, 904–907.
24. Wold, D.J., Haag, R., Rampi, M.A. and Frisbie, C.D. (2002) Distance dependence of electron tunneling through self-assembled monolayers measured by conducting probe atomic force microscopy: unsaturated versus saturated molecular junctions. *J. Phys. Chem. B*, **106**, 2813–2816.
25. Liu, H. *et al.* (2008) Length-dependent conductance of molecular wires and contact resistance in metal–molecule–metal junctions. *ChemPhysChem*, **9**, 1416–1424.
26. Kim, B.-S. *et al.* (2006) Correlation between HOMO alignment and contact resistance in molecular junctions: aromatic thiols versus aromatic isocyanides. *J. Am. Chem. Soc.*, **128**, 4970–4971.
27. Engelkes, V.B., Beebe, J.M. and Frisbie, C.D. (2004) Length-dependent transport in molecular junctions based on SAMs of alkanethiols and alkanedithiols: effect of metal work function and applied bias on tunneling efficiency and contact resistance. *J. Am. Chem. Soc.*, **126**, 14287–14296.
28. Lakshmi, S., Dutta, S. and Pati, S.K. (2008) Molecular electronics: effect of external electric field. *J. Phys. Chem. C*, **112**, 14718–14730.
29. Wang, W., Lee, T. and Reed, M.A. (2003) Mechanism of electron conduction in self-assembled alkanethiol monolayer devices. *Phys. Rev. B*, **68**, 035416.
30. Selzer, Y., Cabassi, M.A., Mayer, T.S. and Allara, D.L. (2004) Thermally activated conduction in molecular junctions. *J. Am. Chem. Soc.*, **126**, 4052–4053.
31. Poot, M. *et al.* (2006) Temperature dependence of three-terminal molecular junctions with sulfur end-functionalized tercyclohexylidenes. *Nano Lett.*, **6**, 1031–1035.
32. Li, X. *et al.* (2007) Thermally activated electron transport in single redox molecules. *J. Am. Chem. Soc.*, **129**, 11535–11542.
33. Choi, S.H., Kim, B.-S. and Frisbie, C.D. (2008) Electrical resistance of long conjugated molecular wires. *Science*, **320**, 1482–1486.
34. Mayor, M. *et al.* (2003) Electric current through a molecular rod? Relevance of the position of the anchor groups. *Angew. Chem. Int. Ed.*, **42**, 5834–5838.
35. Haiss, W. *et al.* (2006) Thermal gating of the single molecule conductance of alkanedithiols. *Faraday Discuss.*, **131**, 253–264.
36. Selzer, Y. *et al.* (2005) Effect of local environment on molecular conduction: isolated molecule versus self-assembled monolayer. *Nano Lett.*, **5**, 61–65.
37. Selzer, Y., Cabassl, M.A., Mayer, T.S. and Allara, D.L. (2004) Temperature effects on conduction through a molecular junction. *Nanotechnology*, **15**, S483–S488.
38. Venkataraman, L. *et al.* (2006) Single-molecule circuits with well-defined molecular conductance. *Nano Lett.*, **6**, 458–462.
39. Venkataraman, L. *et al.* (2007) Electronics and chemistry: varying single-molecule junction conductance using chemical substituents. *Nano Lett.*, **7**, 502–506.
40. Chen, F. *et al.* (2006) Effect of anchoring groups on single-molecule conductance: comparative study of thiol-, amine-, and carboxylic-acid-terminated molecules. *J. Am. Chem. Soc.*, **128**, 15874–15881.
41. Xu, B. and Tao, N.J. (2003) Measurement of single-molecule resistance by repeated formation of molecular junction. *Science*, **301**, 1221–1223.
42. Yasuda, S. *et al.* (2006) Bond fluctuation of S/Se anchoring observed in single-molecule conductance measurements using the point contact method

with scanning tunneling microscopy. *J. Am. Chem. Soc.*, **128**, 7746–7747.
43. Anslyn, E.V. and Dougherty, D.A. (2006) *Modern Physical Organic Chemistry*, Chapter 10, University Science Books, Sausalito.
44. Closs, G.L. and Miller, J.R. (1988) Intramolecular long-distance electron transfer in organic molecules. *Science*, **240**, 440–447.
45. Jordan, K.D. and Paddon-Row, M.N. (1992) Analysis of the interactions responsible for long-range through-bond-mediated electronic coupling between remote chromophores attached to rigid polynorbornyl bridges. *Chem. Rev.*, **92**, 395–410.
46. Marcus, R.A. (1993) Electron transfer reactions in chemistry: theory and experiment (nobel lecture). *Angew. Chem. Int. Ed.*, **32**, 1111–1222.
47. Barbara, P.F., Meyer, T.J. and Ratner, M.A. (1996) Contemporary issues in electron transfer research. *J. Phys. Chem.*, **100**, 13148–13168.
48. Weiss, E.A. *et al.* (2005) Conformationally gated switching between superexchange and hopping within oligo-*p*-phenylene-based molecular wires. *J. Am. Chem. Soc.*, **127**, 11842–11850.
49. Newton, M.D. (1991) Quantum chemical probes of electron-transfer kinetics: the nature of donor–acceptor interactions. *Chem. Rev.*, **91**, 767–792.
50. Lambert, C., Noll, G. and Schelter, J. (2002) Bridge-mediated hopping or superexchange electron-transfer processes in bis(triarylamine) systems. *Nat. Mater.*, **1**, 69–73.
51. Berlin, Y.A. *et al.* (2003) Charge hopping in molecular wires as a sequence of electron-transfer reactions. *J. Phys. Chem. A*, **107**, 3970–3980.
52. Sachs, S.B. *et al.* (1997) Rates of interfacial electron transfer through *p*-conjugated spacers. *J. Am. Chem. Soc.*, **119**, 10563–10564.
53. Curtiss, L.A., Naleway, C.A. and Miller, J.R. (1995) Superexchange pathway calculation of electronic coupling through cyclohexane spacers. *J. Phys. Chem.*, **99**, 1182–1193.
54. Paulson, B. *et al.* (1993) Long-distance electron transfer through rodlike molecules with cubyl spacers. *J. Phys. Chem.*, **97**, 13042–13045.
55. Berlin, Y.A., Grozema, F.C., Siebbeles, L.D.A. and Ratner, M.A. (2008) Charge transfer in donor–bridge–acceptor systems: static disorder, dynamic fluctuations, and complex kinetics. *J. Phys. Chem. C*, **112**, 10988–11000.
56. Nitzan, A. (2001) A relationship between electron-transfer rates and molecular conduction. *J. Phys. Chem. A*, **105**, 2677–2679.
57. Adams, D.M. *et al.* (2003) Charge transfer on the nanoscale: current status. *J. Phys. Chem. B*, **107**, 6668–6697.
58. Zhu, X.-Y. (2004) Charge transport at metal–molecule interfaces: a spectroscopic view. *J. Phys. Chem. B*, **108**, 8778–8793.
59. Zhu, X.Y. (2004) Electronic structure and electron dynamics at molecule–metal interfaces: implications for molecule-based electronics. *Surf. Sci. Rep.*, **56**, 1–83.
60. Park, H. *et al.* (2000) Nanomechanical oscillations in a single-C_{60} transistor. *Nature*, **407**, 57–60.
61. Park, J. *et al.* (2002) Coulomb blockade and the Kondo effect in single-atom transistors. *Nature*, **417**, 722–725.
62. Chae, D.-H. *et al.* (2006) Vibrational excitations in single trimetal-molecule transistors. *Nano Lett.*, **6**, 165–168.
63. Ashcroft, N.W. and Mermin, N.D. (1976) *Solid State Physics*, Chapter 1, Holt Rinehart & Winston, Austin.
64. Wiesendanger, R. (1994) *Scanning Probe Microscopy*, Cambridge University Press, Cambridge.
65. Simmons, J.G. (1963) Generalized formula for the electric tunnel effect between similar electrodes separated by a thin insulating film. *J. Appl. Phys.*, **34**, 1793–1803.
66. Landauer, R. (1981) Can a length of perfect conductor have a resistance? *Phys. Lett. A*, **85**, 91–93.
67. Datta, S. (2004) Electrical resistance: an atomistic view. *Nanotechnology*, **15**, S433–S451.

68. Nitzan, A. (2001) Electron transmission through molecules and molecular interfaces. *Annu. Rev. Phys. Chem.*, **52**, 681–750.
69. Joachim, C. and Ratner, M.A. (2005) Molecular electronics: some views on transport junctions and beyond. *Proc. Natl. Acad. Sci. USA*, **102**, 8801–8808.
70. Segal, D. *et al.* (2000) Electron transfer rates in bridged molecular systems 2. A steady-state analysis of coherent tunneling and thermal transitions. *J. Phys. Chem. B*, **104**, 3817–3829.
71. Segal, D., Nitzan, A., Ratner, M. and Davis, W.B. (2000) Activated conduction in microscopic molecular junctions. *J. Phys. Chem. B*, **104**, 2790–2793.
72. Segal, D. and Nitzan, A. (2001) Steady-state quantum mechanics of thermally relaxing systems. *Chem. Phys.*, **268**, 315–335.
73. Young, R., Ward, J. and Scire, F. (1971) Observation of metal-vacuum-metal tunneling, field emission, and the transition region. *Phys. Rev. Lett.*, **27**, 922–924.
74. Beebe, J.M. *et al.* (2006) Transition from direct tunneling to field emission in metal–molecule–metal junctions. *Phys. Rev. Lett.*, **97**, 026801.
75. Beebe, J.M., Kim, B.-S., Frisbie, C.D. and Kushmerick, J.G. (2008) Measuring relative barrier heights in molecular electronic junctions with transition voltage spectroscopy. *ACS Nano*, **2**, 827–832.
76. Mujica, V., Kemp, M. and Ratner, M.A. (1994) Electron conduction in molecular wires. II. Application to scanning tunneling microscopy. *J. Chem. Phys.*, **101**, 6856–6864.
77. Mujica, V., Kemp, M. and Ratner, M.A. (1994) Electron conduction in molecular wires. I. A scattering formalism. *J. Chem. Phys.*, **101**, 6849–6855.
78. Beebe, J.M., Engelkes, V.B., Miller, L.L. and Frisbie, C.D. (2002) Contact resistance in metal–molecule–metal junctions based on aliphatic SAMs: effects of surface linker and metal work function. *J. Am. Chem. Soc.*, **124**, 11268–11269.
79. Baranovski, S. and Rubel, O. (2006) in *Charge Transport in Disordered Solids with Applications in Electronics*, Chapter 2 (ed. S. Baranovski), John Wiley & Sons, Ltd, New York, NY.
80. Miller, A. and Abrahams, E. (1960) Impurity conduction at low concentrations. *Phys. Rev.*, **120**, 745.
81. Cui, X.D. *et al.* (2001) Reproducible measurement of single-molecule conductivity. *Science*, **294**, 571–574.
82. Yaliraki, S.N., Kemp, M. and Ratner, M.A. (1999) Conductance of molecular wires: influence of molecule–electrode binding. *J. Am. Chem. Soc.*, **121**, 3428–3434.
83. Yaliraki, S.N. and Ratner, M.A. (1998) Molecule–interface coupling effects on electronic transport in molecular wires. *J. Chem. Phys.*, **109**, 5036–5043.
84. Datta, S. (1995) *Electronic Transport in Mesoscopic Systems*, Cambridge University Press, New York, NY.
85. Gadzuk, J.W. and Plummer, E.W. (1971) Field emission energy distribution (FEED). *Rev. Mod. Phys.*, **45**, 487–548.
86. Meisel, K.D., Vocks, H. and Bobbert, P.A. (2005) Polarons in semiconducting polymers: study within an extended holstein model. *Phys. Rev. B*, **71**, 205206.
87. Segal, D. and Nitzan, A. (2002) Conduction in molecular junctions: inelastic effects. *Chem. Phys.*, **281**, 235–256.
88. Zade, S.S. and Bendikov, M. (2008) Study of hopping transport in long oligothiophenes and oligoselenophenes: dependence of reorganization energy on chain length. *Chem. Eur. J.*, **14**, 6734–6741.
89. Efros, A.L. and Shklovskii, B.I. (1975) Coulomb gap and low temperature conductivity of disordered systems. *J. Phys. C: Solid State Phys.*, **8**, L49–L51.
90. Rose, A. (1955) Space-charge-limited currents in solids. *Phys. Rev.*, **97**, 1538.
91. Lampert, M.A. (1956) Simplified theory of space-charge-limited currents in an insulator with traps. *Phys. Rev.*, **103**, 1648.

92. Tuccitto, N. *et al.* (2009) Highly conductive ~-nm-long molecular wires assembled by stepwise incorporation of metal centres. *Nat. Mater.*, **8**, 41–46.
93. Holmlin, R.E. *et al.* (2001) Electron transport through thin organic films in metal-insulator-metal junctions based on self-assembled monolayers. *J. Am. Chem. Soc.*, **123**, 5075–5085.
94. Lafferentz, L. *et al.* (2009) Conductance of a single conjugated polymer as a continuous function of its length. *Science*, **323**, 1193–1197.
95. Rosink, J.J.W.M. *et al.* (2000) Self-assembly of π-conjugated azomethine oligomers by sequential deposition of monomers from solution. *Langmuir*, **16**, 4547–4553.
96. Kelley, T.W., Granstrom, E. and Frisbie, C.D. (1999) Conducting probe atomic force microscopy: a characterization tool for molecular electronics. *Adv. Mater.*, **11**, 261–264.
97. Wold, D.J. and Frisbie, C.D. (2000) Formation of metal–molecule–metal tunnel junctions: microcontacts to alkanethiol monolayers with a conducting AFM tip. *J. Am. Chem. Soc.*, **122**, 2970–2971.
98. Wold, D.J. and Frisbie, C.D. (2001) Fabrication and characterization of metal–molecule-metal junctions by conducting probe atomic force microscopy. *J. Am. Chem. Soc.*, **123**, 5549–5556.
99. Leatherman, G. *et al.* (1999) Carotene as a molecular wire: conducting atomic force microscopy. *J. Phys. Chem. B*, **103**, 4006–4010.
100. Engelkes, V.B. and Frisbie, C.D. (2006) Simultaneous nanoindentation and electron tunneling through alkanethiol self-assembled monolayers. *J. Phys. Chem. B*, **110**, 10011–10020.
101. Engelkes, V.B., Beebe, J.M. and Frisbie, C.D. (2005) Analysis of the causes of variance in resistance measurements on metal–molecule–metal junctions formed by conducting-probe atomic force microscopy. *J. Phys. Chem. B*, **109**, 16801–16810.
102. Giese, B. (2000) Long-distance charge transport in DNA: the hopping mechanism. *Acc. Chem. Res.*, **33**, 631–636.
103. Giese, B. *et al.* (2001) Direct observation of hole transfer through DNA by hopping between adenine bases and by tunnelling. *Nature*, **412**, 318–320.
104. Conwell, E.M. (2005) Charge transport in DNA in solution: the role of polarons. *Proc. Natl. Acad. Sci. USA*, **102**, 8795–8799.
105. Shen, Y., Hosseini, A.R., Wong, M.H. and Malliaras, G.G. (2004) How to make ohmic contacts to organic semiconductors. *ChemPhysChem*, **5**, 16–25.
106. Akaba, R., Tokumaru, K. and Kobayashi, T. (1980) Electronic structure and conformations of *N*-benzylideneanilines. I. Electronic absorption spectral study combined with CNDO/S CI calculations. *Bull. Chem. Soc. Jpn.*, **53**, 1993–2001.
107. Tsuji, T., Takeuchi, H., Egawa, T. and Konaka, S. (2001) Effects of molecular structure on the stability of a thermotropic liquid crystal. Gas electron diffraction study of the molecular structure of phenyl benzoate. *J. Am. Chem. Soc.*, **123**, 6381–6387.
108. Harada, J., Harakawa, M. and Ogawa, K. (2004) Torsional vibration and central bond length of *N*-benzylideneanilines. *Acta Crystallogr. B*, **60**, 578–588.
109. Yu, W. *et al.* (2006) Crystal structure and geometry-optimization study of 4-methyl-3′,5′-dinitro-4′-methyl benzylidene aniline. *J. Mol. Struct.*, **794**, 255–260.
110. Kubatkin, S. *et al.* (2003) Single-electron transistor of a single organic molecule with access to several redox states. *Nature*, **425**, 698–701.
111. Sze, S.M. (1981) *Physics of Semiconductor Devices*, 2nd edn, Chapter 7, John Wiley & Sons, Inc., New York, NY.
112. Morgan, P.W., Kwolek, S.L. and Pletcher, T.C. (1987) Aromatic azomethine polymers and fibers. *Macromolecules*, **20**, 729–739.
113. Yang, C.-J. and Jenekhe, S.A. (1995) Group contribution to molar refraction and refractive index of conjugated polymers. *Chem. Mater.*, **7**, 1276–1285.
114. Liu, C.-L. and Chen, W.-C. (2005) Fluorene-based conjugated

poly(azomethine)s: synthesis, photophysical properties, and theoretical electronic structures. *Macromol. Chem. Phys.*, **206**, 2212–2222.

115. Tsai, F.-C. *et al.* (2005) New thiophene-linked conjugated poly(azomethine)s: theoretical electronic structure, synthesis, and properties. *Macromolecules*, **38**, 1958–1966.
116. Berkesi, O. *et al.* (2003) Hydrogen bonding interactions of benzylidene type schiff bases studied by vibrational spectroscopic and computational methods. *Phys. Chem. Chem. Phys.*, **5**, 2009–2014.
117. Krebs, F.C. and Jorgensen, M. (2002) Controlling the energy levels of conducting polymers. Hydrogen versus fluorine in poly(dialkylterphenylenevinylene)s. *Macromolecules*, **35**, 7200–7206.
118. Anthony, J.E. (2006) Functionalized acenes and heteroacenes for organic electronics. *Chem. Rev.*, **106**, 5028–5048.
119. Yoon, M.-H., Facchetti, A., Stern, C.E. and Marks, T.J. (2006) Fluorocarbon-modified organic semiconductors: molecular architecture, electronic, and crystal structure tuning of arene- versus fluoroarene-thiophene oligomer thin-film properties. *J. Am. Chem. Soc.*, **128**, 5792–5801.
120. Renak, M.L. *et al.* (1999) Fluorinated distyrylbenzene chromophores: effect of fluorine regiochemistry on molecular properties and solid-state organization. *J. Am. Chem. Soc.*, **121**, 7787–7799.
121. Beaujuge, P.M., Ellinger, S. and Reynolds, J.R. (2008) The donor-acceptor approach allows a black-to-transmissive switching polymeric electrochrome. *Nat. Mater.*, **7**, 795–799.
122. Yang, C. *et al.* (2008) Visible-near infrared absorbing dithienylcyclopentadienone-thiophene copolymers for organic thin-film transistors. *J. Am. Chem. Soc.*, **130**, 16524–16526.

Part II
Donor–Bridge–Acceptor Systems

Charge and Exciton Transport through Molecular Wires. Edited by L.D.A. Siebbeles and F.C. Grozema

ISBN: 978-3-527-32501-6

4
Tunneling through Conjugated Bridges in Designed Donor–Bridge–Acceptor Molecules

Bo Albinsson, Mattias P. Eng, and Jerker Mårtensson

4.1
Introduction

The transfer of electrons and excitation energy is of paramount importance for describing many elementary steps in chemical and physical processes of importance for biology, physics, chemistry, and material science. The fundamental nature of these transfer processes is of course the same even if they occur in seemingly different systems such as between redox active proteins in a biological context or in solar cells driving the charge separation that eventually leads to the current production of the cell. Nevertheless, progress and knowledge development are not necessarily readily transferred from one field to another due to the differences in language and nomenclature used in different fields. Hopefully this communication barrier will break down when enough knowledge of the *fundamental* processes themselves will become available. This is one of the reasons for experimentally and theoretically study model systems for electron and energy transfer. The model systems, typically covalently bound donors (D) and acceptors (A), are merely simplified mimics of the much more complicated biological or material science systems but they will, if carefully designed, provide crucial information for understanding the fundamental processes occurring in the complex assemblies. This chapter will concentrate on bridged donor–acceptor systems designed for the systematic study of parameters that influence the rate and efficiency of electron and energy transfer reactions. A particular emphasis will be put on π-conjugated bridges for which there most often is a substantial influence on the rate of transfer. A lot of seminal experimental and theoretical work on donor–acceptor systems connected by saturated covalent bridges has been reported during the 1980s and 1990s and these studies have recently been reviewed [1–3].

4.1.1
Theoretical Background

Some key ideas and mathematical relations are presented in this section and the reader interested in derivations or more thorough discourse of the

Charge and Exciton Transport through Molecular Wires. Edited by L.D.A. Siebbeles and F.C. Grozema

ISBN: 978-3-527-32501-6

underlying theories is referred to the excellent anthology edited by Vincenzo Balzani [4].

4.1.1.1 Molecular Wires

A molecular wire has been defined in different ways throughout the current chemistry, physics, and material science literature. It has either been defined as a molecule that conducts electrical current between two reservoirs of electrons or more broadly as a molecule that enhances the electronic coupling between two terminals (i.e., the donor and acceptor). In this chapter, we will adopt the more broad definition and simply let the wire be a conduit for mediating the electronic coupling between the ends of the wire allowing for long distance electron or energy transfer.

4.1.1.2 Electron and Energy Transfer – the Fundamentals

Electron and energy transfer are related phenomena that can be described by a common theoretical framework. Provided that the electronic coupling is not too large, the Fermi golden rule, Eq. (4.1), predicts the rate of transition between two potential energy surfaces

$$k_{if} = (2\pi/\hbar) V_{if}^2 \mathrm{FCWD} \tag{4.1}$$

In this so-called diabatic (nonadiabatic) approximation the electronic coupling, V_{if}, is defined as the effective electronic Hamiltonian matrix element that couples the initial (Ψ_i) and final (Ψ_f) states

$$V_{if} = \langle \Psi_f | H' | \Psi_i \rangle \tag{4.2}$$

where H' is the operator corresponding to the (small) perturbation mixing the initial and final states. The Franck–Condon weighted density (FCWD) of states accounts for the conservation of energy and describes the influence from the nuclear modes of the system. Its specific form has to be adapted to the transfer reactions studied. For electron transfer (ET), Marcus approximated the involved potential surfaces by simple parabolas with equal force constants (curvature) which, when combined with transition state theory, leads to [5, 6]

$$k_{\mathrm{ET}} = \sqrt{\frac{\pi}{\hbar^2 k_B T \lambda}} V^2 \exp\left(-\frac{(\Delta G^0 + \lambda)^2}{4\lambda k_B T}\right) \tag{4.3}$$

In Eq. (4.3) ΔG^0 is the standard free energy change for the ET reaction (energy displacement of the parabolas) and λ is the reorganization energy defined as the potential energy difference between the reactant and product nuclear configurations for the final electronic state, that is, the energy gained by a nuclear (and solvent) relaxation after vertical excitation from the reactant to the product state. For energy transfer reactions the spectral overlap between the involved states can sometimes be estimated from spectroscopic measurements and the FCWD term of Eq. (4.1) is then evaluated from the spectral overlap integral:

$$J = \int I_D(\nu)\varepsilon_A(\nu)d\nu \tag{4.4}$$

where $I_D(\nu)$ and $\varepsilon_A(\nu)$ are the normalized emission and absorption spectra of the donor and acceptor, respectively. In the case of energy transfer between triplet states either the emission or absorption spectrum or both are difficult to measure and the rates are therefore often estimated by the approximate Marcus equation (Eq. (4.3)) or the slightly more elaborate Marcus–Jortner equation [7, 8]. In a situation where spectroscopic data are available (e.g., for metal complexes), however, one should preferably estimate the FCWD factor from the spectral overlap integral. Singlet energy transfer is usually dominated by the Coulombic interaction between the donor and acceptor and the leading term of the multipole expansion (dipole–dipole) gives the so-called Förster approximation [9, 10]. The Förster equation could be written in a form that clearly shows its relation to the Fermi golden rule [11, 12] but more commonly it is expressed in a way that is related to experimental observables:

$$k_{\mathrm{EET}} = (1/\tau_D)\,(R_0/R_{\mathrm{DA}})^6 \tag{4.5}$$

where τ_D is the excited state lifetime of the donor in absence of the acceptor, R_{DA} the donor–acceptor distance (through space), and R_0 is the critical donor–acceptor distance at which 50% of the donor excited state decay is due to energy transfer. R_0 is easily calculated from the spectroscopic properties (donor emission and acceptor absorption spectra) as

$$R_0^6 = \frac{9000\,(\ln 10)}{128\pi^5 N_A}\,\frac{\kappa^2 \phi_D J_{\text{Förster}}}{n^4} \tag{4.6}$$

where ϕ_D is the emission quantum yield of the donor in the absence of the acceptor, κ is an orientation factor, n is the solvent refractive index, and $J_{\text{Förster}}$ is an overlap integral similar to Eq. (4.4). The orientation factor, κ, can be calculated if the orientations of the interacting transition dipoles are known. This is, however, seldom the case since conformational disorder gives a range of possible values. Methods to average over conformational distributions have been derived [13] but many studies just simply assume that both donor and acceptor are freely rotating dipoles leading to the average, $\kappa^2 = 2/3$. Recently, the expected variation of the orientation factor was experimentally demonstrated in singlet energy transfer between two DNA-base analogs, which were incorporated in the fairly rigid double stranded DNA helix [14].

In parallel to the Coulombic interaction, excitation energy can also be transferred due to a second, purely quantum mechanical interaction, namely the electron exchange interaction. This is how triplet excitation energy normally is transferred and it bears great resemblance to the origin of ET. Excitation energy transfer caused by the exchange interaction is often referred to as Dexter energy transfer. The exchange interaction, irrespectively of mode of transfer, is dealt with in some detail in the next section.

4.1.1.3 Mechanisms for Electron and Energy Transfer

ET between ends of a molecular wire is dictated by either of two mechanisms: the coherent tunneling or incoherent hopping mechanism. The tunneling mechanism is often described by the superexchange model (*vide infra*) which requires the donor and acceptors at either ends of the wire to be energetically well separated from the bridge states [15–17]. One important consequence is that the bridge is never either reduced or oxidized but merely functions as a coupling medium for the transfer process. In contrast, the incoherent hopping mechanism involves real intermediate states that actively transport the electron or hole along the wire. This is why the incoherent mechanism often is called thermally activated hopping [18, 19]. In many real cases the transfer is expected to be governed by a mixture of the two mechanisms and experimentally this has been observed by several groups [20–22].

4.1.1.4 The Distance Dependence of Electron and Energy Transfer Reactions

The coherent tunneling mechanism is, in the absence of a Coulomb interaction, dominated by the exchange interaction that governs the distance dependence of the rates for both electron and triplet energy transfer (and in some cases also singlet energy transfer). For this mechanism the distance dependence of the observed rates is approximately exponential:

$$k = k_0 \exp(-\beta R_{DA}) \tag{4.7}$$

where R_{DA} is the distance between the donor and acceptor (measured along the wire – not through space), k_0 the limiting rate at donor–acceptor contact, and β is the system specific attenuation factor. Numerous experimental and computational studies have verified Eq. (4.7) and β-values for a wide range of molecular wires (molecular bridges) have been determined. In general, β-values are larger for systems comprised of σ-bonds than for systems connected by π-conjugated bridges, which gives more efficient long-range transfer of electrons or excitation energy than in the former systems. This difference in distance dependence is easy to understand from the probability of tunneling as described by Gamow already in 1928 [23, 24]. The quantitative treatment for a particle with mass, m, tunneling through a rectangular barrier with height, $\Delta E_{DB} = E_B - E_D$, and width, R_{DA}, yields

$$k \propto V^2 \propto f(E_D, E_B) \exp\left(-\frac{2}{\hbar}\sqrt{2m\Delta E_{DB}}\, R_{DA}\right) \tag{4.8}$$

which predicts an approximately exponential decay of the tunneling rate with distance, $k \propto \exp(-\beta_{Gamow} R_{DA})$. In Eq. (4.8) the pre-exponential function, $f(E_D, E_B)$, depends on the barrier (bridge) and donor energies, and varies only slightly with the barrier height as long as ΔE_{DB} is large [25]. It is clear from Eq. (4.8) that the distance dependence is a function of the barrier height. If ΔE_{DB} is independent of R_{DA}, a decay parameter can be defined by Eq. (4.9)

$$\beta_{Gamow} = \frac{2\sqrt{2m\Delta E_{DB}}}{\hbar} \tag{4.9}$$

It is instructive to calculate some typical values for this decay parameter; for a single tunneling electron, the Gamow "β-values" are 1.8, 1.0, and 0.46 $Å^{-1}$ for tunneling barriers of 3, 1, and 0.2 eV, respectively, corresponding approximately to the situation for tunneling through vacuum, a σ-bonded bridge, and a π-conjugated bridge, respectively.

Although β-values only apply for mechanisms that are expected to decay exponentially (Eq. (4.7)) they have been used as quality factors also for ET that occur via the incoherent hopping mechanism. As an empirical measure for the attenuation of the rate versus distance this is acceptable but since hopping is not expected to decay exponentially it could be quite confusing when comparing β-values with the purpose of trying to understand the mechanistic differences between hopping and long range tunneling.

4.1.1.5 **Superexchange**

For the coherent tunneling mechanism the influence of the intervening medium has been described by the so-called superexchange model [15–17]. In this model the magnitude of the electronic coupling is given by a first-order perturbation theory treatment as

$$V_{DA} = \frac{V_{DB}V_{BA}}{\Delta}\left(\frac{\nu}{\Delta}\right)^{n-1} \tag{4.10}$$

Equation (4.10) gives the total electronic coupling (V_{DA}) in terms of the electronic coupling of the bridge to the donor and acceptor (V_{DB} and V_{BA}), the interaction, ν, between nearest neighbors of a chain composed of n identical units, and the energy gap, Δ, between the relevant donor and bridge localized states. If the length of the chain connecting the donor and acceptor is directly proportional to n, that is, $R_{DA} = nR_0$, where R_0 is the length of one subunit and the electronic coupling between subunits is small compared to the energy gap ($\nu/\Delta \ll 1$), the distance dependence of the electronic coupling, and therefore also the rate, is exponential. Within this approximation the attenuation factor β is given by Eq. (4.11)

$$\beta = \frac{2}{R_0}\ln\left|\frac{\Delta}{\nu}\right| \tag{4.11}$$

If the bridge is treated as a single chromophore, that is, a single repeating unit, Eq. (4.10) is simplified to $V_{DA} = V_{DB}V_{BA}/\Delta$ that clearly shows the reciprocal energy gap dependence of the electronic coupling. This reciprocal energy gap dependence of both electron and energy transfer reactions has been demonstrated experimentally for sets of π-conjugated bridges [26, 27].

Extensions of the McConnell model has been proposed including the treatment of nonneighbor interactions [28], multiple transfer paths [29, 30], small energy gaps [31–33], and nonhomologous bridges [34]. It may seem as if the distance dependence could be simply predicted by using Eq. (4.11), but the parameters Δ and ν are not available from experiments. This is partly due to the imbedded approximations in the derivation of the McConnell model; Δ is the hypothetical tunneling energy gap (orbital energy difference in McConnell's derivation), which is assumed to be a constant independent on the size of the mediating bridge and ν is the electronic coupling between the subunits of the bridge. For large energy gaps, Δ could be regarded as constant, independent of the size of the molecular bridge and approximately equal to the energy difference between the relevant bridge and donor states. However, Δ has to be regarded as a model parameter for π-conjugated bridges where the energy gap clearly depends on the size of the molecular bridge (wire) and therefore cannot simply be estimated from relaxed bridge and donor states [35]. The coupling between subunits, ν, for π-conjugated bridges is, furthermore, strongly dependent on the conformation of the molecular bridge (and therefore also on temperature) as will be discussed in some detail below.

4.2
Through-Bond Electronic Coupling in π-Conjugated Bridges

If wire-like properties are desirable, then perhaps the most obvious choice of linker between two active components is a π-conjugated bridge. It combines extensive electron delocalization and low-lying orbitals with seeming rigidity and rod like structure, properties that would guarantee a high degree of electronic coupling and a well-defined relative geometry between the active components. Although a large, constant coupling might be of interest, the most intriguing applications will come with the ability to tune the coupling. Here, bridges comprised of series of aromatic subunits that can rotate in and out of conjugation relative to each other will have a prominent position.

In the subsequent section, the performance of π-conjugated bridges, in the first place oligo(phenyleneethynylene) or OPE bridges, will be shown in the context of a selection of transfer processes. The attenuation with increasing donor–acceptor distance and how this is influenced by the relative energy of the low-lying orbitals is emphasized. Temperature as a means to tune the electronic coupling via changes in conformations and conformational dynamics is also highlighted.

4.2.1
Distance and Energy Gap Dependence

4.2.1.1 Singlet Energy Transfer Mediated by π-Bridges

Zinc/free-base porphyrin based donor–bridge–acceptor (D–B–A) systems are ideal for studies of singlet energy transfer (SEET). As required by the theories for excitation energy transfer, there is a substantial overlap between the absorption spectrum of the acceptor, the free-base porphyrin, and the fluorescence spectrum of the zinc porphyrin donor [26]. The magnitudes of the oscillator strengths of the relevant donor and acceptor transitions are considerable, although only weakly allowed, and the donor fluorescence lifetime is sufficiently long to provide the necessary practical requirements for SEET.

Singlet energy transfer is affected by the bridge linking the donor and acceptor together. However, exactly how is often difficult to demarcate because the influence can be both diverse and subtle. Indirectly, the bridge might alter the intrinsic properties, such as transition dipole moments, of the attached donor and/or acceptor [36, 37]. More directly, the Coulomb and exchange interactions, cooperatively governing the overall donor–acceptor interaction, can separately be modulated by the bridge. The transfer process is referred to as through-space or Förster energy transfer (FRET) when the former interaction dictates the transfer, whereas it is referred to as through-bond or Dexter-type transfer when the latter dominates. Increased Coulombic interactions as an effect of the polarizability of the bridge have been advocated to explain experimental transfer rates larger than predicted by the Förster theory [38]. Variations in exchange interactions due to differences in frontier orbital shapes have been used to explain large differences between rates observed in pair wise analogous systems [39, 40]. In line with the latter, variations in the exchange interaction due to superexchange modulation will be discussed below as the major rationale for differences in SEET observed for a series of zinc(II)/free-base diporphyrin systems, see Figure 4.1.

To enable an in-depth examination of the excitation energy transfer dependence on the energy gap, ΔE_{DB}, between relevant states of the donor and the bridge, we took advantage of the fact that the energy levels of zinc porphyrins can efficiently be altered by coordination of axial ligands to the metal center [26, 41]. For a simple measure of the relative energies, see the absorption spectra shown in Figure 4.2.

Coordination of pyridine to the zinc porphyrin lowered the first excited singlet state energy and decreased the spectral overlap between the donor and acceptor significantly, compared to the zinc/free base systems in the absence of pyridine (Figure 4.2). The latter had a profound impact on the contribution of the Förster energy transfer to the overall quenching of the singlet excited state of the donor. Compared to the bicyclo[2.2.2]octane, OB, linked bisporphyrins (Figure 4.1) the D–B–A systems with conjugated bridges show significantly larger energy transfer rates than expected from the Förster theory for excitation energy transfer. The observed rates were 1.5–3.7 times larger, depending on the specific bridge, for the ZnP–XB–H_2P systems. Similarly, the observed rates were larger by a factor ranging from 3.1 to 10.6 for the corresponding pyridine coordinated Zn(py)P–XB–H_2P systems. In harmony with McConnell's superexchange theory (Eq. (4.10)), the larger deviations were found for the bridges whose lowest singlet excited state energies are closer to the lowest singlet excited state of the donor; that is, the smaller the energy difference between the lowest singlet excited state of the bridge and the donor, the larger the energy transfer rate. The superexchange mechanism has been advocated to explain larger singlet energy transfer rates than predicted by the Förster theory also in other more recent D–B–A systems linked by π-conjugated bridges [37, 42].

That a superexchange mechanism for SEET is operative in the ZnP–RB–H_2P and Zn(py)P–RB–H_2P systems was verified by analyzing the deviation of the experimentally observed rates from the rates obtained by the Förster equation (Eq. (4.6)). A bridge-dependent mediation effect – a long range Dexter-type contribution – was defined as the difference between the observed rate and the calculated rate constant according to the Förster theory, that is, $k_{Med} = k_{Exp} - k_{Förster}$. The variation in magnitude of the so-defined rate constants for mediated energy transfer was found to be proportional to the square of the reciprocal energy gap between bridge and the donor ($k_{Med} \propto (1/\Delta E_{DB})^2$) [26, 43], in perfect agreement with McConnell's

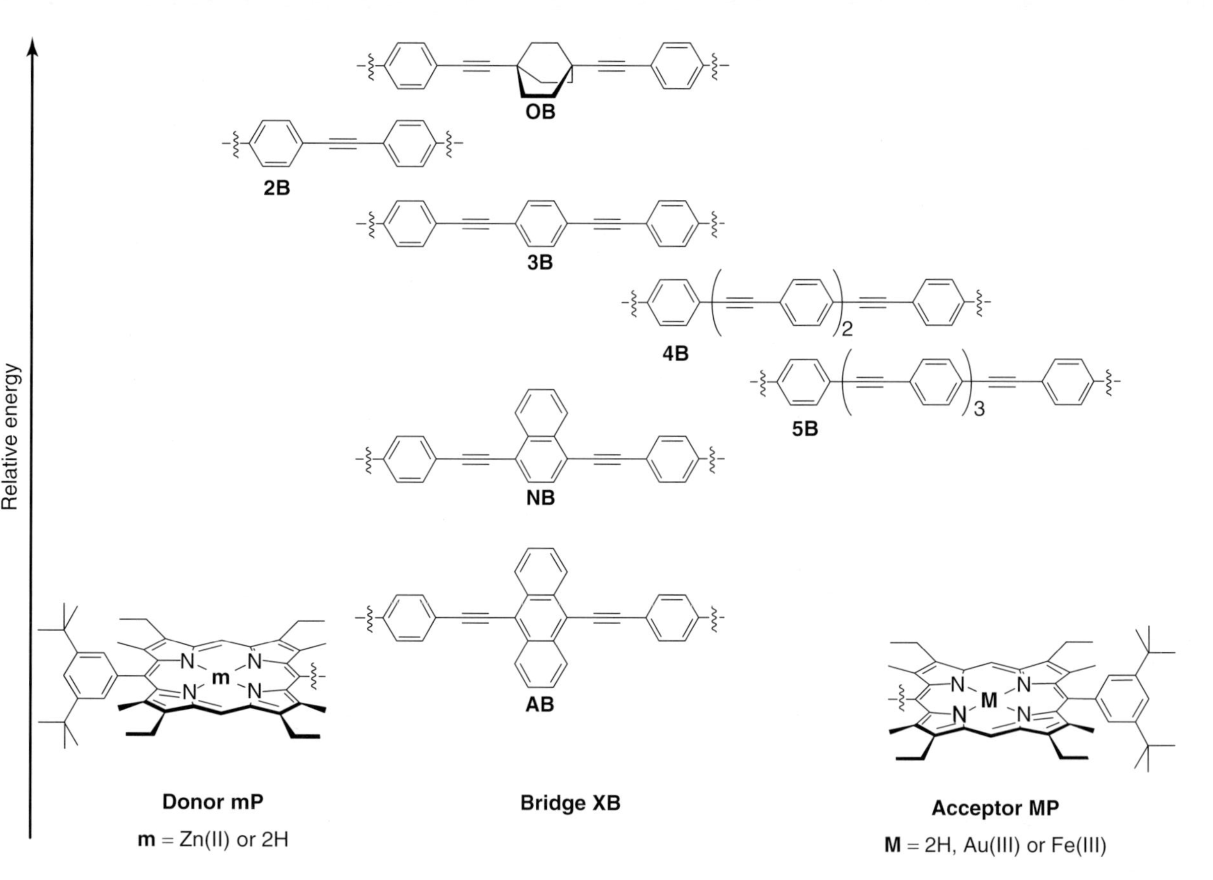

Figure 4.1 The series of D–B–A systems extensively studied in the Albinsson–Mårtensson group. The series are abbreviated mP–XB–MP, where m and M denotes the metalation state of the porphyrins and X the specific bridge structure. The series are divided into two subsets: the energy-gap series mP–RB–MP, with constant D–A distance, and the distance series mP–*n*B–MP. The various central groups in the energy-gap series are denoted by R = O, 3, N, or A. In the series with varying D–A distances, the length of the bridge is indicated by the number *n* that equals the number of phenylene units in the bridge. Any ligands coordinated to the metal is indicated in brackets after the chemical symbol for that metal, for example, Zn(py)P.

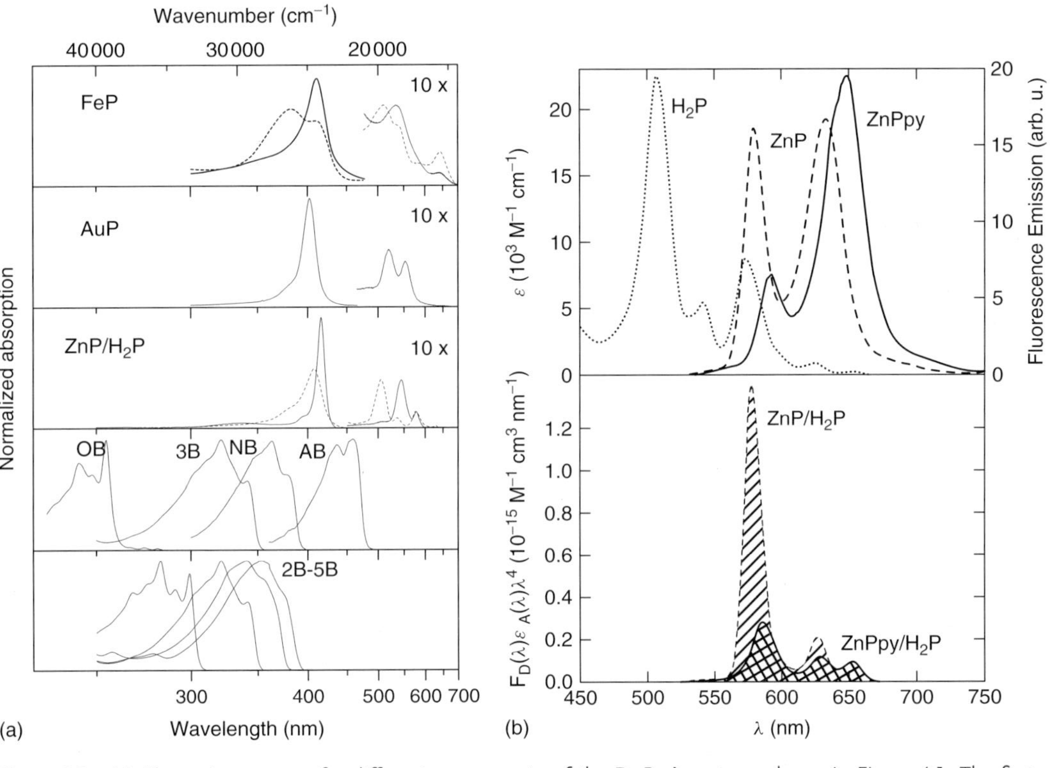

Figure 4.2 (a) Absorption spectra for different components of the D–B–A systems shown in Figure 4.1. The first (top) panel shows the absorption spectra of the iron(III) porphyrins $Fe(Im)_2P$ (——), coordinated with two axial imidazole ligands, and Fe(Cl)P (- - -), coordinated with one axial chloride ligand. The second panel shows the spectrum of the gold(III) porphyrin AuP. The third panel shows the absorption spectra of zinc porphyrin ZnP (——) and the free-base porphyrin H_2P (- - -). The two panels at the bottom show the two subsets of bridges; the energy-gap series **RB** and the length series **nB**. The spectra were recorded in methylene chloride at room temperature. (b) The top panel shows the absorption spectrum of H_2P (· · ·) and the emission spectra of ZnP (- - -) and Zn(py)P (—) in $CHCl_3$ at room temperature. The bottom panel shows the spectral overlap between H_2P and ZnP (///)and between H_2P and Zn(py)P (\\\).

superexchange theory. Notably, although the absolute magnitude decreases, the superexchange contribution to the overall energy transfer rate increases with increasing bridge length. Surprisingly, considering that they belong to different classes – σ type versus fully conjugated – the OB bridge follows the same trend as the other bridges. Thus, it contributes a small but nonzero bridge-mediated electronic coupling of the magnitude predicted by the same reciprocal energy difference relationship as obeyed by the conjugated bridges.

Excitation energy transfer via a hopping/incoherent mechanism is excluded in the topical D–B–A systems because neither of the bridge singlet excited states are thermally attainable at ambient temperature [43]. The lowest singlet excited states of the bridges are all between 4000 and 18 000 cm^{-1} higher in energy than that of the donor. Thus, the lowest singlet excited states of the bridges are only to be considered as virtual states, never occupied during the excitation energy transfer process.

The variation in length of conjugated bridges is usually intimately associated with a variation in the energy of the bridge located electronic states. When analyzed, the distance dependence for excitation energy transfer from a specific donor via a set of conjugated bridges of varying length should therefore include a co-variation of the donor–bridge energy gap, ΔE_{DB}. This hypothesis was tested simply by combining the linear dependence of the rate constant k_{Med} for mediated energy transfer on the reciprocal quadratic energy gap with the approximate exponential decay in transfer rate with donor–acceptor distance suggested by the superexchange theory (combining Eqs. 4.7 and 4.10) [41]:

$$k_{\text{Med}} = \frac{\alpha}{\Delta E_{\text{DB}}^2} \exp\left(-\beta R_{\text{DA}}\right) \tag{4.12}$$

Here α is a new preexponential factor, β an attenuation factor, and R_{DA} the donor–acceptor distance. Applying Eq. (4.12) to the analysis of the data from our series of D–B–A systems with varying bridge lengths and only minor variations in bridge energies gave a value of 43.5 for ln α and an attenuation factor β of 0.25 Å^{-1}. The β-value obtained in this way has, in contrast to most attenuation factors reported in the literature, been corrected for the changes in energy gaps between the donor and the bridges. Using this value for the attenuation factor and the same expression (Eq. (4.12)) in the analysis of the transfer rates obtained from the D–B–A systems with constant donor–acceptor distance but with large variations in bridge energies gave a value of 42.2 for ln α. The close conformity between the two values of ln α strengthens the validity of Eq. (4.12) and emphasizes the caution by which the attenuation factor for SEET should be regarded as a bridge specific parameter. It is to be considered as a system specific parameter that depends on the nature of the bridge as well as the energy levels of the donor and the acceptor with which the bridge is combined.

4.2.1.2 **Triplet Energy Transfer**

Zinc/free-base porphyrin arrays are not only ideally suited for studies of SEET, they also provide excellent systems for exploration of triplet energy transfer (TEET)

[44–47]. In contrast to SEET, triplet excitation energy transfer is not expected to occur over large distances via the Coulomb mechanism because of the spin forbidden transitions involved. Instead, long-range TEET is expected to be mediated by through-bond or superexchange interactions. This close relationship between TEET and ET has not until recently received the attention it deserves.

We found a strong dependence of the observed TEET on the nature and the dynamics of the bridges in our systems. The OB bridge was observed to be insulating, whereas the fully conjugated bridges provided efficient electronic coupling. The TEET rates were first measured at 150 K and were found to decay exponentially with distance in the ZnP–nB–H_2P series with an attenuation factor of 0.45 $Å^{-1}$ [46]. The largest rate ($2.0 \times 10^7\ s^{-1}$) was observed for the system with the shortest 2B bridge and the slowest transfer ($1.5 \times 10^3\ s^{-1}$) was observed in the longest 5B system. In accordance with the superexchange model, a substantial difference in rates between the 3B and the NB systems was observed, see Figure 4.3 [44].

On the one extreme, the triplet lifetime of the zinc porphyrin was, within experimental error, unquenched for the ZnP–OB–H_2P system. On the other extreme, the low laying triplet excited state of the AB bridge alone provide the necessary driving force for efficient quenching of the donor triplet excited state through TEET to the bridge. This latter system was therefore not included in the studies of bridge-mediated TEET. Effectively demonstrated by the absence of quenching in reference compounds ZnP–5B and ZnP–NB, the systems energetically most

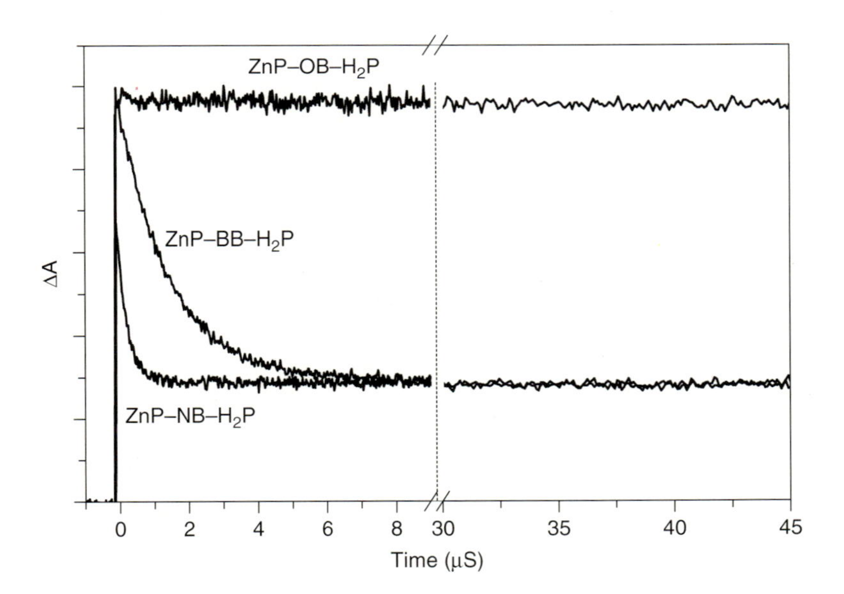

Figure 4.3 Triplet state decay (λ_{pump} = 532 nm, λ_{probe} = 470 nm) for the ZnP–OB–H_2P, ZnP–3B–H_2P, and ZnP–NB–H_2P in 2-MTHF at 150 K. (From Ref. [44].) Please note that the two abbreviations 3B and BB symbolize the same bridge.

similar to the anthracene case, neither of the other bridges provided a sufficiently low laying triplet excited state necessary for thermally activated stepwise TEET.

The triplet excited states $^3ZnP^*$–XB–H_2P were efficiently formed from the singlet excited systems initially formed upon selective photoexcitation of the zinc porphyrin. The quantum yield for intersystem crossing for zinc porphyrins is generally high, usually around 0.9 [48]. The occurrence of triplet excitation energy transfer was then established by monitoring the triplet dynamics of the $^3ZnP^*$- and the $^3H_2P^*$-state at 150 K. Note that the lowest triplet state of both the zinc and the free-base 5,15-aryl-β-octaalkylporphyrin consists of two conformers interrelated via a mother–daughter relationship [45]. This conformational transformation is thermally activated and virtually shut down at 150 K compared to the timescale of TEET, which greatly simplifies the analysis of the transfer kinetics [49].

Except for the OB system, decreased lifetimes for the triplet-excited donors were observed at 470 nm, where the $^3ZnP^*$ state absorption dominates. Corroborating TEET, the corresponding rise-times were observed at a wavelength dominated (434 nm) by the acceptor triplet excited-state absorption [46]. The rate of $^3H_2P^*$ formation due to TEET could also be resolved from kinetic traces recorded at 505 nm where the free-base porphyrin shows a negative absorption (bleaching) in the differential absorption spectrum. A very strong temperature dependence was also observed for the TEET process (*vide infra*) [50].

4.2.1.3 **Electron Transfer**

ET is on energetic grounds a plausible deactivation pathway for singlet excited zinc(II)/gold(III) bisporphyrin arrays in solvents of higher or similar polarity as for $CHCl_3$ and 2-methyltetrahydrofuran (2-MTHF). Some representative architectures of the fair number of elaborate systems based on this donor–acceptor pair that have appeared in the literature can be found in the following Refs. [51–55]. The change in Gibbs free energy for ET from the singlet excited zinc(II)porphyrin to gold(III) porphyrin ranges between −0.43 and −0.95 eV, depending on solvent polarity, for the two series ZnP–nB–AuP^+ and ZnP–RB–AuP^+ [27, 56, 57]. The driving force is calculated from redox potentials using the Born dielectric continuum model based Weller equation (Eq. (4.13)), combined with estimates of the accompanying reorganization energy [58, 59]:

$$\Delta G^\circ = (E_{\mathrm{ox}} - E_{\mathrm{red}}) - \mathrm{E}_{0-0} + \frac{e}{4\pi\varepsilon_0}\left(\frac{1}{\varepsilon_s} - \frac{1}{\varepsilon_{\mathrm{ref}}}\right)\left(\frac{1}{2R_D} + \frac{1}{2R_A}\right) - \frac{e}{4\pi\varepsilon_s\varepsilon_{\mathrm{ref}}R_{\mathrm{DA}}} \quad (4.13)$$

where E_{ox} and E_{red} are the donor and acceptor oxidation and reduction potentials, respectively, determined by cyclic voltammetry [27], ε_s is the dielectric constant of the solvent, and $\varepsilon_{\mathrm{ref}}$ the dielectric constant of the solvent in which the electrochemical measurements were performed. R_D and R_A are the average radii of the donor and acceptor, respectively.

The overall quenching rates in polar solvents were 1–2 orders of magnitude larger than the calculated Förster rate constants for the singlet excited donors in the gold series ZnP–nB–AuP, ZnP–RB–AuP^+, and Zn(py)P–RB–AuP^+, except

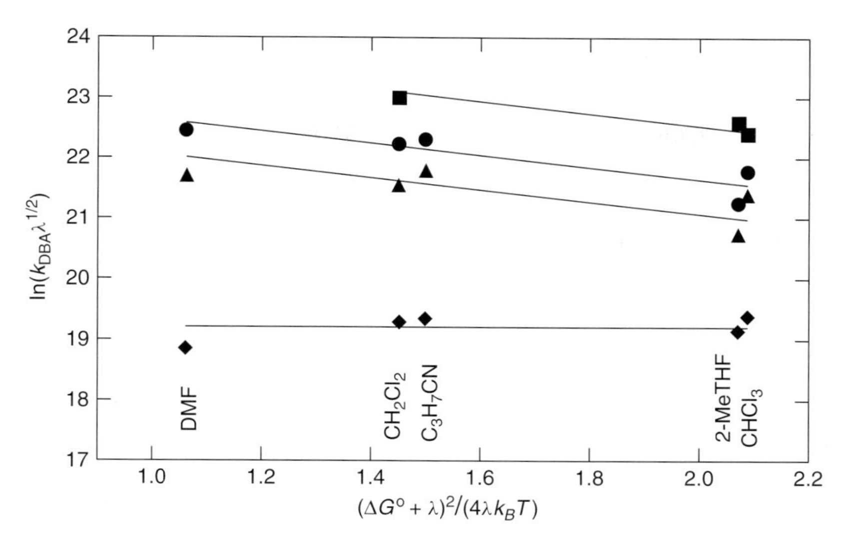

Figure 4.4 Solvent dependence of the quenching rates, k_{DBA}, for the first singlet excited states of the zinc porphyrin in ZnP–RB–AuP^{+}: R = O (♦), R = 3 (▲), R = N (•), and R = A (■). (From Ref. [27].)

for the OB systems [27, 56]. The observed quenching rates for the singlet excited zinc porphyrin donor in the ZnP–RB–AuP^{+} series and its pyridine-coordinated counterpart Zn(py)P–RB–AuP^{+} show the solvent dependence expected for an ET process (Figure 4.4) [27]. A slope of −1 is observed in the plot according to the linearized form of the Marcus equation (Eq. (4.3)). In contrast, a horizontal line (slope 0) is obtained for the OB linked systems indicating that ET is not a viable process for this bridge element.

In solvents polar enough, such as butyronitrile or more polar solvents, the charge-separated ion-pair state ZnP$^{\bullet+}$–AB$^{\bullet-}$ was observed by transient absorption spectroscopy after photoexcitation of the dyad ZnP–AB [20, 60]. None of the other bridges provide the necessary low reduction potential to allow for such charge separation between the zinc porphyrin and the bridge. Thus, sequential and direct superexchange-mediated ET could operate in parallel but was only observed for the AB bridge systems and in sufficiently polar solvents [20]. Sequential ET (incoherent hopping) was not only shown to be strongly solvent polarity dependent but it was also shown to be slowed down, relative to the superexchange pathway, and eventually turned off as the temperature was lowered. Several elegant studies where the sequential/hopping ET pathway has been switched on and off have appeared in the literature [61–63].

ET was confirmed by detection of the zinc-porphyrin radical cation [54, 64, 65]. In line with the lack of solvent dependence, no trace of the zinc-porphyrin radical cation could be detected for the OB linked systems. In fact the quenching rates in these systems were completely or nearly completely accounted for by the calculated Förster rate constants. Bicyclo[2.2.2]octane and its benzo-annulated analogs, in

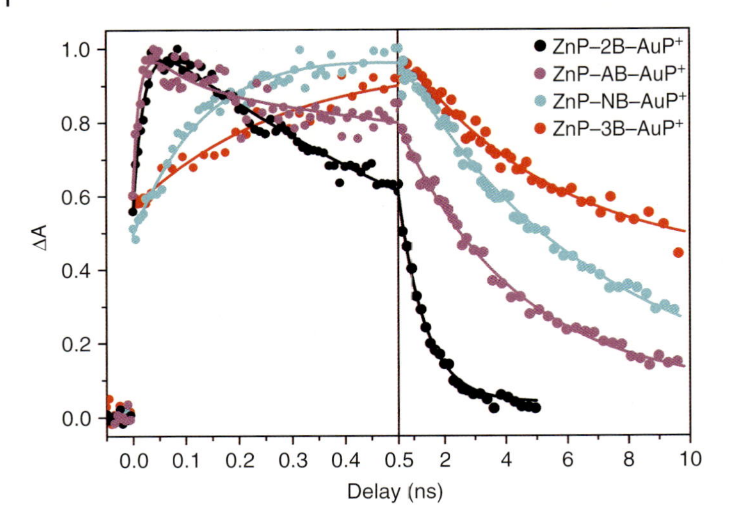

Figure 4.5 Transient absorption monitored at 680 nm (λ_{pump} = 548 nm) shows the initial build up of the zinc porphyrin radical cation and its subsequent decay in the ZnP–2B–AuP$^+$, ZnP–AB–AuP$^+$, ZnP–NB–AuP$^+$, and ZnP–3B–AuP$^+$ systems in DMF at room temperature. (From Ref. [57].)

combination with another donor–acceptor pair, have been thoroughly explored as conduits for ET by Ratner, Wasielewski, and coworkers [66].

The decay of the singlet excited state of zinc porphyrin was monitored at ~470 nm and the formation of the zinc-porphyrin radical cation at ~670 nm by transient absorption measurements (Figure 4.5). The conformity between decay and build-up times was striking for the systems with the shorter bridges. The formation and decay times measured by transient absorption were also in excellent agreement with the ones obtained by time-resolved fluorescence measurements. For the systems with the two longest bridges, 4B and 5B, the signature of the radical cation was partly obscured by the singlet and triplet absorptions as these evolved on the same timescale [56]. However, there was a small differential absorption remaining at delay times longer than 1 ns, indicating the existence of a charge-separated state with considerably longer lifetime in these systems.

An attenuation factor, β, of 0.3 Å^{-1} was obtained for the forward ET process in the ZnP–nB–AuP$^+$ series. It was estimated by simply plotting the logarithm of the difference between the observed quenching rate and the Förster rate constant versus the edge-to-edge donor–acceptor distance. For the energy-gap series ZnP–RB–AuP$^+$ and Zn(Py)P–RB–AuP$^+$, a reasonable fit to a straight line was obtained when the experimentally derived electronic couplings were plotted versus the reciprocal energy differences between the first singlet excited states of the donor and the different bridges [27].

The photogenerated ZnP$^{+\bullet}$–RB–AuP$^{\bullet}$ state was brought back to the ground state by backward ET. These recombination processes were 10–100 times slower than the corresponding forward ET but showed the same characteristic dependences on the bridge structure [57]. That is, the smaller the energy difference between the relevant donor and bridge states or the shorter the donor–acceptor distance the faster the transfer rate. The backward ET process was conveniently studied by monitoring the decay of the zinc-porphyrin radical cation at 680 nm using transient absorption. The rate constant for recombination was calculated as the inverse of the measured zinc-porphyrin radical cation lifetime, because no other deactivation processes competed with the recombination. Excellent signals and relatively fast decay rates were obtained for the systems with high quantum yields for forward ET. The absorption decayed biexponentially with the two lifetimes corresponding to the singlet excited state lifetime, mainly governed by the rate of the forward ET, and the lifetime of the zinc-porphyrin radical cation. For the systems with slow backward ET, the triplet excited state of the zinc porphyrin made a substantial contribution to the transient. This is because intersystem crossing from the singlet excited state of the zinc porphyrin competed successfully with the slow forward ET. The tails of the kinetic traces, at times longer than 10 ns, were therefore fitted to three-exponential expressions. Two lifetimes for the triplet state of the zinc porphyrin [45, 49] and one for the zinc-porphyrin radical cation. Fitting the transfer rates to Eq. (4.7) gave an attenuation factor of $0.4\,\text{Å}^{-1}$.

The driving force for the backward ET was calculated in the same token as for the forward process (Eq. (4.13)). It was found that differences in the driving force between forward and backward ET was not enough to explain the large differences in rates observed for the two processes. The driving forces are very similar, −1.17 and −0.95 eV for the forward and the backward ET, respectively, in $ZnP{-}RB{-}AuP^{+}$ dissolved in dimethylformamide (DMF). Instead, the difference in electronic coupling for the two processes was found to explain the observed trends. The electronic coupling calculated using the Marcus equation (Eq. (4.3)) is 1 order of magnitude larger for the forward ET than for ET in the opposite direction.

Analyzed in the context of superexchange, the virtual state $D^{+\bullet}{-}B^{-\bullet}{-}A$ is invoked to affect the mediation of electronic coupling between the donor and the acceptor, see Figure 4.6. It was found, though, that it was not applicable for the topical systems [27]. Taking the electronic couplings obtained for the 3B and the NB systems as the starting point, extrapolating to infinite energy gap gave a significant and thus unrealistic electronic coupling when energy-gap calculations were based on this virtual state. In addition, a negative energy gap was obtained for the $ZnP{-}AB{-}AuP^{+}$ system even for solvents in which no sequential transfer was observed [20, 60]. The inconsistency was advocated to be caused by the inappropriateness in relating the virtual state in the tunneling process to a property connected to a relaxed state, that is, the experimentally determined reduction potential for the bridge radical-anion, instead of relating it to a vertical Franck–Condon state. Accordingly, the electronic couplings were plotted versus the reciprocal energy difference between the first excited states of the donor and the bridge. A reasonable fit to a straight line was obtained for this plot [27]. Although not a Franck–Condon state, $ZnP{-}RB^{+\bullet}{-}AuP^{\bullet}$ was identified as the virtual state involved in the backward ET process. Estimated in the same way as for the forward

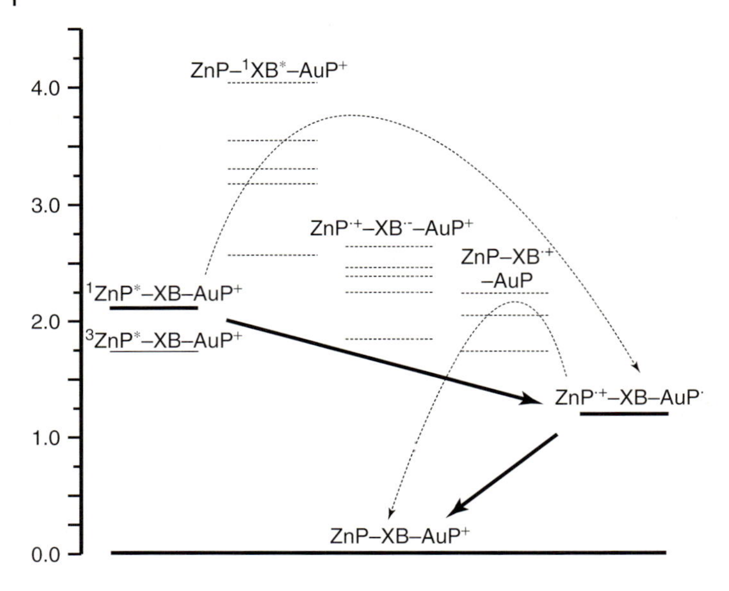

Figure 4.6 A schematic representation of the forward and backward electron transfer processes observed in the ZnP–XB–AuP+ systems. The virtual states potentially involved in coupling the initial states and final states together are shown in the center of the scheme. The pathways described by the dashed arrows indicate the virtual states most likely to be involved.

process, the energy gap differences obtained for the backward ET were found to be 1.5–3.5 times larger than the ones corresponding to the forward ET [57]. The differences in energy gaps corroborated, within the context of the McConnell model, to the difference in damping factors found for the two processes (0.3 vs. 0.4 $Å^{-1}$).

To fully account for the differences in rates for forward and backward ET, not only the energy gap dependence had to be considered but also differences in the donor–bridge electronic coupling V_{DB}. The bridge coupling terms were extracted from the total coupling V_{DA} as the slope ($V_{DB} V_{BA}$, see Eq. (4.10)) obtained by plotting the total coupling versus the inverse energy gap [57]. A significantly smaller slope, that is, smaller bridge coupling terms, was obtained for backward ET compared to forward ET, see Figure 4.7. It is reasonable that this difference can be traced back, at least to some extent, to variations in orbital interactions between the bridges and the different donors. The lowest unoccupied molecular orbital (LUMO) of the zinc porphyrin is of major importance in the forward ET, where the zinc porphyrin acts as an electron donor. Acting instead as an electron acceptor, its highest occupied molecular orbital (HOMO) orbital plays an active role in the ET in the opposite direction. While there is a substantial orbital coefficient at the *meso*-carbon, at the point of attachment of the bridge, in the LUMO there is a node at the same position in the HOMO. In addition, it has been suggested that the electron is initially localized on the porphyrin ring upon reduction of gold(III) porphyrins but subsequent relaxation localizes the electron on the gold atom [67, 68]. This would give markedly different ET pathways for the two processes due to

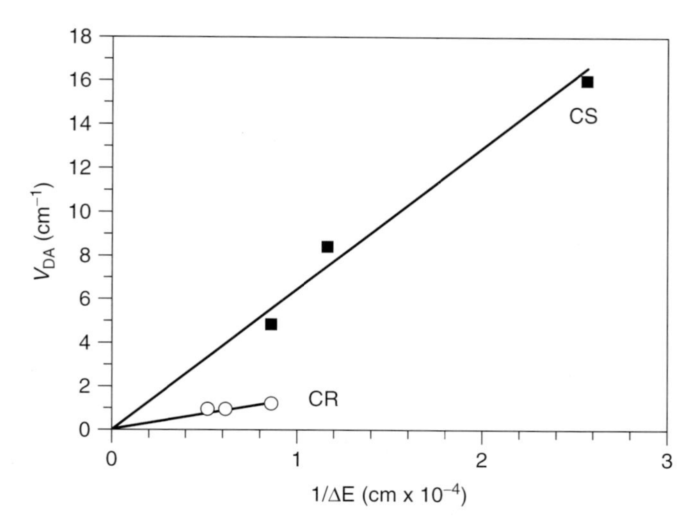

Figure 4.7 Electronic coupling for electron transfer in the "forward" (CS) and "backward" (CR) direction, V_{DA}, versus the inverse energy gap between the first singlet excited state of the donor and the bridge in ZnP–RB–AuP. The lines are linear fits with fixed intercept $V_{DA}(0) = 0$.

alterations in both the transfer distance and in the communication between the bridge and the gold porphyrin/atom. Shorter through-bond distance for ET has been suggested by Lindsey, Bocian, and coworker as the major reason for faster rates for ring-centered redox processes compared to the analogous metal-centered processes observed in monolayers of zinc(II) and iron(III) porphyrins, respectively, on Au(111) [69].

4.2.1.4 **Induced/Enhanced Intersystem Crossing**

In marked contrast to the efficient ET observed in the zinc(II)/gold(III) and free-base/gold(III) porphyrin arrays, porphyrin radical cation formation was never observed for the corresponding zinc(II)/chloroiron(III) and free-base/chloroiron(III) arrays [70, 71]. Exceptions were the anthracene-bridged systems for which the radical cation was observed in very polar solvents. Instead, the high-spin chloroiron(III) porphyrin efficiently induced an enhanced intersystem crossing of the singlet excited zinc(II) porphyrin to its first excited triplet state. A minor fraction (10–40%) of the singlet excited state quenching was accounted for by SEET via the Förster mechanism.

Corroborating the enhanced intersystem crossing hypothesis, a large increase in the transient absorption intensity in the 450–520 nm region, where the ^{3}ZnP* absorption dominates the spectrum, accompanied the singlet excited state quenching. A quenching that was easily monitored, both in steady state and time-resolved fluorescence measurements, because the iron(III)porphyrin is nonfluorescent at ambient temperature. The rate enhancement for intersystem crossing in ZnP–OB–Fe(Cl)P was calculated from transient absorption and time-resolved fluorescence data as the quotient of two experimentally accessible ratios. The numerator – the ratio between the quantum yields for intersystem crossing in the D–B–A system and its reference compound – was obtained as the ratio between the transient absorption intensities at 470 nm for ZnP–OB–Fe(Cl)P and its reference system ZnP–OB. The denominator – the ratio between the fluorescence quantum yields for the D–B–A system and its reference compound – was obtained as the ratio between the singlet lifetimes for the same two compounds. The latter requires, based on the identical ground state absorption spectra this is a reasonable assumption, that the rate constant for fluorescence is the same in both the arrays, ZnP–OB–Fe(Cl)P and ZnP–OB. This analysis gave a factor 2 in rate enhancement, for both the zinc(II) and the free-base porphyrin system. Similar effects have been observed by others in biomimetic system comprising iron(III) porphyrins. A large increase in intersystem crossing (ISC) rate was observed for the excited zinc(II) porphyrin in a set of Hangman biporphyrins with an unusual mixed axial coordination configuration of weak-field and strong-field ligands [72]. In the same vein, enhanced ISC was observed for a deazaflavin chromophore covalently held in the vicinity of an iron porphyrin center [73]. The influence of remote components with an unpaired electron on the excited state dynamics has also been explored in copper(II)/free-base porphyrin systems with respect to its dependence on relative orientation and frontier orbital shapes [74].

A small but clearly discernable dependence of the bridge structure is observed for the rate enhancement, that is, the smaller the energy gap between the zinc(II) or free-base porphyrin and the bridge the larger the rate enhancement. However, a straightforward analysis analogous to the one above is not possible for systems other than the OB linked arrays. This is because the presence of an efficient TEET process decreases the transient triplet absorption and blocks out the effect of the enhanced intersystem crossing in the systems with fully conjugated bridges.

The enhanced intersystem crossing was clearly induced by the high-spin chloroiron(III) porphyrin. The low-spin complexes, the bis(imidazole)iron porphyrin arrays, performed rather differently from the high-spin systems [70]. ET was found to be the major deactivation pathway for these systems, except for the OB linked array. Confirmed by EPR and UV-vis spectroscopy, the spin state of the iron center was readily tuned from high spin ($S = 5/2$) to low spin ($S = 1/2$) by adding imidazole to the solutions of chloroiron(III) porphyrins. The change in spin state was caused by the exchange of the single weak field, axial chloride ligand by two axial imidazole ligands, see Figure 4.8. In contrast to the high-spin systems, no significant transient absorption was observed for the low-spin systems in the 450–520 nm region, where the ^{3}ZnP* absorbs in the transient absorption. Instead, a distinct peak was observed at 680 nm, assigned to the zinc porphyrin radical cation $ZnP^{\bullet +}$, supporting the conclusion that ET was an important deactivation channel in this case. The high-spin complexes show monotonic decaying transients at 680 nm whereas the low-spin complexes, except for the OB system, show an initial increase in transient absorption [70]. The observed build-up times are the same as the single excited state lifetimes for the zinc porphyrins in the 3B and NB bridged systems. The rise time observed for the AB bridged system is much faster

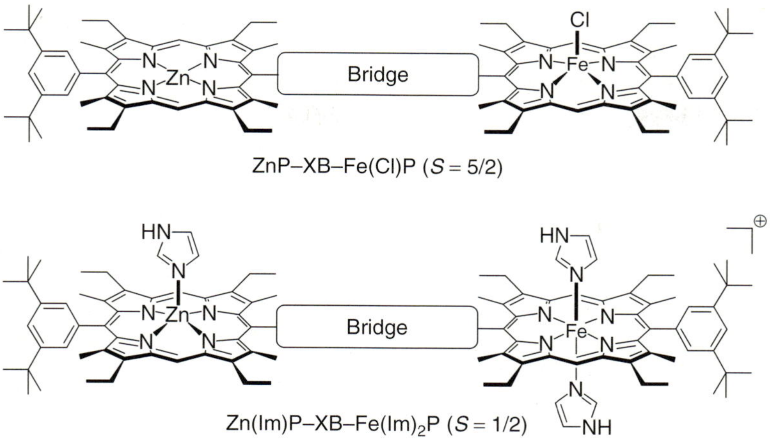

ZnP–XB–Fe(Cl)P ($S = 5/2$)

Zn(Im)P–XB–Fe(Im)$_2$P ($S = 1/2$)

Figure 4.8 The coordination patterns for the high- and the low-spin systems ZnP–XB–Fe(Cl)P and Zn(Im)P–XB–Fe(Im)$_2$P, respectively.

than the fluorescence lifetime for the initially singlet excited zinc porphyrin. This might indicate the presence of an equilibrium between a zinc porphyrin localized singlet excited state and a zinc porphyrin-anthracene charge-separated state, an intermediate on the stepwise ET pathway to Zn(Im)P$^+$–AB–Fe(Im)$_2$P.

Although the high-spin complexes showed a higher degree of average fluorescence quenching, in the range of 67–91% quantum efficiency, the low-spin complexes showed a greater dependence on the bridge structure. The quantum efficiency for fluorescence quenching varied from 23% for the OB linked system to 95% for the anthracene linked one. It was concluded that SEET was the only deactivation accessible for Zn(Im)P–OB–Fe(Im)$_2$P$^+$, because neither the zinc porphyrin radical cation nor enhanced intersystem crossing was observed for this system. Further, the small quenching observed in Zn(Im)P–OB–Fe(Im)$_2$P$^+$ was entirely accounted for by SEET according to the Förster theory. Only 1–20% of the observed quenching could be accounted for by SEET via the dipole–dipole mechanism in the other systems.

Estimations based on the Weller equation (Eq. (4.13)) indicate that the driving force for ET should be approximately 1 eV (2-MTHF and CH_2Cl_2) for both the zinc(II)/chloroiron(III) and the (imidazole)zinc(II)/bis(imidazole)iron(III) porphyrin pairs. Further, the process is estimated to be barrier free. Thus, the driving force alone cannot explain the differences in deactivation of the zinc porphyrin singlet excited state between the high- and low-spin systems. The observed differences must instead arise from differences in the effective electronic coupling.

Changes in both orbital symmetries and orbital energies are expected when the coordination geometry of the iron(III) porphyrin is changed from square pyramidal to octahedral [75, 76]. The resulting changes in electron and spin densities at the point of attachment of the bridge are likely to exert decisive influence on the communication between the iron(III) porphyrin and the rest of the system [77]. The internal interaction between the iron center and the porphyrin ring is also certainly affected by the structural rearrangement from an out-of-plane five-coordinated system to a symmetrical octahedral arrangement [78].

It is interesting to compare the ET rate for the low-spin iron(III) systems with those for the corresponding gold(III) systems, $ZnP{-}XB{-}AuP^+$. Although the driving force is only 0.6 eV (2-MTHF) for the latter, and the reorganizations energies very similar, the observed ET rates are almost the same in both systems. This shows, again, that the electronic coupling has a decisive role in ET processes.

In summary, we have experimentally shown that the structure of the bridge linking a donor and an acceptor clearly and by the same token affects a range of different transfer processes, each one originating from a specific electronic coupling. The donor–acceptor electronic coupling is "mediated" by the bridge and it decreases with distance. The magnitude of the electronic coupling and its distance dependence depend critically on the energy gap between the donor and bridge states relevant for the specific transfer process. The results indicate that the bridge state to which the energy gap should be related is better regarded as a Franck–Condon state rather than a thermally relaxed state. Note, the bridge state is never populated in a superexchange process; it is simply a virtual state reducing the tunneling barrier between the initial and final states. There also seem to be a dependence related to the point of attachment between the donor and the bridge that can be rationalized in terms of frontier orbital theory. In addition, a strong dependence on the bridge conformational dynamics was observed. The effect of this conformational dynamic on the overall electronic coupling will be dealt with in the remainder of this chapter.

4.2.2
Conformational and Temperature Effects on Electron Exchange

In this section, we will explore the effect of molecular conformation on the bridge-mediated through-bond mechanism of electron and energy transfer. Molecular conformation is first shown to control transfer rates through the preparation of donor states with different energetics. The energetics have impact on the transfer rate by changing the driving force (cf. Eq. (4.3)) as well as, via the donor–bridge energy gap, the mixing of donor and bridge states (cf. Eqs. (4.7)–(4.10)). Further, the torsion angles between planes of rigid units within the D–B–A structure are shown to modulate the electronic coupling. The consequential temperature dependence is explored, both experimentally and theoretically. The magnitude of the impact of temperature on the electronic coupling is modeled by taking into account both the Boltzmann distribution of conformations (depends on the energy to rotation) and the conformational dependence of the electronic coupling.

From the semiclassical expression for the ET rate (Eq. (4.3)) it follows that the electronic coupling can be derived from the intercept of a plot of $\ln(kT^{1/2})$ versus T^{-1}. This procedure is often used but assumes that the parameters that enter the expression are temperature independent. As will be discussed thoroughly below, the electronic coupling can sometimes be strongly dependent on temperature. Further, many solvent properties that in one way or the other influence the transfer rates also depend on temperature. Thus, when studying the temperature dependence of intramolecular charge and energy transfer processes it is important to account for these solvent effects. One important factor that needs consideration is how the viscosity of the solvent varies with temperature. This has a large impact on the large amplitude movements, for example, rotations of molecular planes, slowing them down as the solvent viscosity increases. Changes in solvent viscosity can in this way have a significant impact on the effective electronic coupling, which has been shown to be crucially dependent on conformation [30, 35, 45, 46, 50, 79–82]. In addition, both the dielectric constant and refractive index, two important solvent properties that influence both the reorganization energy and the driving force, are functions of both temperature and viscosity. All these solvent-induced effects obscure the effects of molecular parameters, such as energy gaps and conformation, on the studied transfer reactions. Although very important, the solvent-induced effects will not be discussed further. Instead, we will mainly be concerned with the effects of molecular parameters on the intramolecular transfer rate.

To facilitate comparison between experimental and theoretical results it is advantageous to reduce the influence of solvent-induced effects on the parameter of interest. In this respect, studying TEET is very useful since the process does not involve the movement of charge. In marked contrast to ET, the effect of dielectric stabilization and variations in the outer reorganization energies is minimal. Therefore much of what will be presented below are results from theoretical and experimental studies of TEET. As discussed in Section 4.1, from a quantum mechanical perspective, the integrals that govern the electron exchange interactions for ET, hole transfer (HT), and TEET are related. For saturated bridges it has been shown that the electronic coupling elements for the three processes are approximately related according to $|V_{\mathrm{TEET}}| = \mathrm{const}|V_{\mathrm{ET}}||V_{\mathrm{HT}}|$ [83, 84].

Many theoretical and experimental studies have found that the main conformational variables that govern the electronic coupling in D–B–A systems built up by a series of individually planar π-conjugated systems are the dihedral angles between individual units as illustrated in Figure 4.9.

According to Arrhenius the temperature dependence of the rate, k, of an activated process is described by $k = A\exp(-E_a/RT)$, where E_a is the activation energy. In

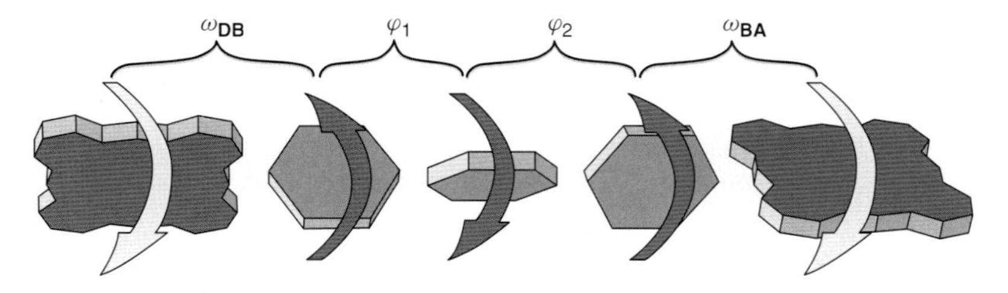

Figure 4.9 A schematic donor–bridge–acceptor molecule built up by units with freedom to rotate relative to each other. The dihedral angles between the subunits are indicated.

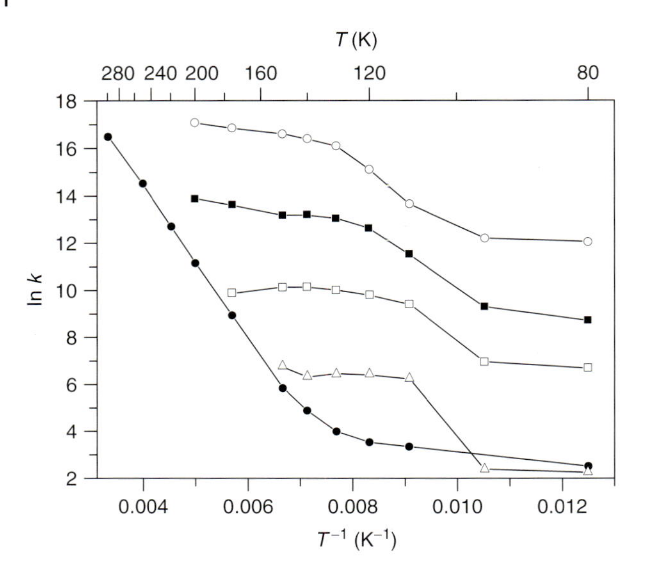

Figure 4.10 The logarithmic rate constant for the relaxation to the ground state of the triplet state of the ZnP–xB reference compounds (•), and the logarithmic rate constants for TEET from ZnP in ZnP–nB–H_2P with $n = 2$(○), 3 (■), 4 (□), and 5 (△).

complex D–B–A systems, built up by several units, each rotation is associated with a unique energy to rotation. It can thus be expected that temperature dependences of transfer rates in this type of systems can be quite complex. If the minimum energy conformation is associated with the highest electronic coupling the transfer rate might even increase with decreasing temperature. This apparent negative activation has been observed for several D–B–A systems in certain temperature intervals [50, 85]. The ZnP–nB–H_2P series shown in Figure 4.1 may serve as an example. An increase in energy transfer rate with decreasing temperature is observed for some of these systems as shown in Figure 4.10 where the logarithmic TEET rate is plotted versus the reciprocal temperature.

4.2.2.1 **Modulating the Transfer Rate by Preparing Different States using Conformational Trapping**

As discussed in the previous sections, the rates of superexchange governed processes depend on the donor–bridge energy gap (cf. Eqs. (4.7)–(4.10)) and the driving force (cf. Eq. (4.3)). Since the energy of a molecule in a particular state is a function of conformation, conformational changes may serve as a way to modulate the transfer rate. This was utilized in a study of the photoinduced processes occurring in a D–B–A molecule where a ferrocene (Fc) donor and a fullerene (C_{60}) acceptor are separated by two porphyrins (P_2), acting as a bridge for charge recombination (Figure 4.11) [86].

In this study the electron donating state was prepared by exciting the bridge (the porphyrin dimer) and an initial ET step from the bridge to the acceptor was

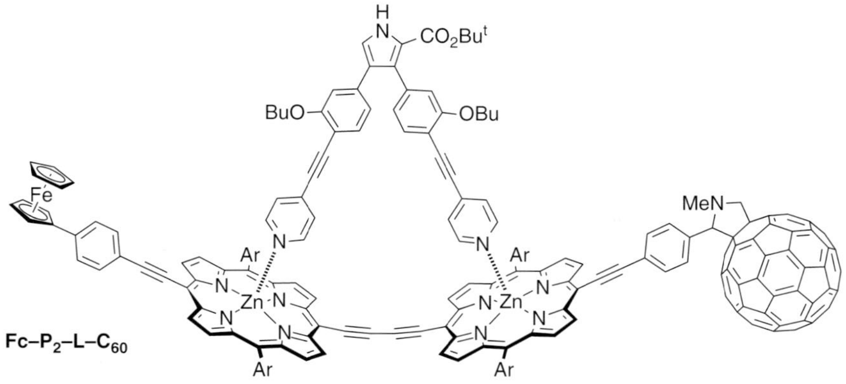

Figure 4.11 The Fc–P_2–C_{60} D–B–A compound shown together with the bound ligand used to planarize the bridge.

realized. To simplify the experimental analysis the initial ET step was studied in a system without the secondary Fc donor. Since the planar and perpendicular conformations of P_2 could be selectively excited, two different conformational states could be prepared [87]. ET occurred independently from each one of these two conformations. Alternatively a bidentate dipyridyl pyrrole ligand could be used to planarize the bridge structure as is illustrated in Figure 4.11. This study found that the transfer rate from P_2 in its perpendicular conformation was approximately 1 order of magnitude faster than that from its planar conformation. This observation was expected because the excited state of the perpendicular conformer is approximately 0.2 eV more energetic and, thus, has a higher the driving force for ET. Further it was discussed that the twisting of the dihedral angle forced the excited state to be localized on the porphyrin closest to the fullerene acceptor and thus decreasing the effective donor (i.e., the bridge) acceptor separation. In the full D–B–A system, after the initial ET step, a subsequent ET step from the secondary donor (Fc) to the bridge (P_2) produces a long-range charge separated state. The recombination of this state is mediated by P_2 in its ground state, which is in this respect acting as a bridge. Thus the recombination rate will be strongly dependent on the porphyrin–porphyrin dihedral angle. The energy required to rotate one of the porphyrin moieties relative to the other is very low in the ground state enabling an almost uniform Boltzmann distribution of conformations [87]. A comparison of the recombination rate for this random distribution to the one in which the ligand planarizes the porphyrin dimer revealed that the rate was an order of magnitude greater for a planar bridge structure in accordance with an, on average, higher electronic coupling.

4.2.2.2 **Conformational Control of Tunneling through Molecular Bridges**

In an elegant series of experimental studies Harriman and co-workers probed the effect of the torsion angle between two phenyl units of a bridge separating a Ru(II)$(tpy)_2$ donor and an Os(II)$(tpy)_2$ acceptor. The average dihedral angle between

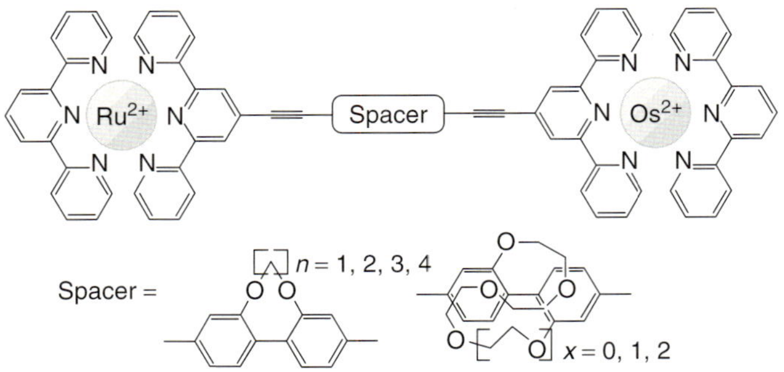

Figure 4.12 Molecular structures of the set of systems used to probe the effect of the dihedral angle in the bridge structure.

the phenyl units was controlled by a covalent linker as shown in Figure 4.12 [79, 81, 88, 89].

The study revealed a pronounced conformational dependence where the electronic coupling is greatly reduced when the phenyl planes are close to orthogonal. In one of the systems a switch in transfer mechanism, from a combination of through-space and superexchange to a combination of hopping and superexchange, was noted when going from low temperatures and rigid media to high temperatures and fluid media [90]. In this series of studies it was, furthermore, experimentally confirmed that the dihedral angle dependent electronic coupling for TEET follows the squared dependence of the overlap integral of two mutually rotating π-orbitals [30, 85]. This is what would be expected for the dependence of the product of the electronic coupling for ET and HT [83, 84]. Thus, for the normalized electronic coupling associated with the dihedral angle, φ, we have

$$V_{\mathrm{TEET}}(\varphi) = V_{\mathrm{ET}}(\varphi) V_{\mathrm{HT}}(\varphi) = \cos^2\varphi \tag{4.14}$$

Note that this only applies to the mediation by π-conjugated orbitals. There is also a small contribution from σ-orbitals that does not show this conformational dependence.

Another, less synthetically demanding, way to control the conformation is to use temperature to vary the Boltzmann distribution of conformations. With this in mind, we have performed a series of parallel theoretical and experimental studies where the conformational impact on TEET in the ZnP–nB–H_2P series (see Figure 4.1) was investigated [33, 50]. Combined with the experimental studies a set of density functional theory (DFT) and time-dependent density functional theory (TD-DFT) calculations were used to derive a model for the Boltzmann averaged electronic coupling. How well the model reproduces the observed trends was evaluated by comparison with the experimental results obtained at different temperatures. The quite good conformity between experiments and model encouraged us to expand the theoretical study and to include a large set of donor and bridge structures with the aim of finding sets of parameters, for both ET and TEET, that would potentially enable a priori predictions of β-values [35].

The quantum chemical calculations were performed using the Gaussian 03 program suite [91] at the B3LYP/6-31G(d) level and have been thoroughly described previously [33, 35, 50]. The calculations include the calculation of the potential energy as a function of dihedral angles between the planes of individual subunits, identified as being the major parameters modulating the electronic coupling in this type of systems [30, 33, 35, 45, 50, 79–82, 85]. There are two sets of dihedral angles relevant for the investigated systems: the dihedral angle between either the donor or the acceptor plane and the plane of the first phenyl unit of the bridge (ω), and a series of dihedral angles between planes defined by neighboring phenyl units of the bridge (φ) (see Figure 4.9). The potential energy as a function of dihedral angle was calculated by changing the angle followed by geometry optimization with respect to all other parameters at each point. To save calculation time the calculations were performed on the building blocks (donor, bridge, or acceptor) instead of the whole D–B–A systems. For the calculations of the electronic coupling, symmetrical model systems – D–nB–D, $n = 2, 3, 4$, and 5 – was used. Again, to reduce calculation time, the systems were stripped of saturated groups, judged to have minor impact on the electronic properties. The electronic couplings for ET and TEET were estimated as half the splitting of the LUMO and LUMO + 1 orbitals and as half the triplet excitation energy difference between the two lowest triplet excited states, respectively. Importantly, for this procedure to be valid the system has to be at the avoided crossing geometry. To facilitate this, the calculations of the electronic coupling were performed on symmetric donor–bridge–donor (D–nB–D, $n = 2, 3, 4$, and 5) systems. Avoided crossing geometries were thus achieved by assuring that the system had a mirror plane or a C_2-rotation axis that forced the wavefunction to be equally distributed on the two donor moieties. The rotations described above were always done in such a way that the symmetry was preserved. Another important prerequisite for the above analysis is the verification that the wavefunctions/orbitals used are the even/odd combination of wavefunctions/orbitals localized on the two donor moieties. The approximation of estimating the electronic coupling through state and orbital splitting is expected to be less valid for shorter distances and when the donor/acceptor and bridge energies are not well separated. In this study we are concerned with systems that operate in the weak to very weak electronic coupling regime where the approximation should be fairly valid.

In order to investigate the conformational dependence of the electronic coupling the angles between the various units were varied in analogy to the procedure described for the potential energy landscape, but without subsequent geometry optimization, and for each of these configurations the electronic coupling was calculated. This procedure allowed mapping of the electronic coupling landscape and the result is shown in Figure 4.13 for ZnP–2B–ZnP that has two dihedral degrees of freedom; ω($\omega = \omega_1 = \omega_2$ due to symmetry) and φ. The figure clearly demonstrates the $\cos^2\varphi$ dependence of the electronic coupling for rotation of the planes of two subunits of the bridge relative to each other. For the dihedral angle, ω, a $\cos^4\omega$ dependence was observed which originates in that, due to symmetry

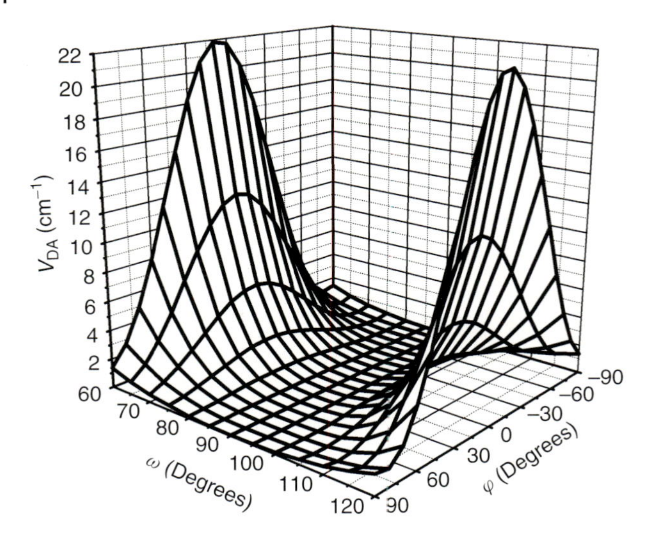

Figure 4.13 The conformational landscape for the calculated electronic coupling of TEET through ZnP–2B–H_2P approximated by the calculated triplet state splitting of a symmetric ZnP–2B–ZnP system.

constraints, both the D–B and B–D angles in the D–B–D system was rotated simultaneously.

This suggests, and was found to be valid for all the systems in the study ($n = 2$–5), that the total electronic coupling can be approximated by a product of one-parameter functions; $V(\omega_1, \varphi_1, \ldots, \varphi_{n-1}, \omega_2) = V(\omega_1)V(\varphi_1)\ldots V(\varphi_{n-1})V(\omega_2)$. Further it was found that, to a good approximation, the total energy for a given conformation could be described by the sum of one-parameter energy functions. This led to that the total Boltzmann averaged electronic coupling can be factorized into one-parameter functions according to

$$\begin{aligned}&\langle V_{\mathrm{DA}}\,(\omega, \varphi_1, \varphi_2, \ldots, \varphi_{n-1})\rangle \\ &\quad = V_x\langle V\,(\omega)\rangle \prod_{m=1}^{n-1} \langle V\,(\varphi_m)\rangle = V_n\langle V\,(\omega)\rangle\langle V(\varphi)\rangle^{n-1}\end{aligned} \tag{4.15}$$

where

$$\langle V(a)\rangle = \frac{\int V(a)e^{-E(a)/RT}da}{\int e^{-E(a)/RT}da} \tag{4.16}$$

where a represents any of the dihedral angles ω or φ. In Figure 4.14 the values of the Boltzmann averaged contributions to the total electronic coupling for the donor–bridge angle (ω) and bridge–bridge angle (φ) are shown. This illustrates the two opposing effects of increasing and decreasing electronic coupling giving transfer rates with the complex temperature dependences shown in Figure 4.10.

The derived model (Eqs. (4.15) and (4.16)) was found to give accurate predictions of the electronic coupling and distance dependence for TEET [33, 50]. Further it

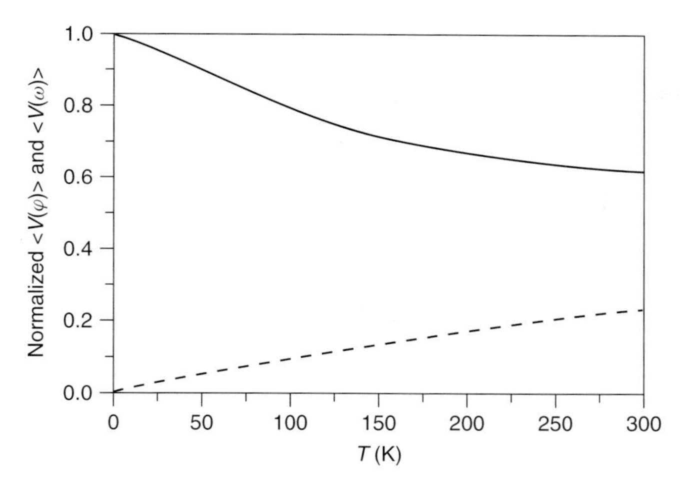

Figure 4.14 The temperature dependence of the Boltzmann averaged weighing factors of the electronic coupling due to donor/acceptor–bridge rotation $\langle V(\omega)\rangle$ (dashed line) and bridge–bridge rotation $\langle V(\varphi)\rangle$ (solid line).

was found that the attenuation factor, β, was independent of the value of ω. This can more easily be visualized if one assumes exponential distance dependence (Eq. (4.7)). It then follows that β can be extracted from the derivative of the logarithmic electronic coupling:

$$\beta = -2\frac{d\ln V_{\mathrm{DA}}}{dR_{\mathrm{DA}}} \tag{4.17}$$

Thus, the conformational dependence of β derived from Eq. (4.15) can be deduced from

$$\ln\langle V_{\mathrm{DA}}\rangle = \ln V_n + \ln\langle V(\omega)\rangle + (n-1)\ln\langle V(\varphi)\rangle \tag{4.18}$$

The first term in Eq. (4.18) is a conformation independent constant, that is, unique for each bridge length and donor–bridge energy gap (ΔE_{DB}). This series (varies with n) of temperature independent constants represent the donor–bridge energy-gap-dependent β-value for a planar bridge structure. The second term is conformation dependent (donor–bridge angle) but independent of bridge length, and will thus not affect the β-value. It will, on the other hand, be a major contribution to the total electronic coupling and, thus, to a large extent govern the temperature dependence of the transfer rates. The last term also depends on conformation and, since n varies with R_{DA}, will, according to the model, be the sole factor governing the conformational dependence of the β-value. Thus, the total β-value can be split into a temperature independent constant that will give the lowest obtainable β-value for a given repeating bridge structure ($\beta_0(\Delta E_{\mathrm{DB}})$), and a temperature-dependent variable that will reflect the average conformation of the bridge $\beta(T)$:

$$\beta_{\mathrm{tot}} = \beta_0\,(\Delta E_{\mathrm{DB}}) + \beta(T) \tag{4.19}$$

As was discussed earlier, the β-value depends on the donor–bridge energy gap. According to Eq. (4.19) this only has impact on the part of the β-value, that is, independent on conformation and/or temperature.

The factor, $\beta(T)$, that governs the temperature dependence of the β-value describes the impact on the electronic coupling due to conformational disorder of the bridge structure. In this factor the electronic coupling for a specific conformation is weighted with its corresponding potential energy. For the OPE bridge structure the energy to rotation approximately follows $E_\varphi \sin^2\varphi$, where E_φ is the barrier height. If the angle dependence of the electronic coupling follows Eq. (4.14) the normalized bridge conformational dependence can be expressed by Eq. (4.20)

$$\langle V(\varphi)\rangle = \frac{\int V(\varphi)\, e^{\frac{-E(\varphi)}{RT}}\, d\varphi}{\int e^{\frac{-E(\varphi)}{RT}}\, d\varphi} = \frac{\int \cos^2\varphi\, e^{\frac{-E_\varphi \sin^2\varphi}{RT}}\, d\varphi}{\int e^{\frac{-E_\varphi \sin^2\varphi}{RT}}\, d\varphi} = \frac{1}{2}\left(1 - \frac{I_1\left(\frac{-E_\varphi}{2RT}\right)}{I_0\left(\frac{-E_\varphi}{2RT}\right)}\right) \tag{4.20}$$

where $I_a(b)$ is the modified Bessel function of the first kind of order a and E_φ is the barrier to rotation of two bridge units in relation to each other. Evaluation of this expression reveals that $\langle V(\varphi)\rangle$ varies from 1 to 0.5 for a planar ($T \to 0$) and randomized ($T \to \infty$) bridge structure, respectively, so that (cf. Eqs. (4.17)–(4.19) and assuming that $R_{DA} = nR_0 + \text{constant}$) $0 \le \beta(T) \le 2\ln 2/R_0$. The same maximum range is expected for all bridge structures with potential energy minima associated with planar conformations. Thus, using this relation, one can easily estimate the maximum effect of rotational disorder on the attenuation factor in any such D–B–A system. Different shapes and heights of the energy barrier to rotation of neighboring units for different bridge structures will give different temperature dependences but *not* affect the maximum range.

For this model to agree with experimental data from solution measurements over a large temperature range it was necessary to take the variation in viscosity of the solvent into account. This was done by introducing an apparent activation energy [92]:

$$E_{app} = E_i + \alpha E_\eta \frac{T^2}{(T - T_0)^2} \tag{4.21}$$

where E_i is the intrinsic energy barrier to rotation, $E_\eta = 1.3469\,\text{kJ/mol}$ and $T_0 = 81$ K for 2-MTHF, and typical α-values are around 0.1 (depending on the sharpness of the viscosity-induced activation barrier).

The effect of using a fitted apparent activation energy instead of a constant value is shown in Figure 4.15. The apparent activation energy can be used to take into account how the hindering of rotational motions can perturb the Boltzmann distribution so that the conformations with higher electronic coupling are given greater weight. This can be visualized by considering the relative probability for continued rotation, $P_{rot}(\varphi)$, compared to the probability for TEET process at each

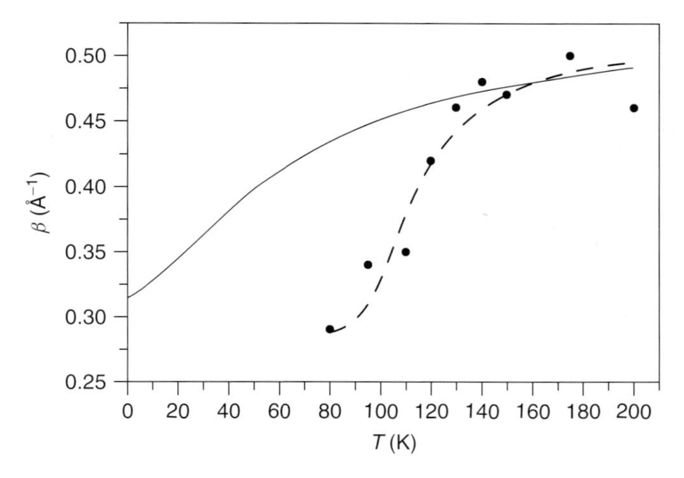

Figure 4.15 The experimentally determined β-values (•) and the simulated β-values (lines) for the case of a temperature independent barrier to rotation ($E(\varphi) = 2.4$ kJ/mol, —) and taking the viscosity into account by using a temperature dependent apparent energy to rotation described by using Eq. (4.21) (- - -).

conformation (see Figure 4.16(b)):

$$P_{\text{rot}}(\varphi) = \frac{k_{\text{rot}}}{k_{\text{rot}} + k|V(\varphi)|^2} = \frac{\frac{k_{\text{rot}}}{k}}{\frac{k_{\text{rot}}}{k} + \cos^4(\varphi)} \tag{4.22}$$

where k_{rot} is the rate of rotation and $k|V(\varphi)|^2 = k_{\text{TEET}}$ is the rate of triplet energy transfer. In Figure 4.16(a) the probability of rotation is plotted for three different rate ratios, k_{rot}/k. The figure shows that if the rate of rotation is much greater than the transfer rate the perturbation of the Boltzmann distribution is not very large because the probability of rotation is almost constant. If the ratio instead is low, the probability of rotation remains high for conformations with low electronic coupling but decreases for the conformations with higher electronic coupling. In this way the effective conformational distribution is shifted toward conformations with higher electronic coupling.

The success of the developed model has encouraged us to extend the theoretical study to include a large set of donors and bridge structures [33, 35]. This work is a tentative step toward building a library of parameters to enable a priori predictions of β-values when only a few parameters of the building blocks are known. The proposed method to achieve this is based on the donor–bridge energy gap dependent minimum attenuation factor $\beta_0(\Delta E_{\text{DB}})$ coupled with the temperature-dependent bridge disorder factor, $\beta(T)$ in accordance with Eq. (4.20). Since the goal is to derive β-values the main focus of this study is on the bridge structures and the appended donors are just a means to tune the donor–bridge energy gap. The studied repeating bridge systems, OPE, oligophenylenevinylene (OPV), oligothiophene (OTP), oligophenylene (OP), and oligofluorene (OF), are

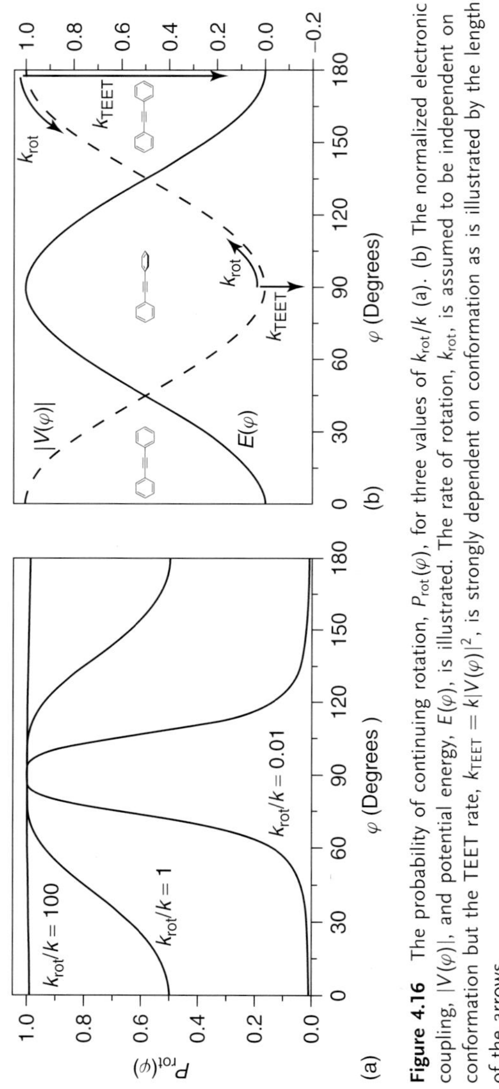

Figure 4.16 The probability of continuing rotation, $P_{rot}(\varphi)$, for three values of k_{rot}/k (a). (b) The normalized electronic coupling, $|V(\varphi)|$, and potential energy, $E(\varphi)$, is illustrated. The rate of rotation, k_{rot}, is assumed to be independent on conformation but the TEET rate, $k_{TEET} = k|V(\varphi)|^2$, is strongly dependent on conformation as is illustrated by the lengths of the arrows.

Table 4.1 The repeating bridge structures and their range (bridge length dependence) of calculated LUMO and triplet state energies.

Bridge structure[a]	$-E$(LUMO) (eV)	$E(T_1)$ (eV)
OP ($n-1$)	0.901–1.653	3.031–2.329
OF (n)	1.330–1.693	2.485–2.188
OPE ($n-1$)	1.252–2.093	2.678–1.993
OPV ($n-1$)	1.357–2.077	2.296–1.606
OTP (S, $n-1$, S)	1.245–2.081	2.323–1.472

[a] $n = 2$–5 in this study.

illustrated in Table 4.1 together with their calculated LUMO and first triplet excited state energies.

The study showed that the bridges can be divided into two types based on their response to temperature changes. On the one hand, the OPE and OPV bridges have planar conformations that are the most energetically favorable. Thus, for these bridge structures, if the temperature is increased, the β-value will also increase in accordance with on an average less planar bridge structure. The OP, OF, and OTP bridge structures, on the other hand, have the lowest energy conformations that are not planar. For these systems increasing the temperature will simultaneously populate conformations associated with both higher and lower electronic coupling. This difference between OPE-type and OP-type bridges is more easily visualized by showing the dependence of the electronic coupling on conformation together with the Boltzmann distribution of conformations for two selected temperatures (Figure 4.17).

From Figure 4.17(a) it can be seen that for OPE-type bridges, when going from low to high temperatures, the distribution moves from conformations with high-electronic coupling ($\varphi = 0^\circ$ or 180°) toward conformations with low-electronic coupling ($\varphi = 90^\circ$). For the OP-type bridges on the other hand, as can be seen from Figure 4.17(b), the distribution moves from conformations with an average electronic coupling ($\varphi \approx 45^\circ$) both toward conformations with higher and lower electronic coupling. Thus the impact of temperature on the latter kind of repeating bridge structure should be much smaller.

In summary, the study firstly derived a set of bridge specific parameters for each repeating unit that allows for a priori determination of $\beta_0(\Delta E_{DB})$. Secondly,

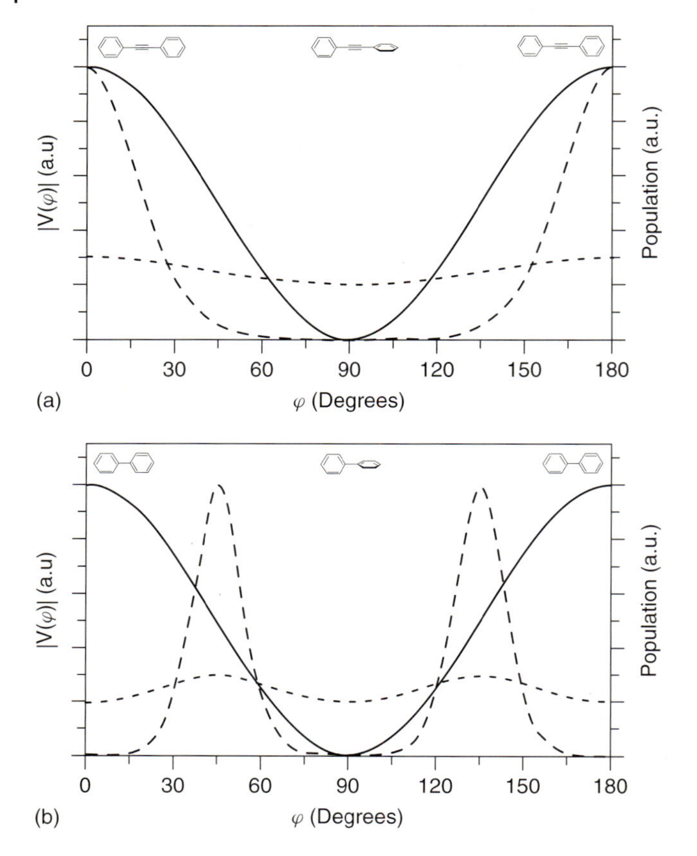

Figure 4.17 The electronic coupling (——) and Boltzmann populations at 20 K (- - -) and 300 K (– – –) for OPE-type bridges (a) and OP-type bridges (b).

the bridge conformational dependence of the electronic coupling was mapped for each bridge system. This allowed for estimations of the factor, $\beta(T)$, which describes the influence of bridge disorder. The total β-value calculated using Eq. (4.19) was shown to agree appreciably with literature values for TEET. For even better agreement with literature values the influence of solvent viscosity on the Boltzmann distribution must be estimated for each bridge system. For ET it was concluded that more elaborate ways of calculating the electronic coupling, rather than from the orbital splitting of symmetric systems, is most probably imperative.

In conclusion we have illustrated how the rates of superexchange mediated processes is controlled by molecular conformation. We have highlighted that it is necessary to take into account that the bridge is a dynamic system. Thus, it is crucial to consider averages over all available conformations when calculating parameters that govern the transfer processes. Further, we have suggested a pathway to create a library of building-block specific parameters that in the long run might enable a priori estimations of transfer rates and β-values.

4.3 Conclusions

The rate of energy and electron transfer reactions has been shown for several systems to be strongly dependent on the mediating power of the intervening medium. The single most important parameter that modulates this dependence is the electronic coupling. This is why we have devoted many years of experimental and computational efforts into finding ways to design systems with predictable electronic coupling. In this chapter we have tried to highlight some of the important molecular properties that can be used to modulate the electronic coupling in D–B–A systems such as: (i) the tunneling barrier height and width and how these interact in determining the distance dependence of electron and energy transfer reactions, (ii) the conformation of the molecular bridge and the related temperature dependence, and (iii) the state (orbital) topology of the interacting donor, bridge, and acceptor components. In complex systems all of these factors might be used to tune the electronic coupling, which, in turn, can make the interpretation of measured rates quite difficult. To say the least, it is very rarely enough to just look at the thermodynamic driving force for ET when analyzing systematic variations in series of different molecules.

References

1. Paddon-Row, M.N. (2003) *Advances in Physical Organic Chemistry*, Elsevier, vol. 38, pp. 1–85.
2. Paddon-Row, M.N. (2003) Superexchange-mediated charge separation and charge recombination in covalently linked donor–bridge–acceptor systems. *Aust. J. Chem.*, **56**, 729–748.
3. Albinsson, B. and Mårtensson, J. (2008) Long-range electron and excitation energy transfer in designed donor–bridge–acceptor systems. *J. Photochem. Photobiol., C. Photochem. Rev.*, **9**, 138–155.
4. Balzani, V. (2001) *Electron Transfer in Chemistry*, Wiley-VCH Verlag GmbH, Weinheim.
5. Marcus, R.A. (1965) On theory of electron-transfer reactions. 6. unified treatment for homogeneous and electrode reactions. *J. Chem. Phys.*, **43**, 679–684.
6. Marcus, R.A. (1956) Theory of oxidation–reduction reactions involving electron transfer.1. *J. Chem. Phys.*, **24**, 966–978.
7. Jortner, J. and Bixon, M. (1988) *J. Chem. Phys.*, **88**, 167–170.
8. Ulstrup, J. and Jortner, J. (1975) Effect of intramolecular quantum modes on free energy relations for electron transfer reactions. *J. Chem. Phys.*, **63**, 4358–4368.
9. Förster, T. (1946) Energiewanderung und Fluoreszenz. *Naturwissenschaften*, **33**, 166–175.
10. Förster, T. (1948) Zwischenmolekulare energiewanderung und fluoreszenz. *Ann. Phys.*, **2**, 55–75.
11. Pullerits, T., Hess, S., Herek, J.L. and Sundstrom, V. (1997) Temperature dependence of excitation transfer in LH2 of *Rhodobacter sphaeroides*. *J. Phys. Chem. B*, **101**, 10560–10567.
12. Dexter, D.L. (1953) A theory of sensitized luminscence in solids. *J. Chem. Phys.*, **21**, 836–850.
13. Dale, R.E., Eisinger, J. and Blumberg, W.E. (1979) Orientational freedom of molecular probes – orientation factor in intra-molecular energy-transfer. *Biophys. J.*, **26**, 161–193.

14. Börjesson, K., Preus, S., El-Sagheer, A.H., Brown, T., Albinsson, B. and Wilhelmsson, L.M. (2009) A nucleic acid base analog FRET-pair facilitating detailed structural measurements in nucleic acid containing systems. *J. Am. Chem. Soc.*, **131**, 4288–4293.
15. Anderson, P.W. (1959) New approach to the theory of superexchange interactions. *Phys. Rev.*, **115**, 2–13.
16. Halpern, J. and Orgel, L.E. (1960) The theory of electron transfer between metal ions in bridged systems. *Discuss. Faraday Soc.*, 32–41.
17. McConnell, H. (1961) Intramolecular charge transfer in aromatic free radicals. *J. Chem. Phys.*, **35**, 508–515.
18. Berlin, Y.A., Burin, A.L. and Ratner, M.A. (2002) *Chem. Phys.*, **275**, 61–74.
19. Berlin, Y.A., Hutchison, G.R., Rempala, P. and Ratner, M.A. (2003) *J. Phys. Chem. A*, **107**, 3970–3980.
20. Winters, M.U., Pettersson, K., Mårtensson, J. and Albinsson, B. (2005) Competition between superexchange-mediated and sequential electron transfer in a bridged donor–acceptor system. *Chem.-Eur. J.*, **11**, 562–573.
21. Paulson, B.P., Miller, J.R., Gan, W.X. and Closs, G. (2005) Superexchange and sequential mechanisms in charge transfer with a mediating state between the donor and acceptor. *J. Am. Chem. Soc.*, **127**, 4860–4868.
22. Weiss, E.A., Wasielewski, M.R. and Ratner, M.A. (2005) *Molecular Wires: from Design to Properties*, vol. **257**, Springer-Verlag, Berlin, pp. 103–133.
23. Gamow, G. (1928) Quantum theory of nuclear disintegration. *Nature*, **122**, 805–806.
24. Gamow, G. (1928) Quantum theory of the atomic nucleus. *Z. Phys. A*, **51**, 204–212.
25. Albinsson, B., Eng, M.P., Pettersson, K. and Winters, M.U. (2007) Electron and energy transfer in donor–acceptor systems with conjugated molecular bridges. *Phys. Chem. Chem. Phys.*, **9**, 5847–5864.
26. Kilså, K., Kajanus, J., Mårtensson, J. and Albinsson, B. (1999) Mediated electronic coupling: singlet energy transfer in porphyrin dimers enhanced by the bridging chromophore. *J. Phys. Chem. B*, **103**, 7329–7339.
27. Kilså, K., Kajanus, J., Macpherson, A.N., Mårtensson, J. and Albinsson, B. (2001) Bridge-dependent electron transfer in porphyrin-based donor–bridge–acceptor systems. *J. Am. Chem. Soc.*, **123**, 3069–3080.
28. Lopez-Castillo, J.M., Filali-Mouhim, A. and Jay-Gerin, J.P. (1993) Superexchange coupling and non-nearest-neighbor interactions in electron transfers. *J. Phys. Chem.*, **97**, 9266–9269.
29. Liang, C.X. and Newton, M.D. (1992) *Ab initio* studies of electron-transfer – pathway analysis of effective transfer integrals. *J. Phys. Chem.*, **96**, 2855–2866.
30. Newton, M.D. (2000) Modeling donor/acceptor interactions: combined roles of theory and computation. *Int. J. Quantum Chem.*, **77**, 255–263.
31. Evenson, J.W. and Karplus, M. (1992) Effective coupling in bridged electron-transfer molecules – computational formulation and examples. *J. Chem. Phys.*, **96**, 5272–5278.
32. Evenson, J.W. and Karplus, M. (1993) Effective coupling in biological electron-transfer – exponential or complex distance dependence. *Science*, **262**, 1247–1249.
33. Eng, M.P. and Albinsson, B. (2006) Non-exponential distance dependence of bridge-mediated electronic coupling. *Angew. Chem. Int. Ed.*, **45**, 5626–5629.
34. Cave, R.J. and Newton, M.D. (1996) Generalization of the Mulliken-Hush treatment for the calculation of electron transfer matrix elements. *Chem. Phys. Lett.*, **249**, 15–19.
35. Eng, M.P. and Albinsson, B. (2009) The dependence of the electronic coupling on energy gap and bridge conformation towards prediction of the distance dependence of electron transfer reactions. *Chem. Phys.*, **357**, 132–139.
36. Yeow, E.K.L., Sintic, P.J., Cabral, N.M., Reek, J.N.H., Crossley, M.J. and Ghiggino, K.P. (2000) Photoinduced energy and electron transfer in bis-porphyrins with quinoxaline Troger's base and biquinoxalinyl spacers. *Phys. Chem. Chem. Phys.*, **2**, 4281–4291.

37. Wallin, S., Hammarström, L., Blart, E. and Odobel, F. (2006) Electronic interactions and energy transfer in oligothiophene-linked bis-porphyrins. *Photochem. Photobiol. Sci.*, **5**, 828–834.
38. Fückel, B., Köhn, A., Harding, M.E., Diezemann, G., Hinze, G., Basché, T. and Gauss, J. (2008) Theoretical investigation of electronic excitation energy transfer in bichromophoric assemblies. *J. Chem. Phys.*, **128**, 074505/1–074501/13.
39. Strachan, J.-P., Gentemann, S., Seth, J., Kalsbeck, W.A., Lindsey, J.S., Holten, D. and Bocian, D.F. (1997) Effects of orbital ordering on electronic communication in multiporphyrin arrays. *J. Am. Chem. Soc.*, **119**, 11191–11201.
40. Cho, H.S., Jeong, D.H., Yoon, M.-C., Kim, Y.H., Kim, Y.R., Kim, D., Jeoung, S.C., Kim, S.K., Aratani, N., Shinmori, H. and Osuka, A. (2001) Excited-state energy transfer processes in phenylene- and biphenylene-linked and directly-linked zinc(II) and free-base hybrid diporphyrins. *J. Phys. Chem. A*, **105**, 4200–4210.
41. Pettersson, K., Kyrychenko, A., Rönnow, E., Ljungdahl, T., Mårtensson, J. and Albinsson, B. (2006) Singlet energy transfer in porphyrin-based donor–bridge–acceptor systems: interaction between bridge length and bridge energy. *J. Phys. Chem. A*, **110**, 310–318.
42. Metivier, R., Nolde, F., Muellen, K. and Basche, T. (2007) Electronic excitation energy transfer between two single molecules embedded in a polymer host. *Phys. Rev. Lett.*, **98**, 047802/1–047802/4.
43. Jensen, K.K., van Berlekom, S.B., Kajanus, J., Mårtensson, J. and Albinsson, B. (1997) Mediated energy transfer in covalently linked porphyrin dimers. *J. Phys. Chem. A*, **101**, 2218–2220.
44. Andreasson, J., Kajanus, J., Mårtensson, J. and Albinsson, B. (2000) Triplet energy transfer in porphyrin dimers: comparison between pi- and sigma-chromophore bridged systems. *J. Am. Chem. Soc.*, **122**, 9844–9845.
45. Andréasson, J., Kyrychenko, A., Mårtensson, J. and Albinsson, B. (2002) Temperature and viscosity dependence of the triplet energy transfer process in porphyrin dimers. *Photochem. Photobiol. Sci.*, **1**, 111–119.
46. Eng, M.P., Ljungdahl, T., Mårtensson, J. and Albinsson, B. (2006) Triplet excitation energy transfer in porphyrin-based donor–bridge–acceptor systems with conjugated bridges of varying length: an experimental and DFT study. *J. Phys. Chem. B*, **110**, 6483–6491.
47. Chen, H.-C., You, Z.-Q. and Hsu, C.-P. (2008) The mediated excitation energy transfer: effects of bridge polarizability. *J. Chem. Phys.*, **129**, 084708/1–084708/10.
48. Murov, S.L., Carmichael, I. and Hug, G.L. (1993) *Handbook of Photochemistry*, Marcel Dekker, New York.
49. Kyrychenko, A., Andreasson, J., Mårtensson, J. and Albinsson, B. (2002) Sterically induced conformational relaxation and structure of meso-diaryloctaalkyl porphyrins in the excited triplet state: experimental and DFT studies. *J. Phys. Chem. B*, **106**, 12613–12622.
50. Eng, M.P., Mårtensson, J. and Albinsson, B. (2008) Temperature dependence of electronic coupling through oligo-*p*-phenyleneethynylene bridges. *Chem.-Eur. J.*, **14**, 2819–2826.
51. Fortage, J., Boixel, J., Blart, E., Becker, H.C. and Odobel, F. (2009) Very fast single-step photoinduced charge separation in zinc porphyrin bridged to a gold porphyrin by a bisethynyl quaterthiophene. *Inorg. Chem. (Washington, DC, United States)*, **48**, 518–526.
52. Ohkubo, K., Sintic, P.J., Tkachenko, N.V., Lemmetyinen, H., Ou, Z., Shao, J., Kadish, K.M., Crossley, M.J. and Fukuzumi, S. (2006) Photoinduced electron-transfer dynamics and long-lived CS states of donor–acceptor linked dyads and a triad containing a gold porphyrin in nonpolar solvents. *Chem. Phys.*, **326**, 3–14.
53. Segawa, H., Takehara, C., Honda, K., Shimidzu, T., Asahi, T. and Mataga, N. (1992) Photoinduced electron-transfer reactions of porphyrin heteroaggregates: energy gap dependence of an intradimer charge recombination process. *J. Phys. Chem.*, **96**, 503–506.

54. Brun, A.M., Harriman, A., Heitz, V. and Sauvage, J.P. (1991) Charge transfer across oblique bisporphyrins: two-center photoactive molecules. *J. Am. Chem. Soc.*, **113**, 8657–8663.
55. Flamigni, L., Dixon, I.M., Collin, J.-P. and Sauvage, J.-P. (2000) A Zn(II) porphyrin-Ir(III) bis-terpyridine-Au(III) porphyrin triad with a charge-separated state in the microsecond range. *Chem. Commun.*, 2479–2480.
56. Pettersson, K., Wiberg, J., Ljungdahl, T., Mårtensson, J. and Albinsson, B. (2006) Interplay between barrier width and height in electron tunneling: photoinduced electron transfer in porphyrin-based donor–bridge–acceptor systems. *J. Phys. Chem. A*, **110**, 319–326.
57. Wiberg, J., Guo, L., Pettersson, K., Nilsson, D., Ljungdahl, T., Mårtensson, J. and Albinsson, B. (2007) Charge recombination versus charge separation in donor–bridge–acceptor systems. *J. Am. Chem. Soc.*, **129**, 155–163.
58. Rehm, D. and Weller, A. (1969) Kinetics and mechanism of electron transfer in fluorescence quenching in acetonitrile. *Ber. Bunsen-Ges.*, **73**, 834–839.
59. Weller, A. (1982) Photoinduced electron transfer in solution: exciplex and radical ion pair formation free enthalpies and their solvent dependence. *Z. Phys. Chem. (Muenchen, Germany)*, **133**, 93–98.
60. Kilså, K., Macpherson, A.N., Gillbro, T., Mårtensson, J. and Albinsson, B. (2001) Control of electron transfer in supramolecular systems. *Spectrochim. Acta. A, Mol. Biomol. Spectrosc.*, **57A**, 2213–2227.
61. Weiss, E.A., Tauber, M.J., Kelley, R.F., Ahrens, M.J., Ratner, M.A. and Wasielewski, M.R. (2005) Conformationally gated switching between superexchange and hopping within oligo-*p*-phenylene-based molecular wires. *J. Am. Chem. Soc.*, **127**, 11842–11850.
62. Yamada, R., Kumazawa, H., Tanaka, S. and Tada, H. (2008) Observation of the transition from tunneling to hopping carrier transport through single oligothiophene molecules. *Mater. Res. Soc. Symp. Proc.*, **1091E**, 1091 - AA05 - 75.
63. Andersson, M., Linke, M., Chambron, J.-C., Davidsson, J., Heitz, V., Hammarström, L. and Sauvage, J.-P. (2002) Long-range electron transfer in porphyrin-containing [2]-rotaxanes: tuning the rate by metal cation coordination. *J. Am. Chem. Soc.*, **124**, 4347–4362.
64. Imahori, H., Hagiwara, K., Aoki, M., Akiyama, T., Taniguchi, S., Okada, T., Shirakawa, M. and Sakata, Y. (1996) Linkage and solvent dependence of photoinduced electron transfer in zincporphyrin-C_{60} dyads. *J. Am. Chem. Soc.*, **118**, 11771–11782.
65. Helms, A., Heiler, D. and McLendon, G. (1992) Electron transfer in bis-porphyrin donor–acceptor compounds with polyphenylene spacers shows a weak distance dependence. *J. Am. Chem. Soc.*, **114**, 6227–6238.
66. Goldsmith, R.H., Vura-Weis, J., Scott, A.M., Borkar, S., Sen, A., Ratner, M.A. and Wasielewski, M.R. (2008) Unexpectedly similar charge transfer rates through benzo-annulated bicyclo[2.2.2]octanes. *J. Am. Chem. Soc.*, **130**, 7659–7669.
67. Ou, Z.P., Kadish, K.M., Wenbo, E., Shao, J.G., Sintic, P.J., Ohkubo, K., Fukuzumi, S. and Crossley, M.J. (2004) Substituent effects on the site of electron transfer during the first reduction for gold(III) porphyrins. *Inorg. Chem.*, **43**, 2078–2086.
68. Kadish, K.M., Wenbo, E., Ou, Z.P., Shao, J.G., Sintic, P.J., Ohkubo, K., Fukuzumi, S. and Crossley, M.J. (2002) Evidence that gold(III) porphyrins are not electrochemically inert: facile generation of gold(II) 5,10,15,20-tetrakis(3,5-di-tert-butylphenyl)porphyrin. *Chem. Commun.*, 356–357.
69. Jiao, J., Schmidt, I., Taniguchi, M., Lindsey, J.S. and Bocian, D.F. (2008) Comparison of electron-transfer rates for metal- versus ring-centered redox processes of porphyrins in monolayers on Au(111). *Langmuir*, **24**, 12047–12053.
70. Pettersson, K., Kilså, K., Mårtensson, J. and Albinsson, B. (2004) Intersystem crossing versus electron transfer in porphyrin-based donor–bridge–acceptor

systems: influence of a paramagnetic species. *J. Am. Chem. Soc.*, **126**, 6710–6719.

71. Kilså, K., Kajanus, J., Larsson, S., Macpherson, A.N., Mårtensson, J. and Albinsson, B. (2001) Enhanced intersystem crossing in donor/acceptor systems based on zinc/iron or free-base/iron porphyrins. *Chem. A Eur. J.*, **7**, 2122–2133.
72. Hodgkiss, J.M., Krivokapic, A. and Nocera, D.G. (2007) Ligand-field dependence of the excited state dynamics of hangman bisporphyrin dyad complexes. *J. Phys. Chem. B*, **111**, 8258–8268.
73. Müller, M.A., Gaplovsky, M., Wirz, J. and Woggon, W.-D. (2006) Synthesis and photophysical properties of a deazaflavin-bridged porphyrinato-iron(III) that mimics the interaction of a deazaflavin inhibitor with the heme-thiolate cofactor of cytochrome P450 3A4. *Helv. Chim. Acta*, **89**, 2987–3001.
74. Asano-Someda, M., Jinmon, A., Toyama, N. and Kaizu, Y. (2001) Orientation effect of enhanced intersystem crossing in copper(II) porphyrin-free base porphyrin dimers. *Inorg. Chim. Acta*, **324**, 347–351.
75. Barkigia, K.M., Chantranupong, L., Smith, K.M. and Fajer, J. (1988) Structural and theoretical models of photosynthetic chromophores. Implications for redox, light-absorption properties and vectorial electron flow. *J. Am. Chem. Soc.*, **110**, 7566–7567.
76. Poveda, L.A., Ferro, V.R., Garcia de la Vega, J.M. and Gonzalez-Jonte, R.H. (2000) Molecular modeling of highly peripheral substituted Mg- and Zn-porphyrins. *Phys. Chem. Chem. Phys.*, **2**, 4147–4156.
77. Yang, S.I., Seth, J., Balasubramanian, T., Kim, D., Lindsey, J.S., Holten, D. and Bocian, D.F. (1999) Interplay of orbital tuning and linker location in controlling electronic communication in porphyrin arrays. *J. Am. Chem. Soc.*, **121**, 4008–4018.
78. Ghosh, A., Halvorsen, I., Nilsen, H.J., Steene, E., Wondimagegn, T., Lie, R., van Caemelbecke, E., Guo, N., Ou, Z. and Kadish, K.M. (2001) Electrochemistry of nickel and copper Î²-octahalogeno-meso-tetraarylporphyrins. Evidence for important role played by saddling-induced metal(dx2-y2)-porphyrin("a2u") orbital interactions. *J. Phys. Chem. B*, **105**, 8120–8124.
79. Benniston, A.C., Harriman, A., Li, P.Y., Patel, P.V. and Sams, C.A. (2005) The effect of torsion angle on the rate of intramolecular triplet energy transfer. *Phys. Chem. Chem. Phys.*, **7**, 3677–3679.
80. Kyrychenko, A. and Albinsson, B. (2002) Conformer-dependent electronic coupling for long-range triplet energy transfer in donor–bridge–acceptor porphyrin dimers. *Chem. Phys. Lett.*, **366**, 291–299.
81. Benniston, A.C. and Harriman, A. (2006) Charge on the move: how electron-transfer dynamics depend on molecular conformation. *Chem. Soc. Rev.*, **35**, 169–179.
82. Filatov, I. and Larsson, S. (2002) Electronic structure and conduction mechanism of donor–bridge–acceptor systems where PPV acts as a molecular wire. *Chem. Phys.*, **284**, 575–591.
83. Closs, G.L., Johnson, M.D., Miller, J.R. and Piotrowiak, P. (1989) A connection between intramolecular long-range electron, hole, and triplet energy transfers. *J. Am. Chem. Soc.*, **111**, 3751–3753.
84. Closs, G.L., Piotrowiak, P., MacInnis, J.M. and Fleming, G.R. (1988) Determination of long-distance intramolecular triplet energy-transfer rates. Quantitative comparison with electron transfer. *J. Am. Chem. Soc.*, **110**, 2652–2653.
85. Davis, W.B., Ratner, M.A. and Wasielewski, M.R. (2001) Conformational gating of long distance electron transfer through wire-like bridges in donor–bridge–acceptor molecules. *J. Am. Chem. Soc.*, **123**, 7877–7886.
86. Winters, M.U., Kärnbratt, J., Blades, H.E., Wilson, C.J., Frampton, M.J., Anderson, H.L. and Albinsson, B. (2007) Control of electron transfer in a conjugated porphyrin dimer by selective excitation of planar and perpendicular conformers. *Chem. A Eur. J.*, **13**, 7385–7394.
87. Winters, M.U., Kärnbratt, J., Eng, M., Wilson, C.J., Anderson, H.L. and

Albinsson, B. (2007) Photophysics of a butadiyne-linked porphyrin dimer: influence of conformational flexibility in the ground and first singlet excited state. *J. Phys. Chem. C*, **111**, 7192–7199.

88. Benniston, A.C. and Harriman, A. (2007) 17th International Symposium on Photochemistry and Photophysics of Coordination Compounds, Dublin, Ireland, pp. 2528–2539.

89. Benniston, A.C., Harriman, A., Li, P., Patel, P.V. and Sams, C.A. (2008) Electron exchange in conformationally restricted donor–spacer–acceptor dyads: angle dependence and involvement of upper-lying excited states. *Chem. A Eur. J.*, **14**, 1710–1717.

90. Benniston, A.C., Harriman, A., Li, P.Y. and Sams, C.A. (2005) Temperature-induced switching of the mechanism for intramolecular energy transfer in a 2,2′: 6′,2″-terpyridine-based Ru(II)-Os(II) trinuclear array. *J. Am. Chem. Soc.*, **127**, 2553–2564.

91. Frisch, M.J., Trucks, G.W., Schlegel, H.B., Scuseria, G.E., Robb, M.A., Cheeseman, J.R., Montgomery, J. A. Jr., Vreven, T., Kudin, K. N., Burant, J.C., Millam, J.M., Iyengar, S.S., Tomasi, J., Barone, V., Mennucci, B., Cossi, M., Scalmani, G., Rega, N., Petersson, G.A., Nakatsuji, H., Hada, M., Ehara, M., Toyota, K., Fukuda, R., Hasegawa, J., Ishida, M., Nakajima, T., Honda, Y., Kitao, O., Nakai, H., Klene, M., Li, X., Knox, J.E., Hratchian, H.P., Cross, J.B., Adamo, C., Jaramillo, J., Gomperts, R., Stratmann, R.E., Yazyev, O., Austin, A.J., Cammi, R., Pomelli, C., Ochterski, J.W., Ayala, P.Y., Morokuma, K., Voth, G.A., Salvador, P., Dannenberg, J.J., Zakrzewski, V.G., Dapprich, S., Daniels, A.D., Strain, M.C., Farkas, O., Malick, D.K., Rabuck, A.D., Raghavachari, K., Foresman, J.B., Ortiz, J.V., Cui, Q., Baboul, A.G., Clifford, S., Cioslowski, J., Stefanov, B.B., Liu, G., Liashenko, A., Piskorz, P., Komaromi, I., Martin, R.L., Fox, D.J., Keith, T., Al-Laham, M.A., Peng, C.Y., Nanayakkara, A., Challacombe, M., Gill, P.M.W., Johnson, B., Chen, W., Wong, M.W., Gonzalez, C. and Pople, J.A. (2003) Gaussian, Inc., Pittsburgh, PA.

92. Brocklehurst, B. and Young, R.N. (1994) Fluorescence anisotropy decays and viscous behavior of 2-methyltetrahydrofuran. *J. Chem. Soc. Faraday Trans.*, **90**, 271–278.

5 Base Pair Sequence and Hole Transfer Through DNA: Rational Design of Molecular Wires

Josh Vura-Weis, Frederick D. Lewis, Mark A. Ratner, and Michael R. Wasielewski

5.1 Introduction

Charge transfer in Donor–Bridge–Acceptor (D–B–A) systems generally occurs via one of two mechanisms: single-step coherent superexchange and multistep incoherent hopping [1]. The rate of superexchange charge transfer is most simply described in Eq. (5.1), where R is the D–A distance and β depends on the energetics and overlap between participating orbitals of D, B, and A.

$$k_{\text{ct}} = k_0 e^{-\beta R} \tag{5.1}$$

In this mechanism, the bridge serves only to mediate charge transfer and does not itself become oxidized or reduced. This mechanism is strongly distance dependent, with $\beta \approx 0.3-1.5$ for most organic materials. In the hopping mechanism, a charge is injected onto the bridge, where it can reversibly hop from one bridge site to another until it reaches a trap site. The rate constant can be described by Eq. (5.2), where N is the number of hopping sites and η has a value between 1 and 2 [2, 3]:

$$k_{\text{ct}} = k_0 N^{-\eta} \tag{5.2}$$

We take the phrase "molecular wire" to mean a molecule that transfers charge rapidly and efficiently (i.e., with high quantum yield) over long distances. The phrase "wire-like behavior" therefore usually refers to the latter mechanism, which shows a much weaker distance dependence and mimics the incoherent nature of charge transfer through macroscopic metallic wires. It should be noted that the *rate* and the *efficiency* of charge transfer are separate quantities, a division not always made in the literature. In photoinduced D–B–A systems, charge recombination from an initial charge-separated state $D^{+\bullet}-B^{-\bullet}-A$ can reduce the probability of full charge separation to $D^{+\bullet}-B-A^{-\bullet}$, even if the few charges that reach A do so rapidly. This distinction is less important for steady-state devices such as DNA transistors [4, 5], where an applied voltage provides the driving force for charge transport and recombination is not a factor.

The idea that the regular, well-defined structure of a DNA base pair stack could function as a molecular wire has a long and controversial history [6–13]. For a

Charge and Exciton Transport through Molecular Wires. Edited by L.D.A. Siebbeles and F.C. Grozema

ISBN: 978-3-527-32501-6

molecule to be considered a molecular wire, one must be able to measure the efficiency and rate of charge transfer over long distances (>20 Å) to show that they are only weakly distance dependent. The use of intercalated donors and acceptors in early studies made accurate determination of the D–A distance difficult and raised the possibility that the observed photophysics might come from minority species in solution. The radical cation and anion states of early donors and acceptors were not spectroscopically observable, so the rates and efficiencies had to be inferred from indirect observations such as either steady-state or time-resolved fluorescence quenching. Other experiments such as strand cleavage provided relative efficiencies of different strand lengths but neither absolute efficiencies nor rates.

The recent development of donor and acceptor moieties with distinct excited state and radical ion spectra allows the accurate determination of charge transfer rates using transient absorption spectroscopy. Because absorption is a stochiometric measurement (the absorption of species X is directly proportional to the concentration of X), this technique allows the measurement of absolute quantum yields. It is also insensitive to small amounts of impurities or minority species. These donors and acceptors may be covalently bound to each end of a stable DNA hairpin to make a series of D–$Base_n$–A molecules with well-defined D–A distances.

Through the use of such techniques, we have conducted a comprehensive examination of hole transfer through double-stranded DNA stacks as a function of distance and base pair sequence. Both the efficiency and rate in poly(A), poly(G), and alternating poly(AT), poly(GC), and poly(GA) sequences (see Figures 5.4(a) and 5.15 for representative structures) are strongly distance dependent and therefore non-wire like. However, charge separation in diblock $A_{2-3}G_n$ sequences proceeds by a fast and moderately efficient hole-hopping mechanism over as many as 9 bp (~34 Å), with distance dependence that may be characterized as wire-like. This ability to transfer charge rapidly and efficiently over long distances may enable the use of DNA for applications ranging from sensors to molecular electronics.

5.2
Spectral Signatures of Charge Transfer

One of the problems in measuring charge transfer through DNA is that the common bases A, T, G, and C have very weak absorption in the visible region, as do their excited states and radical ions. It has therefore been difficult to detect spectroscopically the existence of charge on the bases. The excited state of the adenine isomer 2-aminopurine absorbs in the visible region, and the quenching of this state has been used to monitor charge separation over up to three bases [14]. However, this study could not determine if the charge had left 2-aminopurine via a superexchange or hopping mechanism. Other nonnatural but spectroscopically observable bases such as 1,N^6-ethenoadenine can distort the B-DNA structure, possibly affecting charge transfer rates by altering the orbital overlap between other base pairs in the stack [12, 15].

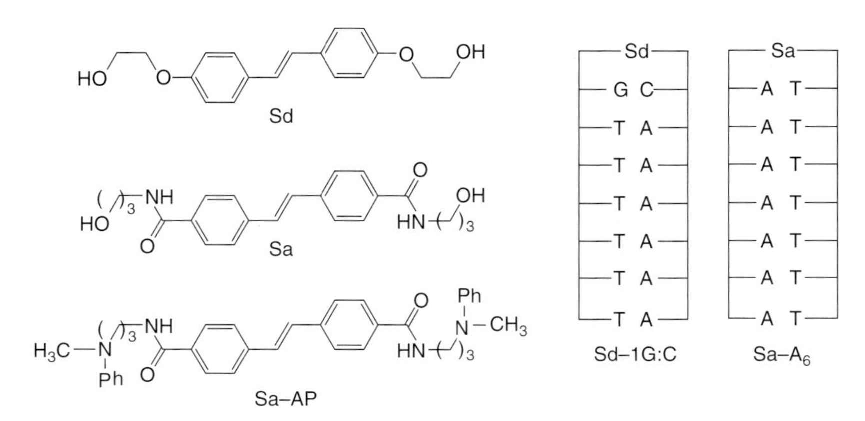

Figure 5.1 Sa and Sd chromophores and model compounds used to characterize their excited states and radical ions.

In our studies, the stilbenediether Sd and stilbenediamide Sa shown in Figure 5.1 were chosen as donor and acceptor chromophores because they have distinct excited state and radical ion spectra in the visible region. This allows us to monitor the creation of both $Sa^{-\bullet}$ and $Sd^{+\bullet}$ and thereby determine the relative timing of charge leaving Sa and arriving at Sd. These stilbenes may be incorporated into DNA hairpins using standard phosphoramidite chemistry to form stable B-DNA structures, as shown by X-ray crystallography [16] and by molecular dynamics simulations [17].

The spectra of excited state Sa^* and radical anion $Sa^{-\bullet}$ were determined by femtosecond transient absorption spectroscopy of the stilbenedi(anilinopropyl)amide model compound Sa–AP in standard 0.1 M NaCl, 10 mM sodium phosphate buffer solution. Neither AP nor $AP^{+\bullet}$ absorb strongly in the visible region, so all features may be assigned to transient states of Sa.

The Sa^* excited state is formed within 1 ps after photoexcitation of Sa–AP at 350 nm, leading to stimulated emission at 380 nm and an absorptive band at 575 nm (Figure 5.2(a)) [18]. These features are nearly identical to those seen after photoexcitation of Sa in methanol. Charge separation in Sa–AP occurs within a few picoseconds, causing the stimulated emission to disappear and the 575-nm band to change shape, losing its red tail and growing a shoulder at 525 nm. This shoulder, which gives a 525/575 nm absorption ratio of ~0.4, is the key to distinguishing the radical anion $Sa^{-\bullet}$ from the excited state Sa^*. It is common for the first excited state and radical anion of organic chromophores to have similar absorption spectra, and this band shape change was in fact overlooked in early Sa–DNA investigations [19]. The wide spectral window used in this study (350–750 nm) allowed observation of the stimulated emission decay and conclusive assignment of the $Sa^{-\bullet}$ radical anion spectrum. The $Sa^{-\bullet}$ feature decays with a time constant of 20 ps as the charges recombine to the ground state.

The excited state and radical cation spectra of the electron donor Sd were identified through transient absorption of hairpin Sd-1G:C (Figure 5.2(b)) [20].

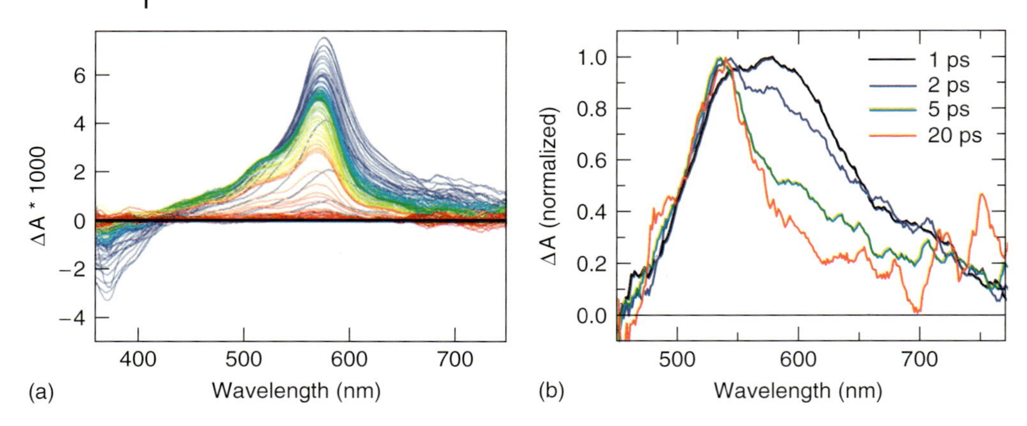

Figure 5.2 Transient spectra of model compounds. (a) Sa–AP from −0.1 ps (blue lines) to 150 ps (red lines) after excitation at 350 nm. (b) Sd-1G:C from 1 to 200 ps after excitation at 340 nm.

Photoexcitation at 340 nm gives the excited state Sd^*, which absorbs in a broad band from ~500 to 700 nm. $Sd^{+\bullet}$, which is formed in 3 ps, is characterized by a narrower peak at 525 nm. The $Sd^{+\bullet}$ absorption is similar to that of 4,4-dimethoxystilbene in acetonitrile ($\lambda_{max} = 340$ nm), and $C^{-\bullet}$ does not absorb in this region. In this model complex, the $Sd^{+\bullet}$–$C^{-\bullet}$ radical ion pair recombines to the ground state in 40 ps.

5.3 Charge Injection into A-Tracts

The hairpin Sa–A_6 (Figure 5.1) has a substantially higher fluorescence quantum yield and longer excited state lifetime ($\Phi_f = 0.38, \tau = 2.0$ ns) than Sa itself ($\Phi_f = 0.11, \tau = 0.28$ ns), likely because of the increase in rigidity induced by the DNA structure [19]. Because charge separation generally quenches excited states, this was originally taken as evidence that charge injection from Sa to A did not occur. The presumed energetics of charge injection supported this conclusion. Sa has an excited state energy of 3.35 eV and a reduction potential of −1.91 eV vs SCE, whereas isolated A has an oxidation potential of 1.69 eV [21]. The free energy of charge separation was estimated using Weller's equation [22] ($\Delta G_{cs} = -E_s - E_{rdn} + E_{ox} - 0.1$, where E_s is the Sa^* excited state energy and 0.1 is a solvent screening factor) to be +0.15 eV. Because charge injection onto A appears to be endothermic, any charge separation in Sa–A_n-Donor hairpins was originally thought to occur through superexchange (Figure 5.4), precluding molecular wire behavior.

However, a reinvestigation of Sa–A_n transient absorption spectroscopy using the 350–750 nm spectral window shows that hole injection from Sa^* into the A-tract does in fact occur [18]. The transient absorption of Sa–A_6 displays the same spectral features as that of Sa–AP: initial stimulated emission at 380 nm and Sa* absorption at 575 nm, then decay of stimulated emission accompanied by a rise in the shoulder at 525 nm, with the 525/575 nm ratio again reaching a maximum value of ~0.4 (Figure 5.3). These processes occur with time constants of 20–30 ps. Time-resolved fluorescence with higher time resolution than previous experiments gave a triple-exponential decay times of 45 ps (71%), 0.86 ns (8%), and 2.2 ns (21%). The short component is near the instrument response time, and is therefore in good agreement with the 20–30 ps time constant obtained from transient absorption. The similarity of the Sa–AP and Sa–A_6 transient spectra, along with the multiexponential fluorescence, indicate that Sa–A_6 undergoes reversible charge injection in ~30 ps to form $Sa^{-\bullet} - (A_n)^{+\bullet}$. Note that while the 525/575 nm rise can be fit quite well to a single exponential, it is in fact the combination of several rates. The rate of this rise should therefore be treated as only an approximate value for the hole injection rate constant $k_i = \tau_i^{-1}$. The 0.86 and 2.2 ns fluorescence decay times, which match the biexponential decay of the $Sa^{-\bullet}$ feature in transient absorption, are attributed to recombination to Sa* from charge-separated states with one or more A:T base pairs separating the radical ions. This type of reversible charge injection has been previously observed in contact radical ion pairs and D–B–A charge-separated states [23–25].

For hole injection to occur, the process must be exothermic or only weakly (within k_bT) endothermic. Two factors may lower the energy of the $Sa^{-\bullet}–A^{+\bullet}$ radical ion pair. The 0.1 mM NaCl aqueous buffer used in these experiments is highly polar, stabilizing the charges. Further stabilization may arise from π-stacking interactions between Sa and A. The reversibility of hole injection from Sa into A-tracts suggests that the Sa* and $Sa^{-\bullet}–A^{+\bullet}$ are nearly isoergodic. This ability of Sa to inject charge into A-tracts opens the possibility of hopping as the mechanism for long-range hole transfer in these systems.

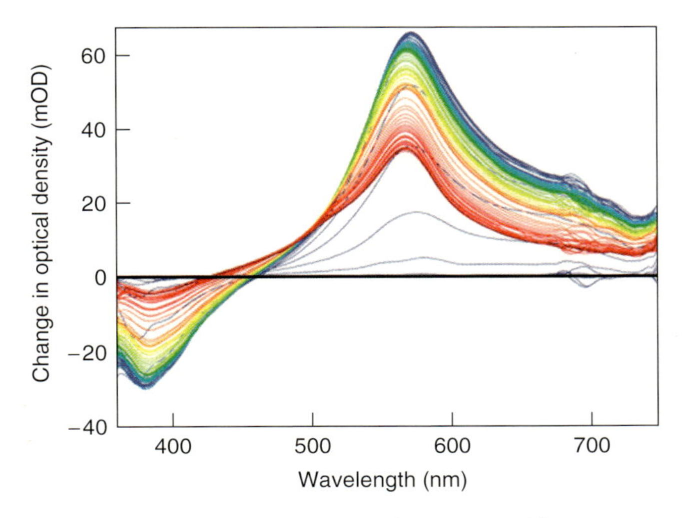

Figure 5.3 Transient spectra of Sa–A_6 from −0.1 ps (blue lines) to 150 ps (red lines) after excitation at 350 nm.

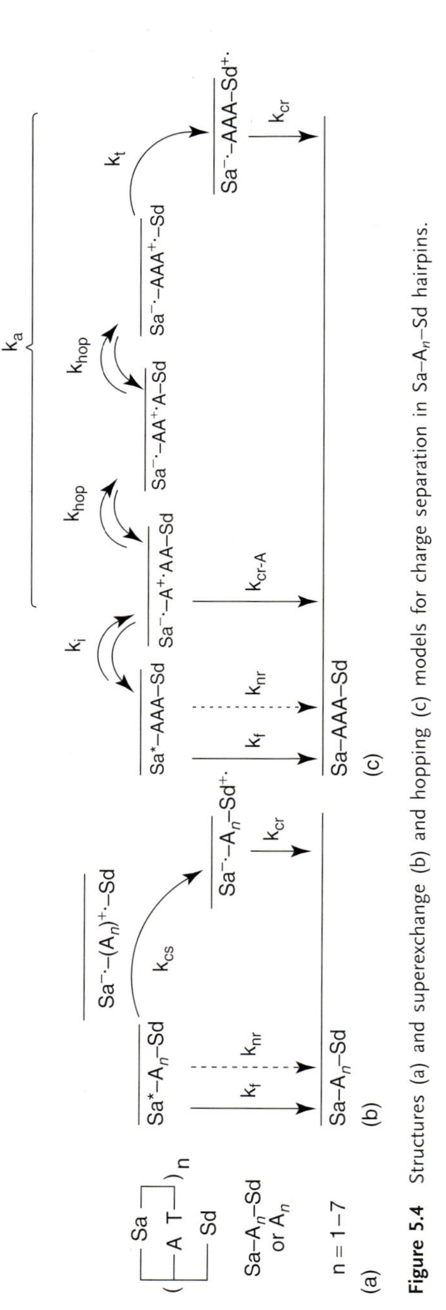

Figure 5.4 Structures (a) and superexchange (b) and hopping (c) models for charge separation in Sa–A_n–Sd hairpins.

5.4
Crossover from Superexchange to Hopping in Sa–A_n–Sd

The Sa/Sd acceptor/donor chromophores were incorporated into Sa–A_n–Sd ($n = 1 - 7$) hairpins (Figure 5.4(a)) and studied using transient absorption spectroscopy and time-resolved fluorescence [18, 26]. *Note: Sa–Bases–Sd sequences will be abbreviated wherever possible as Bases, unless a transient species such as* $Sa^{-\bullet}$*–Bases–*$Sd^{+\bullet}$ *is shown. Sequences without both Sa and Sd such as Sa–*A_6 *or Sa–*A_n*–Sa will always be named in full.* The transient spectrum of A_5 shown in Figure 5.5 is representative of the longer hairpins ($n \geq 3$). As in Sa–A_6, hole injection in ~40 ps is shown by the loss of Sa* stimulated emission and the rise of the 525/575 nm ratio to ~0.4. Fluorescence decay at 390 nm also shows a fast component of 40–100 ps. Hole arrival at Sd to form the fully charge separated state $Sa^{-\bullet}-A_5-Sd^{+\bullet}$ is shown by the rise of a distinct transient absorption peak at 525 nm, increasing the 525/575 ratio to about 1.0 with a time constant of 4.5 ns.

The hole injection and arrival rates ($k = \tau^{-1}$) of these sequences are shown in Figure 5.6(b). The decay of Sa* and formation of $Sd^{+\bullet}$ occur simultaneously in 5 ps for $n = 1$. This is significantly faster than the hole injection time measured for Sa–A_6 or longer A_n sequences, indicating that in A_1 charge separation proceeds via the single-step superexchange mechanism (Figure 5.4(b)). The injection and arrival rates for A_2 are similar, suggesting that both mechanisms may be operative for this sequence. For sequences with $n \geq 3$, the injection time is much faster than the arrival time, consistent with a hopping model in which the charge resides on the A_n bridge before being trapped at Sd (Figure 5.4(c)).

The hole arrival rate is strongly distance dependent at short distances, possibly due to coulombic attraction between the hole and $Sa^{-\bullet}$ (see Section 5.9.2). After 4 bp the distance dependence weakens, with each additional A adding ~2 ns to the arrival time. Charge recombination from $Sa^{-\bullet}-A_n-Sd^{+\bullet}$ to the ground state is

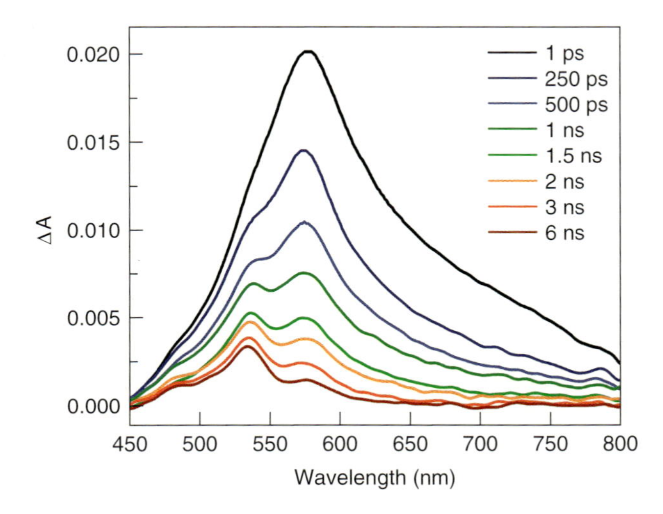

Figure 5.5 Transient spectra of A_5 from 1 ps (ΔA/2.5) to 6 ns.

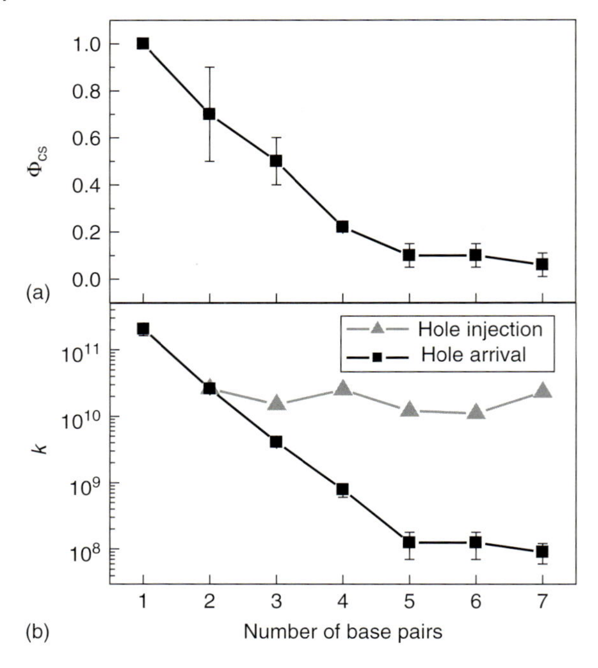

Figure 5.6 Quantum yield of charge separation (a) and rates of hole injection and arrival (b) for Sa–A_n–Sd sequences.

2–4 orders of magnitude slower than hole arrival, as determined by nanosecond transient absorption.

The quantum yield of charge separation Φ_{cs} can be determined from the strength of the $Sa^{-\bullet}/Sd^{+\bullet}$ absorption. Based on its extremely fast charge separation rate, Φ_{cs} for A_1 is assumed to be 1. Quantum yields of the longer sequences were estimated by comparing the integrated intensity of the $Sa^{-\bullet}-A_{n>1}-Sd^{+\bullet}$ spectra after full charge separation (typically $3\tau_a$) with that of A_1. As shown in Table 5.1, Φ_{cs} drops sharply for $n = 1-4$ but then levels off at around 10%. A similar leveling-off behavior was seen after 3 bp in the strand cleavage studies of Giese *et al.* [27]. The quantum yield is determined by the balance of several rates: hole injection k_i, hole transport along the A tract k_{hop}, hole trapping at Sd k_t, and charge recombination to the ground state from the contact ion pair (e.g., $Sa^{-\bullet}-A^{+\bullet}AA-Sd$) k_{cr-A}. Recombination from more distant adenines such as $Sa^{-\bullet}-AA^{+\bullet}A-Sd$ is considered to be much slower than from the contact ion pair due to the exponential distance dependence of tunneling and is neglected. Fluorescence k_f or internal conversion k_{nr} from Sa* is not considered to be a significant factor due to the rapid hole injection. In these sequences the rate-limiting step appears to transport along the A tract, and recombination from $Sa^{-\bullet}-(A_n)^{+\bullet}-Sd$ before the hole becomes trapped on Sd is the primary factor limiting long-distance charge transfer efficiency. Several theories for the mechanism of hole transport along the A tract will be discussed in Section 5.9.

Table 5.1 Charge separation quantum yields, hole injection times, hole arrival times, hole arrival times, and charge recombination times for Sa–A_n–Sd sequences[a].

Sequence	Φ_{cs}	τ_i (ps)	τ_a (ps)	τ_{cr} (ns)
A_1	1.0	–	5	0.42
A_2	0.80	38	39	8.3
A_3	0.50	67	250	230
A_4	0.23	40	1300	8300
A_5	0.10	83	4500	3.0×10^4
A_6	0.09	91	9000	1.5×10^5
A_7	0.06	43	12000	7.0×10^5

[a] Φ_{cs} and τ values in this and subsequent tables are the average of two to three experiments. Errors are generally 5–20%, and are displayed on the accompanying figures and listed in the cited references.

5.5 Symmetry Breaking in Sa–A_n–Sa

The ability of Sa–A_n–Sd to support both superexchange and hopping mechanisms suggests that one could select a desired mechanism by tuning the properties of the electron donor. In fact, Sa itself can be used as a donor in symmetric Sa–A_n–Sa dumbbells [28] (Figure 5.7(b)). Based on electrochemistry of Sa in DMF, the energy of the broken symmetry charge-separated state $Sa^{-\bullet}-A_n-Sa^{+\bullet}$ is predicted to be 0.15 eV above the excited state Sa*. However, as seen for hole injection into A, π-stacking interactions and solvation from the aqueous buffer can stabilize the charge-separated state and make it energetically accessible. Fluorescence quantum yields Φ_f for Sa–A_{1-5}–Sa are independent of length and nearly identical to that of Sa–A_6 (0.32 ± 0.02 vs. 0.38). As in Sa–A_6, fluorescence decay at 390 nm is fit to a triple exponential, though each component is approximately twice as long. The high fluorescence quantum yield suggests that emission or internal conversion

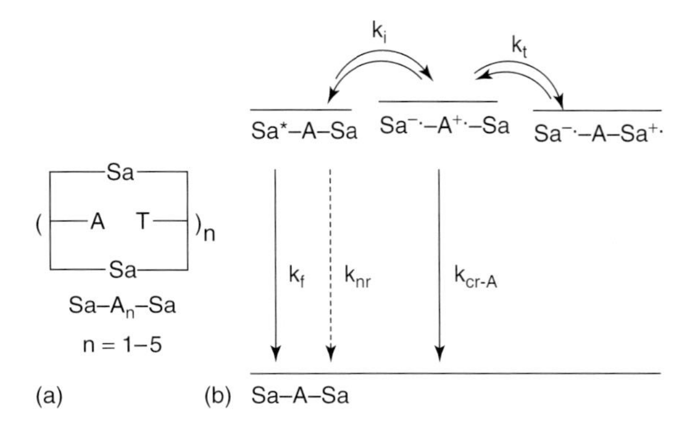

Figure 5.7 Structures of Sa–A_n–Sa dumbbells (a) and hopping scheme for Sa–A_1–Sa (b).

from Sa* is the only relaxation pathway available, since charge separation as seen in the Sa–A_n–Sd hairpins would add another pathway (charge recombination to the ground state from $Sa^{-\bullet}–A_n–Sa^{+\bullet}$) and lower Φ_f.

However, transient absorption spectroscopy of Sa–A_n–Sa shows the same ~40 ps rise of the 525 nm shoulder as in Sa–A_6 and Sa–A_n–Sd, indicating hole injection to form $Sa^{-\bullet} – (A_n)^{+\bullet}–$ Sa. Hole injection times are similar for all Sa–A_n–Sa and A_n sequences, suggesting that the hole is initially located on only the first A, rather than delocalized over two or more adjacent bases.

At longer times, a new peak appears at 505 nm, and the 505/575 nm intensity ratio rises to ~1. This peak is attributed to $Sa^{+\bullet}$, based on similar absorption of other stilbene cation radicals [29] and semi-empirical calculations using the Zerner's Intermediate Neglect of Differential Overlap (ZINDO) method which show the absorption of $Sa^{+\bullet}$ to be blue-shifted from that of $Sa^{-\bullet}$ (λ_{max} = 460 and 520 nm, respectively). The slow component of the 505/575 nm rise is therefore indicative of the formation of $Sa^{-\bullet}–A_n–Sa^{+\bullet}$, and can be fit to a single exponential for Sa–A_{1-2}–Sa. Hole arrival for Sa–A_{3-5}–Sa is incomplete on the 6-ns instrument timescale, so arrival times for the longer sequences were determined using an initial slope method [30]. Hole injection and arrival times are shown in Table 5.2, with rates plotted in Figure 5.8.

Table 5.2 Hole injection (τ_i) and arrival (τ_a) times for Sa–A_n–Sa sequences.

Sequence	τ_i (ps)	τ_a (ps)
Sa–A_1–Sa	38	110
Sa–A_2–Sa	40	530
Sa–A_3–Sa	36	3200
Sa–A_4–Sa	35	4800
Sa–A_5–Sa	32	6600

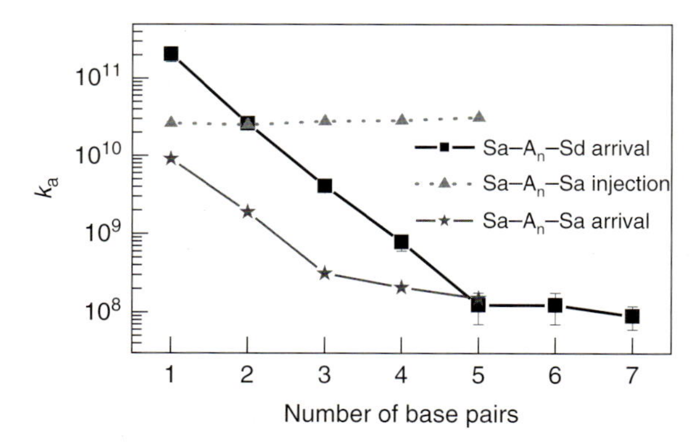

Figure 5.8 Hole injection and arrival rates for Sa–A_n–Sa sequences versus arrival rates for Sa–A_n–Sd.

The short Sa–A_n–Sa sequences exhibit very different photophysics than their analogs in the Sa–A_n–Sd series. Sa–A_1–Sn undergoes single-step superexchange in 5 ps, whereas charge separation in Sa–A_1–Sa is a two-step hopping process (Figure 5.7(b)). This is a consequence of the difference in the driving force ΔG_{cs} between the two systems. Sa–A_n–Sd has $\Delta G_{cs} \approx -0.5$, whereas ΔG_{cs} for Sa–A_n–Sa ≈ 0, based on the Weller equation. As charge separation is expected to be in the Marcus normal region, this loss of driving force slows the single-step reaction enough that it is not competitive with hole injection onto A. Full charge separation is slower for Sa–A_n–Sa than for Sa–A_n–Sd, especially at shorter distances (Figure 5.8). Since the charge injection and A-tract hopping are identical, this must be caused by slower charge trapping on Sa, again as a consequence of lower driving force for hole transfer. The Sa trapping rate k_t may be estimated as $10^{10}\ s^{-1}$ from the hole arrival time for Sa–A_1–Sa, where A-tract transport is not necessary. Hole arrival rates in the two series converge at $n = 5$, indicating that at long distances hole hopping to the end of the A tract becomes the rate-limiting step.

Whereas the fluorescence of short Sa–A_n–Sd sequences is nearly quenched due to irreversible hole trapping at Sd, the length-independence of the Sa–A_n–Sa fluorescence quantum yield indicates that Sa is a weak and reversible hole trap. Fluorescence decay as well as decay of the transient absorption signal indicate that charge recombination takes >1 ns for Sa–A_{1-2}–Sa and 5–10 ns for Sa–A_{3-5}–Sa. Superexchange would be slow and strongly distance dependent over such long distances, as shown by the ~30 μs recombination time for $Sa^{-\bullet}$–A_5–$Sd^{+\bullet}$. Recombination in Sa–A_n–Sa thus occurs through hole detrapping and hopping back through the A tract. Based on the weak distance dependence, detrapping is likely the rate determining step. Charge recombination via multistep hopping has not been observed in other DNA-based systems, which usually have deeper hole traps.

We note that the Sa–A_n–Sa series underscores the dangers of using steady-state measurements to characterize possibly complex time-dependent phenomena, a practice often found in the DNA charge transfer literature. Based on fluorescence data alone, these sequences appear to be photochemically inert, and only through femtosecond transient absorption measurements are the charge transfer events made apparent.

5.6 Influence of a Single G on Charge Transport

In the poly-A sequences above, all hopping steps are roughly isoenergetic except for irreversible trapping on Sd. Replacement of an A:T base pair with G:C creates an intermediate hole trap, approximately 0.3 eV below A but 0.2 eV above Sd. To determine the effect of a single G on hole transfer, the five possible di- and trinucleotide sequences along with the longer sequences A_3G [31] and $A_{2-4}GA$ (Figure 5.9) were synthesized and characterized using femtosecond transient absorption spectroscopy [32]. Charge recombination rates from Sa–$A_{2-4}G$ (without an Sd hole trap) were determined using femtosecond and nanosecond transient

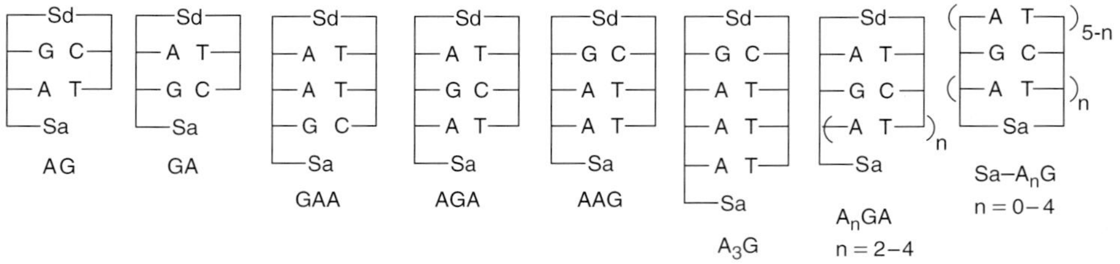

Figure 5.9 Structure of single-G sequences.

Table 5.3 Quantum yields of charge separation and hole arrival times for single-G hairpins.

Sequence	Φ_{cs}	τ_a (ps)	Sequence	Φ_{cs}	τ_a (ns)
AG	0.7	32	A_2GA	0.29	0.45
GA	0.2	80	A_3GA	0.34	1.0
AAG	0.52	120	A_4GA	0.12	5.9
AGA	0.16	190	–	–	–
GAA	0.07	210	–	–	–

absorption [19]. The values obtained for τ_a and Φ_{cs} are given in Table 5.3. Location of G at the end of the A-tract lowers τ_a but has little effect on Φ_{cs} for sequences with the same total number of base pairs N (e.g., A_2G vs. A_3). However, there is a large decrease in Φ_{cs} when G is at the beginning of the A-tract in sequences GA_{1-2} and a modest decrease for G in the middle (AGA vs. A_3).

The hole arrival rates and Φ_{cs} for the A_nGA sequences are shown in Figure 5.10, along with corresponding values for A_n sequences with the same total number of base pairs. For all but the shortest sequence GA, hole arrival is faster for A_nGA. Values of Φ_{cs} rise from $n = 0$ to 4 before decreasing, unlike the monotonic decrease for A_n. At four total base pairs and longer, charge separation is more efficient in A_nGA. This effect is most pronounced for A_3GA, which is ~4.5× faster and 3× more efficient than A_5.

These observations may be understood in terms of the mechanisms shown in Figure 5.11. Whereas charge separation in AG (a) is progressively downhill, hole transfer from $Sa^{-\bullet}{-}G^{+\bullet}A{-}Sd$ to $Sa^{-\bullet}{-}GA^{+\bullet}{-}Sd$ (b) is energetically unfavorable, and the hole is more likely to recombine with $Sa^{-\bullet}$ than to tunnel through the ~0.3 eV adenine barrier to reach Sd. In the longer A_nGA sequences, G serves as a resting point in the overall charge separation process [33, 34]. Once the cation reaches G, it must either tunnel forward to Sd or backward to recombine with $Sa^{-\bullet}$. In the most efficient sequences $A_{2-3}GA$, the initial A tract is short enough for the hole to reach G efficiently, and from there it is easier to tunnel forward through one adenine than backward through two or three. Similar manipulation of site energies has been achieved by Kawai *et al.* [35], who used the intermediate oxidation potential of the nonnatural base 7-deazaadenine to promote rapid long-distance charge separation.

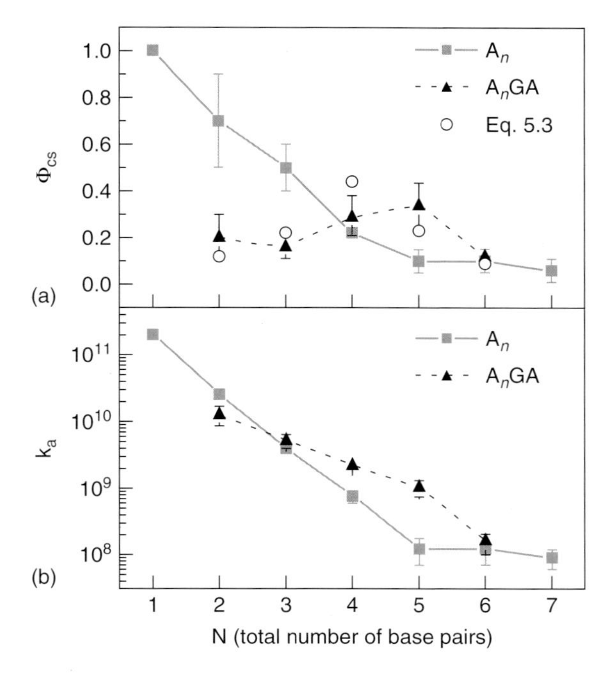

Figure 5.10 Charge separation quantum yields (a), and rate constants for hole arrival (b), A_n and A_nGA. Note that the *x*-axis is the total number of base pairs *N*, not the subscript *n* which indicates the length of the A-tract in a particular sequence. In (a), the calculated values for Φ_{cs} in A_nGA hairpins (see text) are displayed with open circles.

The A_nGA results may be understood more quantitatively with the use of Eq. (5.3), where Φ_G and Φ_t are the efficiencies of hole arrival at G and hole transport from G to Sd, respectively. k_t and k_{cr-G} are the rates of hole transport from G to Sd and the rate of charge recombination from $G^{+\bullet}$, respectively.

$$\Phi_{cs} = \Phi_G \Phi_t = \Phi_G[k_t/(k_t + k_{cr-G})] \quad (5.3)$$

Charge recombination from $Sd^{+\bullet}$ is neglected, as is much slower than the nanosecond timescale of these measurements. Φ_G is a complex function of the reversible injection and A → A hopping rates (k_i and k_{hop}), fluorescence and nonradiative decay rates of Sa* (k_f and k_{nr}), and charge recombination from the contact ion pair (k_{cr-A}). As in Section 5.4, recombination from more distant adenines is neglected. Although Φ_G cannot be measured directly, it is assumed to be similar to the quantum yield of charge separation to Sd over the same A tract, because both G and Sd are deep hole traps. The hole trapping rate k_t was estimated as $5 \times 10^9\ s^{-1}$ from AGA. Calculated values of Φ_{cs} using Eq. (5.3) are shown as white circles in Figure 5.10, and are in good agreement with the experimental values, providing support for the stepwise mechanism.

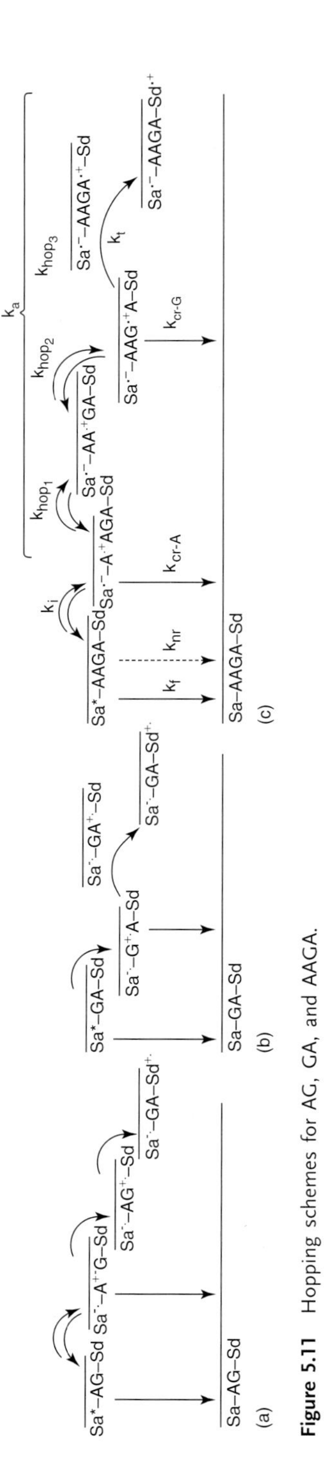

Figure 5.11 Hopping schemes for AG, GA, and AAGA.

5.7 Molecular Wire Behavior in Sa–$A_{2-3}G_{1-7}$–Sd

The increase in quantum yield from judicious placement of a single G suggests that properly constructed base pair sequences could in fact transport charge rapidly and efficiently over long distances, realizing the long-standing goal of a DNA molecular wire. The quantum yields of A_2GA and A_3GA are higher than that of the shorter AGA sequence, indicating that a "charge recombination blocking layer" of two or more adenines is particularly helpful. With this design principle in mind, the rate and efficiency of charge separation in diblock A_mG_n sequences (Figure 5.12) was studied using femtosecond transient absorption spectroscopy. Although long poly-G sequences often form G-quadruplexes, the Sa–Sd hairpin structures are quite stable and retain the usual B-DNA geometry, as shown by circular dichroism [31].

The hole arrival rates and charge separation quantum yields for $A_{0-4}G_{1-7}$ are shown in Figure 5.13, along with those of Sa–A_n–Sd. Arrival times and yields are also given in Table 5.4. Values of k_a for G_{1-3} are similar to those for A_{1-3}. However, quantum yields for G_{2-3} are much smaller than for A_{2-3}. A simplified hopping diagram for poly(A) or poly(G) sequences is shown in Figure 5.14(a). As in previously mentioned sequences, k_a and Φ_{cs} are determined by the competition between the rate constants for hole arrival k_{tB} and charge recombination k_{cr-B} (Eq. (5.4)). Note that $k_a \approx k_{tB}$ when $k_{tB} \gg k_{cr-B}$. As charge recombination tends to occur in the Marcus inverted region, this process is expected to be faster for G_n than for A_n because of the smaller energy gap for recombination from the $Sa^{-\bullet}/G^{+\bullet}$ versus $Sa^{-\bullet}/A^{+\bullet}$ radical ion pair [19]. Faster charge recombination results in slightly smaller values of τ_a but much lower values of Φ_{cs} in G_3 versus A_3 sequences.

$$\Phi_{cs} = k_{tB}/(k_{tB} + k_{cr-B}) \; \tau_a = k_a^{-1} = (k_{tB} + k_{cr-B})^{-1} \tag{5.4}$$

The energy level diagram for charge separation in diblock sequences is shown in Figure 5.14(b). In this case the quantum yield of charge separation is the product of two yields: that of transport through the A tract Φ_{csA} to form $Sa^{-\bullet}–A_m(G_n)^{+\bullet}–Sd$ and that of transport through the G tract Φ_{csG} to form $Sa^{-\bullet}–A_mG_n–Sd^{+\bullet}$ (Eq. (5.5)).

$$\Phi_{cs} = \Phi_{csA}\Phi_{csB} = [k_{tA}/(k_{tA} + k_{cr-A})][k_{tG}/k_{tG} + k_{cr-G})] \tag{5.5}$$

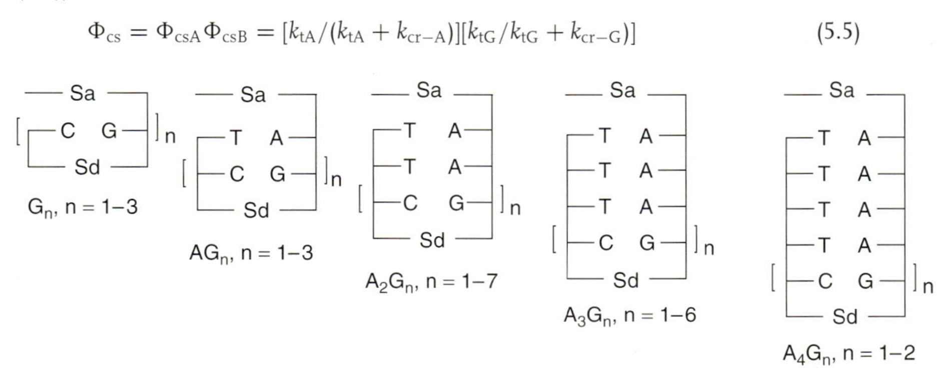

Figure 5.12 Diblock A_mG_n sequences.

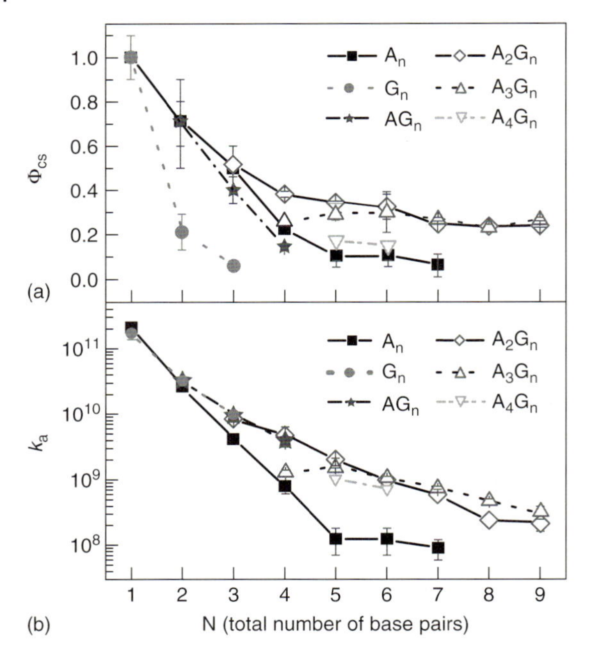

Figure 5.13 (a) Charge separation quantum yields and (b) rate constants for hole arrival in A_mG_n diblock sequences.

Table 5.4 Charge separation quantum yields and hole arrival times for diblock A_mG_n sequences.

N	G_n	Φ_{cs}	τ_{cs} (ps)	AG_n	Φ_{cs}	τ_{cs} (ps)	A_2G_n	Φ_{cs}	τ_{cs} (ps)	A_3G_n	Φ_{cs}	τ_{cs} (ps)	A_4G_n	Φ_{cs}	τ_{cs} (ps)
1	G	1.0	6	–	–	–	–	–	–	–	–	–	–	–	–
2	G_2	0.21	31	AG	0.70	32	–	–	–	–	–	–	–	–	–
3	G_3	0.06	103	AG_2	0.40	104	A_2G	0.52	120	–	–	–	–	–	–
4	–	–	–	AG_3	0.15	253	A_2G_2	0.38	223	A_3G	0.26	764	–	–	–
5	–	–	–	–	–	–	A_2G_3	0.34	490	A_3G_2	0.29	633	A_4G	0.16	945
6	–	–	–	–	–	–	A_2G_4	0.32	1000	A_3G_3	0.30	910	A_4G_2	0.15	1400
7	–	–	–	–	–	–	A_2G_5	0.25	1740	A_3G_4	0.27	1330	–	–	–
8	–	–	–	–	–	–	A_2G_6	0.24	4150	A_3G_5	0.23	2130	–	–	–
9	–	–	–	–	–	–	A_2G_7	0.24	4620	A_3G_6	0.27	3140	–	–	–

Lengthening the A tract produces two competing effects. Hole transport slows (smaller k_{tA}), reducing Φ_{csA}. However, recombination from the more distant G tract also slows (smaller k_{cr-G}), increasing Φ_{csG}. As seen in the A_nGA series, there is an optimal balance at two to three adenines. Charge recombination from Sa–AG hairpins without an Sd hole trap is rapid ($\sim 1 \times 10^{10}$ s^{-1}) [36], which is reflected in the low Φ_{cs} values for AG_n (Figure 5.13(a)). On the other hand, values of Φ_{cs} for A_2G_n

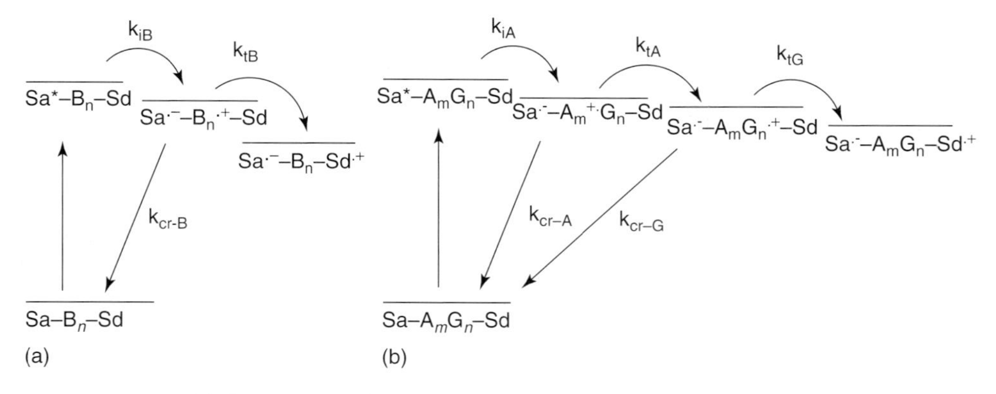

Figure 5.14 Simplified mechanisms for charge separation in (a) homo(purine) sequences (B = adenine or guanine) and (b) in A_mG_n diblock sequences. Within longer blocks, the positive charge could move by either a localized hopping or delocalized polaron mechanism.

and A_3G_n are substantially higher than for A_n sequences with the same total number of base pairs N (~0.25 vs. 0.06 for $N = 7$). Efficiency then drops ~50% for the A_4G_n series, reflecting the low efficiency of hole transport through four adenines. Charge separation rates in A_mG_n sequences are faster and less distance dependent than in A_n, consistent with calculations of stronger electronic coupling between G and G than A and A [15]. From the longer diblock systems we estimate a transport time of 0.5–1 ns/bp through G tracts, compared with ~2 ns/bp in A tracts.

The charge separation efficiencies of A_2G_n and A_3G_n are the highest yet reported for sequences of six or more base pairs. In addition, they are nearly distance independent from $N = 5$ to 9, suggesting that moderately efficient charge separation could be observed for even longer G tracts. This high quantum yield has recently allowed us to measure hole arrival times for sequences as long as A_3G_{19} [37]. It is interesting to note that improvement in Φ_{cs} over A_n sequences does not begin until there are two Gs ($A_2G \approx A_3$ but $A_2G_2 > A_4$). In addition, both k_{cs} and Φ_{cs} rise for A_3G_n from $n = 1$ to 3. This may be evidence for hole delocalization over two or more Gs, as suggested in the polaron model (see Section 5.9).

5.8
Charge Transfer through Alternating Sequences

Hole transport though alternating sequences such as poly(AT) or poly(GC) (Figure 5.15) is expected to be slower than the corresponding homogeneous sequences. If the holes are considered to be localized, then electronic coupling either over the T/C barrier or diagonally to the complementary DNA strand will be smaller than to an adjacent A/G on the same strand. If a delocalized hole model is used for poly(A) or poly(G), then alternating sequences would force charge localization and slow charge transport.

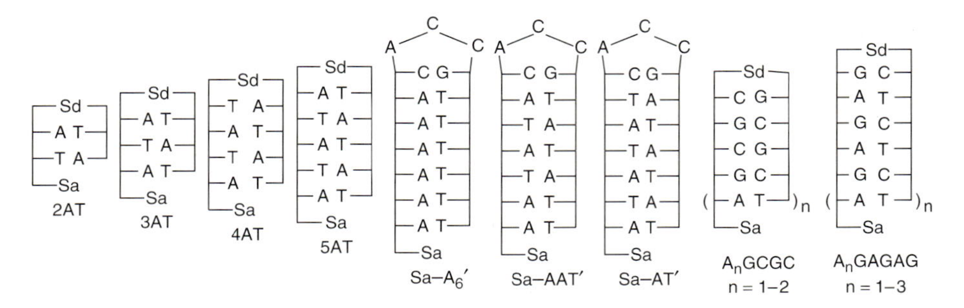

Figure 5.15 Alternating poly(AT), poly(GC), and poly(GA) sequences.

Table 5.5 Fluorescence quantum yields, hole injection times, hole arrival times, and hole trapping quantum yields[a].

Sequence	Φ_f	τ_i (ps)	τ_a (ns)	Φ_{cs}
2AT	0.047	–	0.09	0.5
3AT	0.21	–	1.9	0.12
4AT	0.25	–	6	0.03
5AT	0.26	–	6	0.02
Sa–A6	0.24	50	–	–
Sa–AT′	0.28	110	–	–
Sa–AAT′	0.24	58	–	–

[a] τ_i was not measured for nAT.

The sequences in Figure 5.15 were studied using femtosecond transient absorption and steady-state spectroscopy to probe the relative rate and efficiency of hole transport through alternating versus homogeneous sequences [31, 38]. Excitation of Sa–A_6', Sa–AT′, and Sa–AAT′ yields the familiar Sa* peak at 575 nm followed by hole injection and the rise of the 525/575 nm ratio. Injection in Sa–A6′ and Sa–AAT′ occurs in the usual 30–50 ps, whereas injection into Sa–AT′ takes about twice as long (Table 5.5). This may be evidence for cation delocalization in A tracts, which would increase the driving force and therefore the rate of charge injection. Note, however, that this effect was not seen in the Sa–A_1–Sa versus Sa–A_2–Sa sequences.

Hole transfer through alternating AT sequences was found to be 2–10 times slower and significantly less efficient than through homogeneous A tracts, as shown in Figure 5.16. The charge separation quantum yield of 5AT and 6AT approaches zero, and fluorescence quantum yields for these longer sequences match those of Sa–AT′ within experimental error, confirming that very little charge reaches the irreversible $Sd^{+\bullet}$ trap. Takada *et al.* observed a ~4× decrease in Φ_{cs} between A_5 and 5AT using a different donor/acceptor combination [39], consistent with our results. Other studies based on strand cleavage have found smaller penalties for

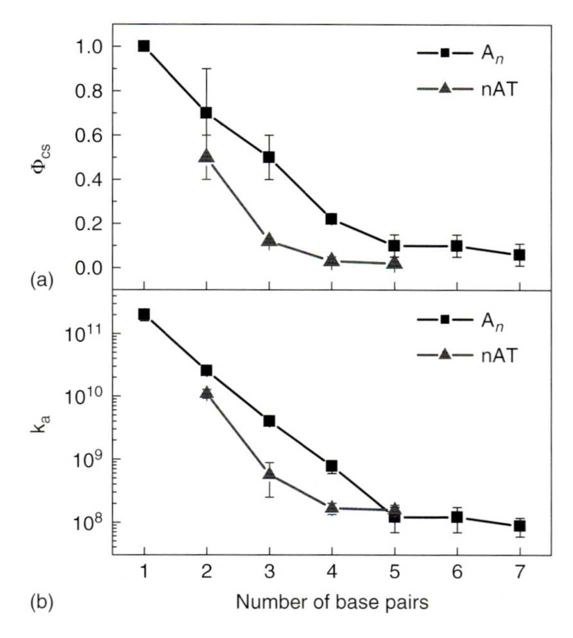

Figure 5.16 (a) Charge separation quantum yields and (b) rate constants for hole arrival in nAT versus A_n sequences.

alternating sequences [27], though those results may not be comparable since they are based on charge shift reactions where the hole does not have to overcome Coulomb attraction to the acceptor radical anion.

Sequences A_{1-3}GAGAG and A_{1-2}GCGC have also been studied using transient absorption, but their hole arrival times were too slow to measure on the 6 ns instrument timescale, and Φ_{cs} is estimated to be <0.05. As equal-length A_2G_{4-7} sequences have charge separation quantum yields of about 0.25 and arrival times from 1 to 5 ns, it is clear that alternating GA or GC sequences severely hinder charge migration.

5.9 Theoretical Descriptions of Charge Transfer through DNA

This section will briefly discuss the prevailing models of charge transfer through DNA, including treatments of the Sa–A_n–Sd system. The reader is encouraged to consult Refs. [9, 10] for several detailed reviews.

5.9.1 General Models of DNA Charge Transfer

A general consensus has emerged that hole transfer from G to G across an A tract (i.e., GA_nG) occurs through superexchange for short sequences ($n < 3–4$)

[3, 40, 41]. A charge may be transferred over long distances (up to ~200 Å) via a series of such short hops. Hole transfer through longer A tracts may occur via population of the A bridge, although the details of this process are subject to debate. In the simplest formulation for A_n blocks with $n \gtrsim 4$, each individual A is a distinct hole resting site, and the charge hops from A to A until it is trapped at G or another lower energy site [33, 41, 42]. Alternatively, the hole could delocalize over more than one base, with geometric distortions of the DNA stack and solvent reorganization stabilizing the charge. Both "small polaron" [43, 44] ($n = 2-3$) and "large polaron" [45] ($n = 3-7$) formulations have been proposed. In the latter case, transport would occur via a polaron drift model as opposed to incoherent hopping. Other calculations indicate that solvation effects localize the hole on a single base [46, 47]. Ion gating has also been implicated as a possible limiting factor for hole hopping rates [44, 48].

The size of a polaron in DNA is determined by the relative strengths of the electronic coupling between bases and dephasing/relaxation caused by dynamic motion of the DNA stack [43].

Combined quantum mechanics/molecular dynamics (QM/MD) simulations of poly(A) and poly(G) sequences have shown large fluctuations in electronic coupling due to thermal motions of the flexible DNA molecule [49–51]. These fluctuations cannot be linked to a single vibrational mode but are influenced by many degrees of freedom such as twisting, buckling, shearing, and stretching motions [52]. In both A–A and G–G coupling, the standard deviation of the coupling is larger than its mean value. Site energies also fluctuate widely because of motion of water and counterions [53], leading to transient configurations where $A^{+\bullet}$ is lower in energy than $G^{+\bullet}$. These results suggest that the Condon approximation is not strictly appropriate in DNA systems [34, 54], though non-Condon contributions to the hole transfer rate are expected to be small [55].

5.9.2 Sa–Sd Hairpins

Several computational studies have been performed on the Sa–A_n–Sd hairpins discussed in Section 5.4. QM/MD simulations have shown that the structure and electronic coupling of Sa–A_{2-3}–Sd hairpins is very similar to that of $(GA_{2-3}G)_n$ sequences, indicating that the model systems are representative of charge transfer through native DNA [52]. A similar method was used to show that the strong distance dependence of CT through short Sa–A_n–Sd hairpins is partly caused by the Coulomb attraction between $Sa^{-\bullet}$ and the hole on the A_n bridge [51]. After hole injection into the A-tract, most of the charge remains localized on the first adenine, although fluctuations in site energies allow slight population of more distant bridge sites. This method was also used to rationalize the influence of a single G on hole arrival time at Sd [56]. The coupling between the frontier molecular orbitals of various chromophores and adenine has been investigated at the Intermediate Neglect of Differential Overlap/Singles Configuration Interaction (INDO/SCI) level. Sd in particular was found to exhibit several tunneling pathways, with significant coupling between the HOMO-1 levels of Sd and A [57]. Finally, rates of charge recombination from $Sa^{-\bullet}-A_n-Sd^{+\bullet}$ are proposed to be controlled by counterion binding to $Sd^{+\bullet}$ [58].

5.10
Conclusion

Early investigations into wire-like behavior in DNA were hampered by structural ambiguity and indirect measurements of the charge transfer process. The study of synthetic hairpins with chromophore linkers using ultrafast transient absorption spectroscopy has made possible the accurate determination of rates and absolute quantum yields of charge transfer through well-defined base pair DNA strands. We have undertaken a systematic study of hole transfer as a function of base pair sequence and have shown that while in most cases the rate and yield of charge transfer is strongly distance dependent, rationally designed mixed A and G sequences can transfer holes rapidly and efficiently over long distances. The balance of charge separation, hopping, and recombination rates in these systems is crucial to their wire-like behavior, and provides guidance for the design of organic conductive materials such as sensors, photovoltaics, and molecular electronic devices.

Acknowledgments

The authors would like to thank Prof. Thorsten Fiebig, Dr. Boiko Cohen, and Dr. Pierre Daublain for their efforts in developing the methods and interpretation discussed above. Dr. Yuri Berlin is thanked for helpful suggestions on the manuscript. JVW was supported by an NSF Graduate Research Fellowship during the early stages of this project. MAR thanks the chemistry divisions of NSF and ONR for support. FDL and MRW were supported by the Chemical Sciences, Geosciences, and Biosciences Division, DOE Office of Basic Energy Sciences, grants no. DE-FG02-96ER14604 and DE-FG02-99ER14999 respectively.

References

1. Barbara, P.F., Meyer, T.J., and Ratner, M.A. (1996) Contemporary issues in electron transfer research. *J. Phys. Chem.*, **100**, 13148–13168.
2. Bixon, M., Giese, B., Wessely, S., Langenbacher, T., Michel-Beyerle, M.E., and Jortner, J. (1999) Long-range charge hopping in DNA. *Proc. Natl. Acad. Sci. USA*, **96**, 11713–11716.
3. Senthilkumar, K., Grozema, F.C., Guerra, C.F., Bickelhaupt, F.M., Lewis, F.D., Berlin, Y.A., Ratner, M.A., and Siebbeles, L.D.A. (2005) Absolute rates of hole transfer in DNA. *J. Am. Chem. Soc.*, **127**, 14894–14903.
4. Yoo, K.H., Ha, D.H., Lee, J.O., Park, J.W., Kim, J., Kim, J.J., Lee, H.Y., Kawai, T., and Choi, H.Y. (2001) Electrical conduction through poly(dA)-poly(dT) and poly(dG)-poly(dC) DNA molecules. *Phys. Rev. Lett.*, **87**, 198102.
5. Porath, D., Cuniberti, G., and Di Felice, R. (2004) *Long-Range Charge Transfer in DNA II*, vol. 237, Springer Verlag, pp. 183–227.

6. Murphy, C.J., Arkin, M.R., Jenkins, Y., Ghatlia, N.D., Bossmann, S.H., Turro, N.J., and Barton, J.K. (1993) Long-range photoinduced electron-transfer through a DNA helix. *Science*, **262**, 1025–1029.
7. Turro, N.J. and Barton, J.K. (1998) Paradigms, supermolecules, electron transfer and chemistry at a distance. What's the problem? The science or the paradigm? *J. Biol. Inorg. Chem.*, **3**, 201–209.
8. Priyadarshy, S., Risser, S.M., and Beratan, D.N. (1998) DNA-mediated electron transfer. *J. Biol. Inorg. Chem.*, **3**, 196–200.
9. Schuster, G.B. (2004) *Long-Range Charge Transfer in DNA I and II*, Springer, Berlin
10. Wagenknecht, H.-A. and Gray, H. (2005) *Charge Transfer in DNA: From Mechanism to Application*, Wiley-VCH Verlag GmbH, Weinheim
11. Meggers, E., Michel-Beyerle, M.E., and Giese, B. (1998) Sequence dependent long range hole transport in DNA. *J. Am. Chem. Soc.*, **120**, 12950–12955.
12. Kelley, S.O. and Barton, J.K. (1999) Electron transfer between bases in double helical DNA. *Science*, **283**, 375–381.
13. O'Neilnl, M.A. and Barton, J.K. (2004) *Long-Range Charge Transfer in DNA I*, vol. 236, Springer Verlag, pp. 67–115.
14. Wan, C.Z., Fiebig, T., Schiemann, O., Barton, J.K., and Zewail, A.H. (2000) Femtosecond direct observation of charge transfer between bases in DNA. *Proc. Natl. Acad. Sci. USA*, **97**, 14052–14055.
15. Rosch, N. and Voityuk, A.A. (2004) *Long-Range Charge Transfer in DNA II*, vol. 237, Springer Verlag, pp. 37–72.
16. Lewis, F.D., Liu, X.Y., Wu, Y.S., Miller, S.E., Wasielewski, M.R., Letsinger, R.L., Sanishvili, R., Joachimiak, A., Tereshko, V., and Egli, M. (1999) Structure and photoinduced electron transfer in exceptionally stable synthetic DNA hairpins with stilbenediether linkers. *J. Am. Chem. Soc.*, **121**, 9905–9906.
17. McCullagh, M., Zhang, L., Karaba, A.H., Zhu, H., Schatz, G.C., and Lewis, F.D. (2008) Effect of loop distortion on the stability and structural dynamics of DNA hairpin and dumbbell conjugates. *J. Phys. Chem. B*, **112**, 11415–11421.
18. Lewis, F.D., Zhu, H.H., Daublain, P., Fiebig, T., Raytchev, M., Wang, Q., and Shafirovich, V. (2006) Crossover from superexchange to hopping as the mechanism for photoinduced charge transfer in DNA hairpin conjugates. *J. Am. Chem. Soc.*, **128**, 791–800.
19. Lewis, F.D., Wu, T.F., Liu, X.Y., Letsinger, R.L., Greenfield, S.R., Miller, S.E., and Wasielewski, M.R. (2000) Dynamics of photoinduced charge separation and charge recombination in synthetic DNA hairpins with stilbenedicarboxamide linkers. *J. Am. Chem. Soc.*, **122**, 2889–2902.
20. Lewis, F.D., Liu, X.Y., Miller, S.E., Hayes, R.T., and Wasielewski, M.R. (2002) Dynamics of electron injection in DNA hairpins. *J. Am. Chem. Soc.*, **124**, 11280–11281.
21. Seidel, C.A.M., Schulz, A., and Sauer, M.H.M. (1996) Nucleobase-specific quenching of fluorescent dyes. 1. Nucleobase one-electron redox potentials and their correlation with static and dynamic quenching efficiencies. *J. Phys. Chem.*, **100**, 5541–5553.
22. Weller, A. (1982) Photoinduced electron transfer in solution: exciplex and radical ion pair formation free enthalpies and their solvent dependence. *Z. Phys. Chem.*, **131**, 93–98.
23. Gould, I.R. and Farid, S. (1996) Dynamics of bimolecular photoinduced electron-transfer reactions. *Acc. Chem. Res.*, **29**, 522–528.
24. Gould, I.R., Young, R.H., Mueller, L.J., Albrecht, A.C., and Farid, S. (1994) Electronic-structures of exciplexes and excited charge-transfer complexes. *J. Am. Chem. Soc.*, **116**, 8188–8199.
25. Lor, M., Thielemans, J., Viaene, L., Cotlet, M., Hofkens, J., Weil, T., Hampel, C., Mullen, K., Verhoeven, J.W., Van der Auweraer, M., and De Schryver, F.C. (2002) Photoinduced electron transfer in a rigid first generation triphenylamine core dendrimer substituted with a peryleneimide acceptor. *J. Am. Chem. Soc.*, **124**, 9918–9925.
26. Lewis, F.D., Zhu, H.H., Daublain, P., Cohen, B., and Wasielewski, M.R. (2006)

Hole mobility in DNA A tracts. *Angew. Chem. Int. Ed.*, **45**, 7982–7985.

27. Giese, B., Amaudrut, J., Kohler, A.K., Spormann, M., and Wessely, S. (2001) Direct observation of hole transfer through DNA by hopping between adenine bases and by tunnelling. *Nature*, **412**, 318–320.
28. Lewis, F.D., Daublain, P., Zhang, L., Cohen, B., Vura-Weis, J., Wasielewski, M.R., Shafirovich, V., Wang, Q., Raytchev, M., and Fiebig, T. (2008) Reversible bridge-mediated excited-state symmetry breaking in stilbene-linked DNA dumbbells. *J. Phys. Chem. B*, **112**, 3838–3843.
29. Samori, S., Hara, M., Tojo, S., Fujitsuka, M., and Majima, T. (2006) Important factors for the formation of radical cation of stilbene and substituted stilbenes during resonant two-photon ionization with a 266- or 355-nm laser. *J. Photochem. Photobiol. A Chem.*, **179**, 115–124.
30. Dougherty, E.V.A. and Dennis, A. (2005) *Modern Physical Organic Chemistry*, University of the Sciences, Philadelphia
31. Vura-Weis, J., Wasielewski, M.R., Thazhathveetil, A.K., and Lewis, F.D. (2009) Efficient charge transport in DNA diblock oligomers. *J. Am. Chem. Soc.*, **131**, 9722–9727.
32. Lewis, F.D., Daublain, P., Cohen, B., Vura-Weis, J., and Wasielewski, M.R. (2008) The influence of guanine on DNA hole transport efficiency. *Angew. Chem. Int. Ed.*, **47**, 3798–3800.
33. Berlin, Y.A., Burin, A.L., and Ratner, M.A. (2002) Elementary steps for charge transport in DNA: thermal activation vs. tunneling. *Chem. Phys.*, **275**, 61–74.
34. Berlin, Y.A., Kurnikov, I.V., Beratan, D., Ratner, M.A., and Burin, A.L. (2004) *Long-Range Charge Transfer in DNA II*, vol. 237, Springer Verlag, pp. 1–36.
35. Kawai, K., Kodera, H., Osakada, Y., and Majima, T. (2009) Sequence-independent and rapid long-range charge transfer through DNA. *Nat. Chem.*, **1**, 156–159.
36. Lewis, F.D., Wu, T.F., Zhang, Y.F., Letsinger, R.L., Greenfield, S.R., and Wasielewski, M.R. (1997) Distance-dependent electron transfer in DNA hairpins. *Science*, **277**, 673–676.
37. Micky Conron, S.M., Thazhathveetil, A.K., Wasielewski, M.R., Burin, A.L., and Lewis, F.D. (2010) Direct measurement of the dynamics of hole hopping in extended DNA G-Tracts. An unbiased random walk. *J. Am. Chem. Soc.*, **132**, 14388–14390.
38. Lewis, F.D., Daublain, P., Cohen, B., Vura-Weis, J., Shafirovich, V., and Wasielewski, M.R. (2007) Dynamics and efficiency of DNA hole transport via alternating AT versus poly(A) sequences. *J. Am. Chem. Soc.*, **129**, 15130–15131.
39. Takada, T., Kawai, K., Cai, X.C., Sugimoto, A., Fujitsuka, M., and Majima, T. (2004) Charge separation in DNA via consecutive adenine hopping. *J. Am. Chem. Soc.*, **126**, 1125–1129.
40. Bixon, M. and Jortner, J. (2001) Charge transport in DNA via thermally induced hopping. *J. Am. Chem. Soc.*, **123**, 12556–12567.
41. Jortner, J., Bixon, M., Voityuk, A.A., and Rosch, N. (2002) Superexchange mediated charge hopping in DNA. *J. Phys. Chem. A*, **106**, 7599–7606.
42. Giese, B. (2004) *Long-Range Charge Transfer in DNA I*, vol. 236, Springer Verlag, pp. 27–44.
43. Kurnikov, I.V., Tong, G.S.M., Madrid, M., and Beratan, D.N. (2002) Hole size and energetics in double helical DNA: competition between quantum delocalization and solvation localization. *J. Phys. Chem. B*, **106**, 7–10.
44. Schuster, G.B. and Landman, U. (2004) *Long-Range Charge Transfer in DNA I*, vol. 236, Springer Verlag, pp. 139–161.
45. Conwell, E. (2004) *Long-Range Charge Transfer in DNA II*, vol. 237, Springer Verlag, pp. 73–101.
46. Voityuk, A.A. (2005) Charge transfer in DNA: hole charge is confined to a single base pair due to solvation effects. *J. Chem. Phys.*, **122**, 204904.
47. Burin, A.L. and Uskov, D.B. (2008) Strong localization of positive charge in DNA induced by its interaction with environment. *J. Chem. Phys.*, **129**, 025101.

48. Barnett, R.N., Cleveland, C.L., Joy, A., Landman, U., and Schuster, G.B. (2001) Charge migration in DNA: ion-gated transport. *Science*, **294**, 567–571.
49. Troisi, A. and Orlandi, G. (2002) Hole migration in DNA: a theoretical analysis of the role of structural fluctuations. *J. Phys. Chem. B*, **106**, 2093–2101.
50. Voityuk, A.A. (2008) Electronic couplings and on-site energies for hole transfer in DNA: systematic quantum mechanical/molecular dynamic study. *J. Chem. Phys.*, **128**, 115101.
51. Grozema, F.C., Tonzani, S., Berlin, Y.A., Schatz, G.C., Siebbeles, L.D.A., and Ratner, M.A. (2008) Effect of structural dynamics on charge transfer in DNA hairpins. *J. Am. Chem. Soc.*, **130**, 5157–5166.
52. Siriwong, K. and Voityuk, A.A. (2008) Pi stack structure and hole transfer couplings in DNA hairpins and DNA. A combined QM/MD study. *J. Phys. Chem. B*, **112**, 8181–8187.
53. Voityuk, A.A., Siriwong, K., and Rosch, N. (2004) Environmental fluctuations facilitate electron–hole transfer from guanine to adenine in DNA pi stacks. *Angew. Chem. Int. Ed.*, **43**, 624–627.
54. Berlin, Y.A., Grozema, F.C., Siebbeles, L.D.A., and Ratner, M.A. (2008) Charge transfer in donor–bridge–acceptor systems: static disorder, dynamic fluctuations, and complex kinetics. *J. Phys. Chem. C*, **112**, 10988–11000.
55. Troisi, A., Nitzan, A., and Ratner, M.A. (2003) A rate constant expression for charge transfer through fluctuating bridges. *J. Chem. Phys.*, **119**, 5782–5788.
56. Grozema, F.C., Tonzani, S., Berlin, Y.A., Schatz, G.C., Siebbeles, L.D.A., and Ratner, M.A. (2009) Effect of GC base pairs on charge transfer through DNA hairpins: the importance of electrostatic interactions. *J. Am. Chem. Soc.*, **131**, 14204–11420.
57. Beljonne, D., Pourtois, G., Ratner, M.A., and Bredas, J.L. (2003) Pathways for photoinduced charge separation in DNA hairpins. *J. Am. Chem. Soc.*, **125**, 14510–14517.
58. Blaustein, G.S., Dernas, B., Lewis, F.D., and Burin, A.L. (2009) Charge recombination in DNA hairpins controlled by counterions. *J. Am. Chem. Soc.*, **131**, 400–401.

6
Charge Transport through Molecules: Organic Nanocables for Molecular Electronics

Mateusz Wielopolski, Dirk M. Guldi, Timothy Clark, and Nazario Martín

6.1
Introduction

A key motivation in developing so-called molecular electronics is the fundamental concept that individual molecules or supramolecular assemblies of nanometer scale may perform the functions of electronic devices.

Using molecular building blocks to develop electronic circuits requires the design and synthesis of specific and suitably functionalized molecular structures intended to imitate the components of an electronic circuit. One of the simplest of these components is a wire. Not surprisingly, the design of "molecular wires" has received the attention of a significant part of the scientific community in recent years [1, 2]. Despite its apparent simplicity, the definition of the term "molecular wire" is still rather controversial. In a broad sense, we employ the term "molecular wire" here to refer to any molecular structure that mediates charges between two chemical species such as, for instance, donor and acceptor moieties. In this chapter, we will focus our attention on assemblies in which photo- or redox-active organic molecules serve as donors and acceptors, which are in turn covalently connected by π-conjugated bridges acting as "molecular wires."

Molecular wires have been studied under a variety of different experimental conditions. Their molecular structures and the nature of the donors and acceptors to which they have been connected determine the precise conditions. In fact, the rational design of chemical structures and the use of modern organic synthesis prompt to a wide variety of such donor–wire–acceptor systems that can exhibit different conduction mechanisms. A variety of detection methods for the electron flow has been developed: for example, fast electrochemistry, especially for self-assembled monolayers (SAMs) containing redox-active groups [3–5], conductance measurements in metal–wire–metal junctions ("break junctions") [6–10] or ultrafast spectroscopy for photoinduced electron-transfer (ET) processes, typically used for donor–bridge–acceptor (DBA) systems [11–13].

The present chapter focuses on the last method, the investigation of photoinduced electron- and energy-transfer reactions in organic DBA conjugates (Figure 6.1).

Charge and Exciton Transport through Molecular Wires. Edited by L.D.A. Siebbeles and F.C. Grozema

ISBN: 978-3-527-32501-6

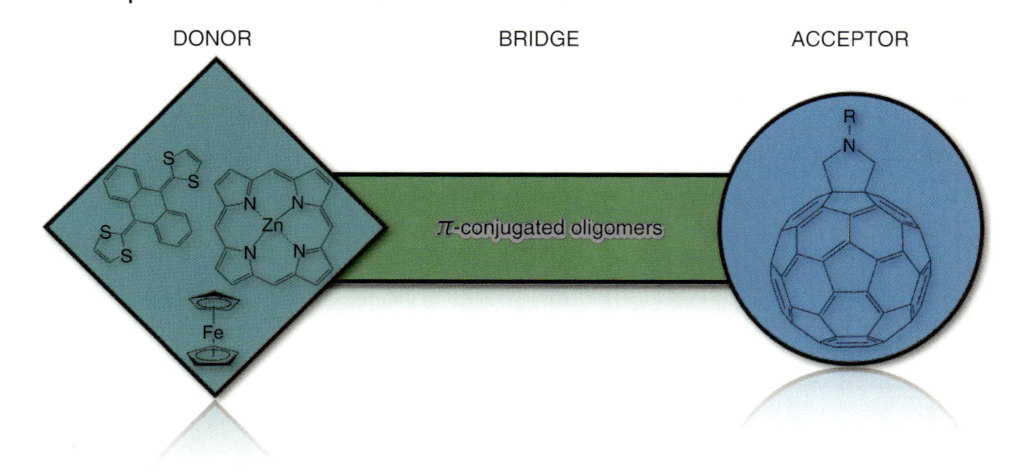

Figure 6.1 Schematic representation of Donor(D)–Bridge(B)–Acceptor(A) conjugates.

In the following, we contrast two different mechanisms; energy transfer and ET, where light excitation powers either of the involved processes. Accordingly, three possible sites for the initial local excitation can be discussed: (i) the donor, (ii) the bridge, and (iii) the acceptor. Figure 6.2 shows one possibility to demonstrate the two different deactivation mechanisms of light excitation, energy transfer versus ET, schematically. In this example, the donor is excited, which is most likely to result in ET, whereas exciting the acceptor or the bridge may lead to hole transfer. Electronic energy transfer and ET reactions exhibit some notable similarities, which renders distinguishing between them complicated. In most cases, the ET

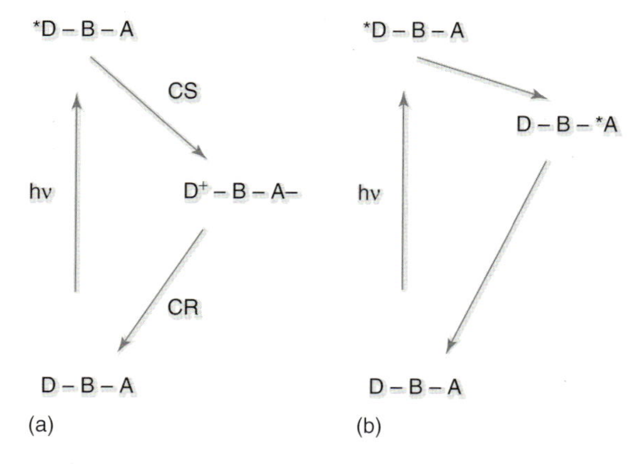

Figure 6.2 Schematic representation of photoinduced electron transfer (a) and electronic energy transfer (b) in DBA systems; CS = charge separation and CR = charge recombination; the position of the local excitation is indicated by an asterisk.

mechanism follows a chronological sequence of excitation, then energy transfer and finally ET. Here we will discuss the mechanistic aspects related to the role of the bridge in DBA conjugates.

6.2 Theoretical Concepts

6.2.1 Definition of the "Molecular Wire" Term

Charge-transfer (CT) and energy-transfer processes are characterized by many different parameters. The molecular building blocks of a system do not govern such processes independently of each other. Instead, the entire supramolecular structure should be analyzed in its entirety. Only such an integrated approach allows understanding the interplay between the components. In particular, close inspection reveals that the energetic relationships between the energy levels of donor, bridge, and acceptor govern the energy-transfer/CT properties of these systems. However, the molecular bridges connecting the donor with the acceptor play the key role in these transport processes. In the following, we survey these processes with particular emphasis on the function of the bridges, or, in other words, on their molecular wire properties. Definitions of the term "molecular wire" range from the description of a specific behavior of a system to a simple reference to the structural parameters of the molecules examined. Thus, finding a precise definition is difficult. In 1998 Emberly and Kirczenow [14] described a molecular wire as "a molecule between two reservoirs of electrons." Nitzan and Ratner, on the other hand, defined a molecular wire as "a molecule that conducts electrical current between two electrodes" [15]. The most appropriate definition for our topic, is the limited one by Wasielewski, who classifies a molecular wire as "a device that conducts in a regime, wherein the distance dependence (of ET) may be very weak" [13].

Regardless of the terminology used, molecules must meet some basic criteria to act as molecular wires. Desirable features that make molecules potential candidates for the electronic components are [16]:

1) Structural tunability, given by a large number of synthetic possibilities that allow closer control and more flexibility than given by metallic and semimetallic conductors.
2) An energetic gradient among the molecular orbitals (MOs) (or more properly CT states) that permits a unidirectional movement of charge carriers within the molecule.
3) Predisposed electron density distribution in the molecules that determines whether charge transport occurs with spin preservation or with a change in spin state.

We can now proceed with a definition of wire-like transport along molecular bridges. Energy and charge transport phenomena along a molecular wire are closely related, despite the fact that energy transfer is represented as a motion of electron/hole pairs (Frenkel exciton or boson) and CT as a motion of a single hole or electron (a fermion). For transport processes, molecular wires are often considered as Hückel-like or tight-binding systems. A thorough quantum mechanical examination of Hückel-like systems can be found elsewhere [17–19]. Within this framework, wire-like transport may be defined as a motion that is assisted by the molecular bridge. From the mechanistic point of view, one may compare theoretical results with experimental data for molecular-wire-like behavior in transferring electronic charge and/or energy. Intramolecular ET rate constants characterize the charge transport in DBA conjugates. In electronic transport junctions, on the other hand, it is convenient to use the word "conductance."

Here, we will focus on molecules as individual wires. It is nevertheless important to point out that electron and energy transport in higher dimensional molecular materials (i.e., polymers, ordered self-assembled layers, molecular crystals, etc.) is closely interconnected. The mechanisms of CT that will be discussed here, coherent tunneling and thermal hopping, are present in both individual wire systems and molecular materials.

6.2.2 Mechanisms of Charge Transfer through Molecular Wires

Intrinsically, long-distance CT is a nonadiabatic process. The rate of CT is determined by a combination of strongly distance-dependent tunneling and weakly distance-dependent incoherent transport events [20, 21]. The tunneling obeys a *superexchange* mechanism, where electrons or holes are transferred from donor to acceptor through an energetically isolated bridge. In this case, the bridge orbitals are considered solely as a coupling medium [22–24]. On the other hand, incoherent or sequential CT involves real intermediate states that couple to internal nuclear motions of the bridge and the surrounding medium. Such states are energetically accessible and may change their geometry [25]. This mechanism is defined as thermally activated *incoherent hopping* [13, 26]. The attenuation factor β is used to describe the quality of a molecular wire, because it describes the decay of the CT rate constant k as a function of distance, r_{DA}, as defined in Eq. (6.1)

$$k = k_0 \exp^{-\beta r_{DA}} \tag{6.1}$$

where k_0 is the inherent rate.

The attenuation factor β is also used to determine whether a DBA system should be considered to display wire-like behavior at all. Equation (6.1) predicts that a small β would allow charges to be transferred over larger distances. Importantly, such a definition of β only applies to exponentially decaying processes. In addition, the limit of very small β describes band transport, the so-called π-electron pathway, along which the electrons can travel coherently [27]. Shallow distance dependence may also result from a series of short-range tunneling events. This is equivalent

Table 6.1 Selected β-values for bridge units in solution and through self-assembled monolayers (SAMs) of organic thiols on the surfaces of metal electrodes (Ag, Au, and Hg).

Typical β-values for selected bridge units	
Alkanes [28–34]	0.6–1.2 $Å^{-1}$
Oligophenylenes [35–39]	0.32–0.66 $Å^{-1}$
Oligo(phenyleneethynylene)s oPEs and oligo(phenylenevinylene)s (oPVs) [3–5, 40, 41]	0.01–0.5 $Å^{-1}$
Oligoenes [4, 42, 43]	0.04–0.2 $Å^{-1}$
Oligoynes [44–46]	0.04–0.17 $Å^{-1}$
Vacuum [47]	2.0–5.0 $Å^{-1}$

to incoherent hopping, which, per se, does not follow an exponential decay with increasing distance. Implicit is that β is an intrinsic factor. This holds for the entire wire system rather than only for the molecule that is providing the linkage between donor and acceptor. In this respect, there is no fundamental difference between DBA systems and metallic contacts linked directly to a bridge.

Interestingly, β-values for identical bridge units can vary significantly. β reflects, for example, the sensitivity of the transport process to the surrounding environment. Distance dependence is one of these parameters. Typical β-values for bridge units in solution and through SAMs of organic thiols on the surfaces of metal electrodes (Ag, Au, and Hg) are listed in Table 6.1.

6.2.2.1 **Superexchange Charge-Transfer Mechanism in Molecular Wires**

Having established the framework for describing molecular-wire behavior, it is necessary to analyze transport phenomena as a function of molecular-wire properties. First the superexchange mechanism, which is considered to be the main mechanism for efficient ET within the photosynthetic reaction center [48, 49], and has been studied in various biomimetic systems [50, 51] should be discussed.

In superexchange CT reactions, the bridge never accommodates any charge. Those states that the molecule occupies between the time that the electron leaves the donor and arrives at the acceptor are called virtual excitations. The superexchange electronic coupling, t_{DA}, reflects the transmission probability for electrons to reach the acceptor [20, 52]. It depends on the individual resonance integrals between molecular subunits and the energy gap between donor and acceptor and bridge [53]. The model assumes an exponential dependence of t_{DA} on the donor–acceptor distance, r_{DA}. The decay is once more expressed by a β-parameter, which is designated β^{SE} for superexchange:

$$\beta^{SE} = -2\left(\ln\left|\frac{t}{\Delta}\right|\right)\Big/ r \qquad (6.2)$$

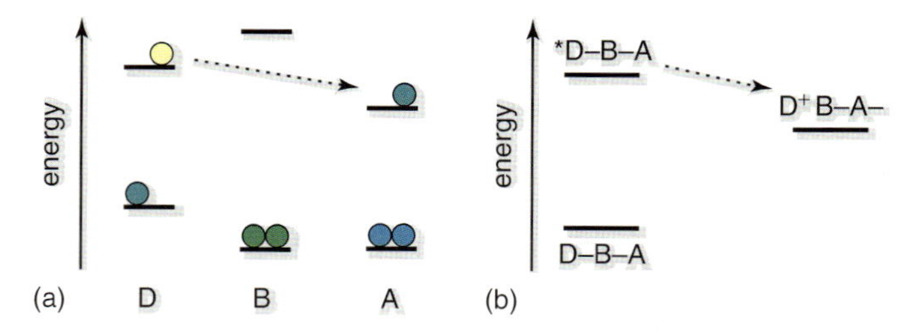

Figure 6.3 Schematic representation of orbital (a) and state (b) energy diagrams of the superexchange mechanism of photoinduced electron transfer in DBA conjugates.

Here t is the nearest-neighbor transfer integral, Δ the energy of the bridge orbitals relative to the donor/acceptor, and r the width of one subunit. This expression has been validated under various experimental conditions [54–56]. Furthermore, it opens up the possibility of calculating the probability of CT via superexchange according to the chemical composition of the bridge (see Figure 6.3).

6.2.2.2 **Sequential Charge-Transfer Mechanism in Molecular Wires**

The superexchange coupling decays exponentially with increasing donor–acceptor distance. Consequently, the probability of superexchange transfer can become very small. In the special case that the donor and the acceptor energy levels lie within $k_B T$ in resonance, an incoherent term dominates the CT rate constant. Several conditions must be fulfilled for sequential CT. Following the definition of Jortner *et al.* [57], sequential CT takes place when (i) near-resonant charge injection is present, (ii) vibronic overlap exists between ion-pair states that are formed while charges move from one bridge site to another, and (iii) vibronic overlap between the ion-pair state given by the terminal bridge site and the ultimate acceptor is significant. In other words, if the energy levels of the bridge approach the levels of the donor, such as represented in Figure 6.4, sequential CT occurs, namely, donor-to-bridge and bridge-to-acceptor. In this case, charge hopping along the bridge is considered as the rate-determining step. Importantly, the distance dependence under such circumstances is influenced by the nature of diffusing

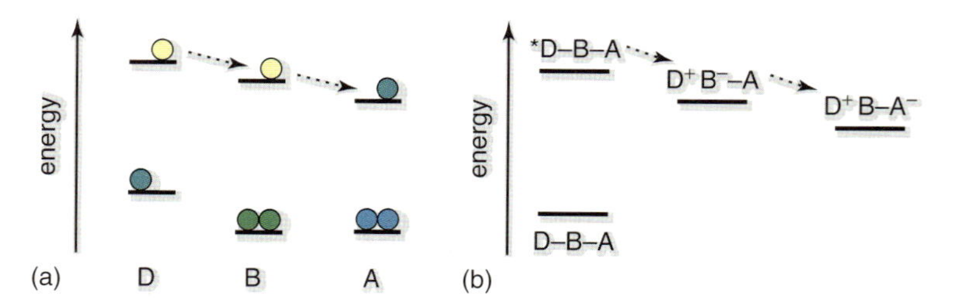

Figure 6.4 Schematic representation of orbital (a) and state (b) energy diagrams of the charge-hopping mechanism of photoinduced electron transfer in DBA conjugates.

charges from the donor to the acceptor. An adequate amount of free energy is thus required to transfer the charges from donor to acceptor. Therefore, the hopping rate may be approximated by

$$k = k_{\mathrm{hop}} \left(\frac{r_{\mathrm{DA}}}{r_{\mathrm{U}}} \right)^{-2} = k_{\mathrm{hop}} N^{-2} \tag{6.3}$$

Here, r_{DA} is the donor–acceptor distance, r_{U} is the length of one modular bridge unit and, therefore, N the number of the units. It can be easily understood that the obtained distance dependence of the CT rates in the hopping regime is relatively weak. The CT rate scales with r_{DA}^{-2} rather than exponentially as found for the superexchange mechanism. Further, it depends on the individual tunneling probabilities between the individual hopping steps. Thus, in the hopping regime the bridge is directly involved in the CT process [27, 57, 58]. As a consequence, long-distance incoherent transfer is, on one hand, more efficient than the coherent analog. On the other hand, it requires potent electron donors and highly conjugated bridges. If these conditions are given, very low β-values characterize the shallow distance dependence of the CT processes [14, 59].

Obviously at intermediate distances, competition arises between the superexchange and sequential mechanisms. This problem has been treated extensively theoretically in order to be able to understand both cases [60–62].

Without going into details, some conclusions can be reached. Firstly, large and positive energy gaps between the initial and intermediate states favor superexchange mechanisms. They are seemingly activationless. Secondly, competition between the superexchange mechanism and sequential mechanism commences when the energy gap approaches the magnitude of the reorganization energies or electronic couplings. Since in any real system the increase of the bridge length causes a more or less efficient lowering of the bridge energy levels, it affects the electronic coupling between the bridge and the donor/acceptor sites. This, in turn, results in a change of the exponential distance dependence of the ET rate. Generally speaking, increasing the bridge length may lead to a switch of mechanism from superexchange to hopping. Thus, when the bridge levels approach the levels of the donor or fall even below, a change in mechanism will be reflected in a sudden increase of the CT rate and simultaneously a decrease of the distance dependence.

We now discuss the impact of electronic and structural parameters on the CT mechanism.

6.2.3 Parameters for Controlling the Charge-Transfer Mechanism

6.2.3.1 Energy Matching

One of the key steps in CT reactions in DBA systems is charge injection into the bridge. The evaluation of charge injection energies is crucial when examining the transport properties. To a first approximation, charge injection energies can be estimated from the energies of the frontier orbitals of the molecular bridge relative to those of the donor and acceptor or the Fermi energy of the metallic contacts. This common description is, however, deceptive. To be more accurate, one should account for the multielectron state energies of the donor (or the Fermi level) and of the bridge cation and bridge anion. However, the frontier orbital energies are often a reasonable estimate. In particular, electron transport across metal–molecule–metal junctions depends strongly on the position of the Fermi level of the metal electrodes relative to the LUMO and HOMO of the molecular bridge.

Applying this concept onto DBA systems, leads to two probable situations:

- When the energetic difference between the orbitals of the bridge and the orbitals of the donor (or the Fermi level) is large, electron transport occurs via superexchange tunneling, that is, tunneling mediated by interactions between donor and acceptor and unoccupied orbitals of the organic bridges that separate them.
- If the orbitals of the donor approach the energy of the unoccupied orbitals of the molecular bridge, resonant ET may take place – either by hopping or resonant tunneling. In this case, the conduction of electrons will occur through the MOs of the bridge.

Fan *et al.* [63], were able to demonstrate these two mechanisms by means of shear-force-based scanning probe microscopy (SPM) experiments. These emphasize the importance of energy matching in determining the transport mechanism. SPM was complemented by current–voltage ($I-V$) and current–distance ($I-d$) measurements. In these investigations, ET processes were tested across SAMs of hexadecanethiol and nitro-substituted compounds, including oligo(phenyleneethynylene) (oPE) on gold surfaces. In these simple experiments, the hexadecanethiol SAMs gave rise to an exponential current increase as the distance was decreased. Large β-values that ranged from 1.3 to 1.4 $Å^{-1}$ resulted. In the low-bias regime, β was almost independent of the tip bias. Such observations suggest a superexchange mechanism for transport through SAMs. The mismatch between molecular energy levels of oPEs and the Fermi level of gold can be considered to be responsible for this trend.

Substituting the oPE structures with nitro groups changed the results. The MOs of the nitro-substituted oPEs are close in energy to the Fermi level. In this case, the current depends only weakly on the distance with a low β-value. Reversible peak-shaped $I-V$ characteristics indicate that part of the conduction mechanism involves resonant tunneling. The β-values for the oPE molecules remained low (i.e., around 0.15 $Å^{-1}$). A possible rationale involves that MOs incorporate interactions with the contacts and give rise to resonant hopping events when applying voltage [64, 65].

Spectrophotoelectrochemistry, for instance, allows the occupied and unoccupied energy levels to be investigated directly and characterized spectroscopically in order to determine their relative positions. These results allow the products of a photoreaction to be identified.

6.2.3.2 Electronic Coupling

The electronic coupling between the components of molecular or metallic DBA systems is an important parameter that influences both incoherent and superexchange electron transport between donors and acceptors. Therefore, it has become increasingly important to develop computational techniques that allow us to calculate the electronic coupling.

In general, most ET reactions occur in the nonadiabatic regime. The reactions are accompanied by a change in the electronic configuration in a thermally fluctuating environment. For such processes, the electronic coupling between donor and acceptor, also called electron transfer integral, is of central importance [21, 47], since it determines the reaction rate and its dependence on the nature, arrangement, and structural fluctuations of the DBA sites. In the weak-coupling limit, the rate of the CT reaction is given by the Fermi golden rule:

$$k_{i \to f} = \frac{2\pi}{\hbar} \left| V_{if} \right|^2 \delta \left(E_i - E_f \right) \tag{6.4}$$

in which V_{if} is the electronic coupling factor describing the transition between two electronic states (i and f). Then, the rate of such processes is related to the density of states and the electronic coupling factor, V_{if}, which is determined by the electronic nature of the molecules or fragments involved. In organic materials it is common to use the coupling between neighboring units for the description of the ET properties [66–68]. Furthermore, due to the close relation of the electrical conductance of molecular wires with the ET rate constant [15, 69], computational methods which are capable of treating ET processes may be applied to study molecular wires. Particularly, since it is difficult to obtain the electronic coupling from experimental data, quantum mechanical calculations often offer the only possibility to acquire this parameter [67, 70, 71].

In the weak-coupling regime the rate of ET can be described in terms of the Marcus [48, 52, 72, 73] theory:

$$k_{\mathrm{ET}} = \frac{2\pi}{\hbar} \left| V_{if} \right|^2 \frac{1}{\sqrt{4\pi\lambda k_B T}} \exp - \frac{\left(\Delta G^0 + \lambda \right)^2}{4\pi\lambda k_B T} \tag{6.5}$$

where λ is the reorganization energy and ΔG^0 is the standard Gibbs free energy change. Similar to Eq. (6.4), the ET rate is proportional to the electronic coupling amplitude squared. This promotes us to define the electronic coupling as the off-diagonal Hamiltonian matrix elements between the initial and final diabatic states. Then, the coupling becomes essentially the interaction of the two MOs where the electron occupancy is changed.

Most calculation methods employ a two-state model in order to estimate the matrix element. This model assumes that when the system moves toward the transition state, only the donor and acceptor states will mix with each other. Any interactions with other intermediate states is neglected; thus in a DBA system, the coupling between the reactant, *DBA, and the product, D^+BA^-, may be considerably facilitated by the virtual bridge state, D^+B^-A.

If the energy gap between the reactant state, $^*DBA$, and the bridge state, D^+B^-A, is significantly large, the latter state will not be reached throughout the CT process. Such a situation, for instance, would favor a superexchange CT mechanism.

Many strategies for the calculation of the couplings between the different states in ET reactions have been developed. They can be classified into (i) energy-gap-based approaches, (ii) direct calculations of the off-diagonal matrix elements, or (iii) the use of an additional operator to describe the extent of charge localization and to calculate the coupling value. A detailed description of the methods would certainly extend the scope of this chapter. That is why we will shortly introduce only the most popular approach.

Commonly, all quantum chemical calculation methods take into account the entire DBA system and, therefore, consider the contributions of all virtual bridge states. As a consequence, interference of different ET pathways is appropriately treated and the donor–acceptor electronic coupling, V_{DA}, estimates become accurate. When the donor–bridge energy gap decreases, bridge states may mix with the adiabatic states of interest. In such a situation, the simple two-state model becomes inappropriate. To account for this deficiency and obtain correct values for V_{DA}, a multistate treatment is required. In other words, one has to consider several additional adiabatic states. Various multistate approaches have been developed. The most common of such is the Generalized Mulliken–Hush (GMH) approach [74, 75]. It evaluates electronic couplings on the basis of quantum chemical calculations of ET processes. Solving the Schrödinger equation yields a set of eigenstates, with the coupling being the Hamiltonian element in a charge-localized space. GMH and its variants use additional operators to define this charge-localized space and calculate the off-diagonal Hamiltonian matrix element. An important advantage of the GMH method is that it is capable of treating systems where more than two adiabatic states are used for the description of the diabatic states.

The fragment charge difference (FCD) approach [76] is very similar to the GMH method. In FCD a charge difference operator is employed. The molecule is partitioned into two fragments, one for the donor and the other for the acceptor. Thereby, a matrix for the charge difference is obtained from which the electronic coupling can be calculated.

While the two-state GMH scheme to calculate electronic couplings is widely used, the multistate scheme still does not enjoy broad application. However, for some DBA systems the two-state model fails, especially in the case of three-state systems, that is, when two diabatic states (ground and locally excited state) are localized on the donor and a single CT state is localized on the acceptor. Nevertheless, a rather technical reason opposes the use of the multistate approach: many of the standard quantum chemical programs do not provide transition dipole moments between excited states, while the corresponding values for the transition between the ground and excited states are usually available. Because of that, the dipole moment matrix required in the GMH scheme cannot be constructed for a system where three or more states must be considered.

6.2.3.3 Temperature

Finally, the influence of different temperatures on the CT mechanism must be considered. In particular, the temperature plays a key role in understanding the mechanistic change between coherent tunneling (superexchange) and incoherent charge hopping. Typically, the low temperature regime is dominated by superexchange because of high-energy gaps for charge injection into the bridge. The electronic levels of the bridge sites couple only weakly to the donor and acceptor and are usually separated by a reasonably high-energy gap from the donor and acceptor frontier orbital levels. The CT exhibits only weak distance dependence. Increasing the temperature, on the other hand, induces thermal injection of the charge carriers into the bridge. The bridge eigenstates become accessible and therefore one expects to observe localized hopping if the vibronic coupling is strong. Thus, upon raising the temperature, an abrupt mechanism change from fairly distance-dependent coherent tunneling to charge hopping with a far smaller distance dependence is likely to occur. Often a sudden increase in rate constants accompanies a transition from low to high temperatures.

Notably, the temperature dependence also reflects the importance of π-conjugation. In saturated organic molecules (alkanes) high injection gaps are present. Extended π-conjugation, on the other hand, leads to stronger electronic coupling between the bridge and the donor/acceptor sites and therefore to lower energy gaps for charge injection from either the donor or the acceptor, even at low temperatures [16].

6.2.3.4 Specific Aspects of Photoinduced Electron Transfer in Organic π-conjugated Systems

Photoinduced CT reactions have been investigated extensively for more than 15 years. It is of fundamental interest to understand the photophysics and photochemistry of excited states in organic molecules, especially in the fast-developing field of photovoltaics [77]. Photosynthetic energy conversion in green plants serves as the ultimate prototype for solar energy conversion [78]. As outlined above, the basic description of photoinduced CT between a donor and an acceptor considers several different steps. For the following discussions, we will assume the following reaction sequence:

Step 1: $D + A \rightarrow D^* + A$ (photoexcitation of D)
Step 2: $D^* + A \rightarrow (D - A)^*$ (excitation delocalized between D and A)
Step 3: $(D - A)^* \rightarrow (D^{\text{TM}+} - A^{\text{TM}-})^*$ (partial CT)
Step 4: $(D^{\text{TM}+} - A^{\text{TM}-})^* \rightarrow (D^{\bullet+} - A^{\bullet-})$ (formation of a radical ion pair)
Step 5: $(D^{\bullet+} - A^{\bullet-}) \rightarrow D^{\bullet+} + A^{\bullet-}$ (complete charge separation)

Alternatively, hole transfer from an excited acceptor to a donor can be formulated. In general, A and D should be either covalently linked (intramolecular CT) or spatially very close (intermolecular CT). At this point, we will focus on intramolecular CT, because the systems investigated contain molecular bridges that link donors and acceptors.

Importantly, Step 2 requires that the electronic wavefunctions of A and D are significantly coupled. In our case, the coupling is provided by the molecular bridge.

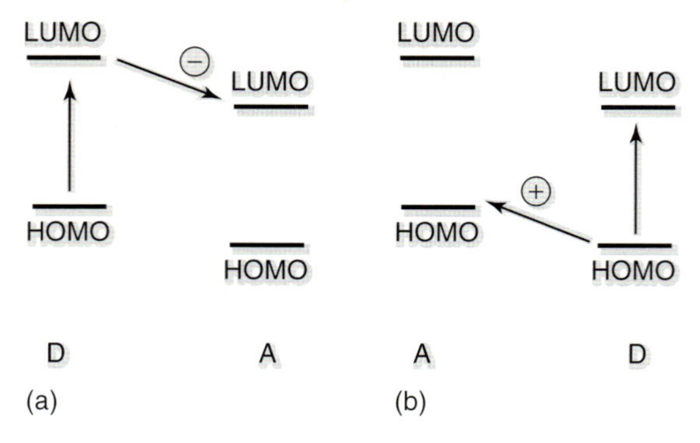

Figure 6.5 Energy-level alignment for electron transfer from D to A (a) and for hole transfer from D to A (b).

On the other hand, a relatively small distance between the donor and acceptor sites would also meet this requirement [79, 80].

In contrast, Step 4 requires the energy levels of the participating components to be aligned. Only if the offset between the LUMOs (for ET) and the HOMOs (for hole transfer) of donor and acceptor is large enough to overcome Coulombic attractions between charges charge separation is induced. The energetic difference between initial and final states represents the driving force for the CT reaction. This is shown schematically in Figure 6.5. When donor and acceptor are covalently linked, the CT reaction is then mediated by the transport properties of the intervening linker. In general, the energetics and kinetics of a photoinduced charge separation are described by the theories of Marcus and Jortner, which take the orientation and the distance between the donors and the acceptors into account [81–83].

In some cases, the CT state might be metastable because of fast delocalization of charge carriers on one or both molecules [84].

Electron transfer profoundly affects chemical reactivity by inverting normal electron densities in electron donor/acceptor pairs and therefore activating previously inaccessible reaction modes. To understand the relaxation pathways for photoinduced CT reactions in solution fully, solvent effects have to be taken into account. This is typically done in terms of the classical Marcus theory for ET reactions in solution, which considers the underlying driving forces ($-G^{\circ}$) and the corresponding reorganization energies ($\lfloor$). The basic principles have been widely discussed in the literature [72, 73, 85, 86].

6.3
Charge Transport along π-conjugated Bridges in C_{60}-Containing Donor–Bridge–Acceptor Conjugates

Recently, systems capable of charge and energy transport have gained tremendous scientific interest because of their potential use in the electronics industry. Thus, a variety of potential molecular wire systems has been designed, synthesized, and tested.

π-Conjugation clearly plays a vital role. In this context, benzenes [87, 88], *p*-phenylacetylenes [89, 90], and porphyrin arrays [91–93] have been intensively studied in terms of their molecular structures. More complex structures such as modified proteins and peptides [94] are now at the forefront of research, especially in the field of molecular wires. The methods used to investigate molecular wires are similarly diverse. Organic SAMs have been the basic building block for the majority of organic devices [95, 96]. Various types of metal–insulator–metal (MIM) junctions [97] including single layers of molecules between aluminum and titanium/aluminum [98, 99], gold and other metal contacts [87], mechanically controllable break junctions [100], silicon adlayers [101], electrochemical break junctions [102], and SAMs sandwiched between two mercury electrodes [100, 103] have been investigated using conduction measurements. At the same time, the different experimental setups allowed to test the performance of molecular wires in an extensive set of different transport environments.

It is possible to characterize both the response coefficients (rate constant or conductance) and the mechanisms for such transfers by restricting the transfer to either within a given molecule (intramolecular ET or energy transfer kinetics) or across a given molecule (single molecule transport junction conductance). As it was shown, the dominant mechanisms range from coherent tunneling (i.e., large injection energy gaps, low temperatures, and short bridges) to incoherent hopping (i.e., small injection energy gaps, higher temperatures, and longer wires). The relative importance of these two can be understood from energetic (the injection barrier versus thermal energy) and temporal (Landauer/Buttiker contact time versus vibrational period) considerations.

This chapter focuses on intramolecular ET processes. In particular, π-conjugated systems and the influence of their structures on ET in solution will be examined. Moreover, features of single molecules rather than properties of ensembles or higher architectures will be pointed out. Therefore, highly diluted solutions are used to rationalize the performance of such ET systems on the single-molecule scale. The thrust is to survey, highlight, and compare the methods currently used to probe molecular systems or molecular fragments in the context of transporting charges efficiently from one end to the other along the molecular framework. The molecular wires are thus covalently connected to electron donors and electron acceptors. In principle, a certain degree of charge delocalization from the donor to the acceptor can occur in the ground state. Nevertheless, the main focus will be on electron donor–acceptor structures, in which excitation by light powers the promotion of electrons from the donor to the acceptor to give a spatially separated radical ion pair. This topic has already been covered widely in the literature on nonfullerene-containing systems, so that herein the description will be limited to examples in which fullerenes play an important role. The presence of an electron donor linked to a fullerene does not necessarily guarantee the formation of charge-separated species. Energy transduction or even simultaneous or sequential energy and ET processes compete with each other.

6.3.1 C_{60}–WIRE–exTTF

Since DBA architectures have proved to be a suitable platform for probing ET processes along the wire-like bridges at the molecular level, several different wire-like bridges combined with various donors have been tested. In all of the examples discussed below, C_{60} serve as electron acceptors. The delocalization of charges within the spherical carbon framework of the fullerene together with its rigid structure offers unique opportunities for stabilizing charged entities. The rigid framework means that fullerenes possess small reorganization energies in ET reactions. This represents a significant advantage for generating long-lived charge-separated states [104].

The first examples contain extended tetrathiafulvalene (exTTF) donors, which are covalently linked to (i) oligo(*para*-phenylenevinylene) (oPPV), (ii) oligo(*para*-phenyleneethynylene) (oPPE), and (iii) oligofluorene (oFL) oligomers of varying length. exTTF is a particularly interesting electron–donor molecule and has been used extensively to prepare electrically conducting and superconducting molecular materials [105, 106]. Despite its highly distorted structure, which deviates significantly from planarity, electrically conducting CT complexes have been prepared by reacting these donors with strong electron acceptors such as fullerenes. These π-extended TTF derivatives release two electrons simultaneously (i.e., in a single quasireversible ET process).

C_{60}–WIRE–exTTF donor–acceptor ensembles offer the great advantage of linking the oligomeric bridge to exTTF without compromising the π-conjugation. The C_{60} derivative *N*-methylfulleropyrolidine (see Figure 6.1), where a pyrolidine ring is linked to the framework of the fullerene, was utilized as the electron accepting part. Particularly, extended π-conjugation is ensured between the anthracenoid part of the *ex*TTF-donor, the oligomeric bridge (*p*-phenylenevinylenes, *p*-phenyleneethynylenes, or oligofluorenes), and the pyrolidine ring of the C_{60} derivative.

6.3.1.1 C_{60}–oPPV–exTTF

A series of C_{60}–oPPV–exTTF [107] donor–acceptor arrays (see Figure 6.6) incorporating π-conjugated oPPV wires of different length between the π-extended exTTF as electron donor and C_{60} as electron acceptor has been prepared by multistep convergent synthesis. The electronic interactions between the three electroactive species present in **1a, b** were investigated by UV–visible spectroscopy and cyclic voltammetry (CV). Although the C_{60} units are connected to the exTTF donors through a π-conjugated oPPV framework, no significant electronic interactions were observed in the ground state. Interestingly, photoinduced ET processes over distances of up to 50 Å afford highly stabilized radical ion pairs. The measured lifetimes for the photogenerated charge-separated states were in the range of hundreds of nanoseconds (~500 ns) in benzonitrile, regardless of the wire length (i.e., from the monomer to the pentamer). A 10-fold lifetime (4.35 ms) was observed for the heptamer-containing array **1b**. This difference in lifetime has been accounted for by

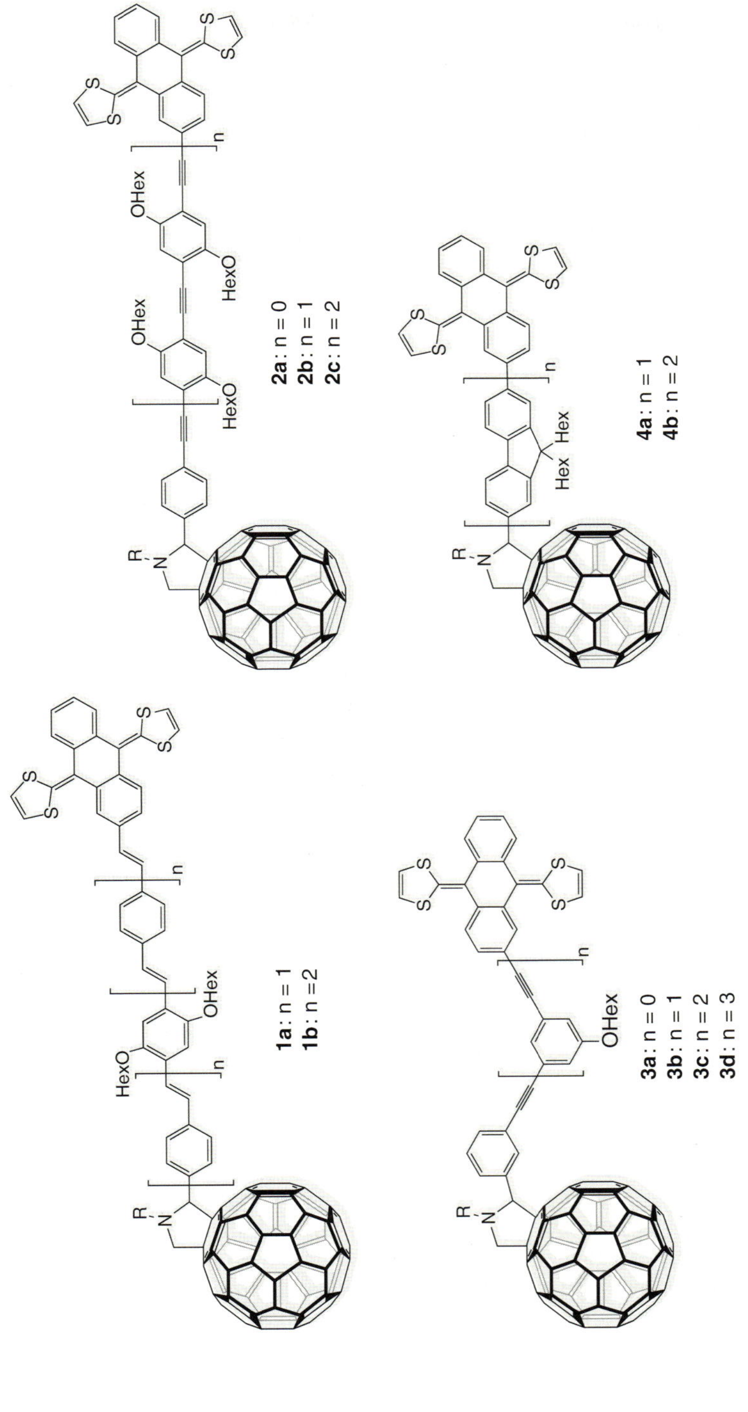
R
N
HexO
OHex
n
S
S
1a: n = 1
1b: n = 2
OHex
HexO
2a: n = 0
2b: n = 1
2c: n = 2
OHex
3a: n = 0
3b: n = 1
3c: n = 2
3d: n = 3
Hex
Hex
4a: n = 1
4b: n = 2

Figure 6.6 Examples of C_{60}–WIRE–exTTF DBA conjugates.

the loss of extended π-conjugation throughout the oPPV moiety. The bridge loses its rigidity with increasing bridge length and torsional flexibility twists the dihedral angles between the two terminal phenyl rings away from planarity. The molecular structures of **1a** and **1b** were optimized by semiempirical calculations at the PM3 level and density functional theory (DFT) calculations at the HF/6-31G(d) and B3-P86/6-31G(d) levels, which all confirmed the increasing deviation from planarity of the oPPV moiety when going from the pentamer to the heptamer-containing array. Plotting the ET kinetics as a function of donor–acceptor distance led to linear dependencies in THF and benzonitrile. An extraordinary small attenuation factor ($\beta = 0.01 \pm 0.005\,\text{Å}^{-1}$) was determined from the slope of these plots. This is mainly because of the energy matching between the C_{60} HOMOs and the long oPPV chains. Of equal importance is the strong electronic coupling between the donor (exTTF) and acceptor (C_{60}), which is realized through the *para*-coupling of the oPPVs to the exTTF donor. In turn, donor–acceptor coupling constants V in $C_{60}{}^{-\bullet}$–oPPV–exTTF$^{\bullet+}$ are about 5.5 cm^{-1}. This assists the rather weakly distance-dependent CT reactions.

6.3.1.2 C_{60}–oPPE–exTTF

Bearing in mind the influence of rotational flexibility on CT reactions described above, the focus will be shifted to very similar molecular wire architectures, oPPEs. Replacing the double bonds by triple bonds adds additional rigidity to the π-system compared with oPPVs. A series of C_{60}–oPPE–exTTF [108] conjugates was tested in order to investigate the influence of such a structural modification on the ET properties of the oPPEs. Three donor–acceptor arrays containing monomeric, dimeric, and trimeric π-conjugated oPPEs linking π-extended exTTF as electron donor and C_{60} as electron acceptor were synthesized (Figure 6.6).

Transient absorption spectroscopy confirmed the presence of spectral signatures of the one-electron oxidized exTTF$^{\bullet+}$ radical cation and the one-electron reduced $C_{60}{}^{\bullet-}$ radical anion. Photoexcitation is followed by a rapid deactivation of the singlet-excited state of the oPPE moiety, which generates the radical ion pair state $C_{60}{}^{\bullet-}$–oPPE–exTTF$^{\bullet+}$, which is apparently lower in energy than the corresponding triplet state of C_{60}. These findings hold, however, only for the monomer **2a** and dimer **2b**, while in the trimer **2c** no radical ion pair formation was observed. The dependence of the charge separation and charge recombination dynamics in THF on the donor–acceptor distance gave an attenuation factor (β) for the oPPE bridges of $0.2 \pm 0.05\,\text{Å}^{-1}$. Extrapolating the linear relationship to the center-to-center distance of the trimer gives a charge separation rate that would be significantly slower than the intrinsic deactivation of the C_{60} singlet excited state. This, in turn, helps to rationalize the lack of ET in the trimer. A schematic representation of the possible reaction pathways, which is also appropriate for the C_{60}–oPPV–exTTF systems shown above, is given in the energy diagram of Figure 6.7.

Geometry optimizations at the B3LYP/6-31G(d) and B3PW91/6-31G(d) levels of DFT, which were supplemented by corresponding single-point calculations with 6-311G(d,p) basis sets gave no significant rotational barriers (i.e., < 2 kcal mol^{-1}) for the phenyl rings in the oPPE-bridge. The planar conformation of the bridge is

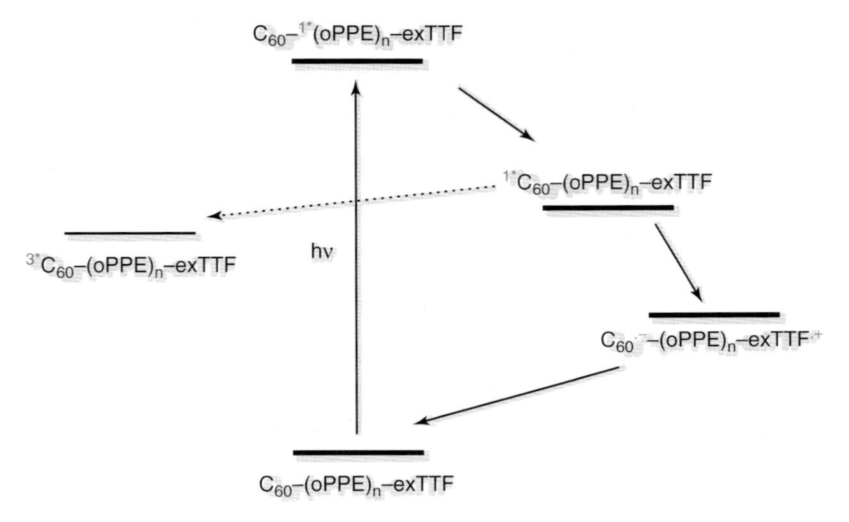

Figure 6.7 Schematic illustration of the reaction pathways in photoexcited **2a–b**.

energetically slightly favored. Unlike the situation found in the oPPV systems, the electronic coupling between the donor and acceptor units seems not to be affected by the dihedral angles of the terminal benzenes. Inspecting the frontier orbitals of both the oPPV and oPPE systems shows, however, that the HOMO in C_{60}–oPPV–exTTF (**1**) reaches farther into the oPPV-bridge than in the C_{60}–oPPE–exTTF (**2**), where it is completely localized on exTTF. Thus, charge injection into the bridge is more facile in **1** than in **2**. In addition, local electron affinity mappings show a homogeneous distribution of electron affinity throughout the whole bridge in C_{60}–oPPV–exTTF, whereas in the oPPE systems local maxima were found on the phenyl rings and minima on the triple bonds (Figure 6.8). Similarly, C_{60}–oPPV–exTTF reveals a homogeneous distribution of the local electron affinity. These features point to the polarizing character of the triple bonds and their shorter bond length. In comparison to the pure double-bond character of the oPPV bridges, the bond length alternation caused by the presence of triple bonds in **2a–c** disrupts the extended π-conjugation and strongly influences the CT path.

Finally, the properties of *meta*-connected isomers oligo(*meta*-phenyleneethynylene)s (oMPEs) shall be discussed. A series of C_{60}–oMPE–exTTF systems (**3a–d**) with up to four oligomeric bridge units was synthesized [109]. The structures are shown in Figure 6.6. The photophysical characteristics of these systems and their aggregation phenomena at high concentrations were investigated. The aggregation properties turned out to be relevant for the photophysical studies. Differences in electronic properties between the *para*- and *meta*-connected systems are mainly governed by the more compact structure and the loss of *para*-conjugation of the latter.

In the low-concentration regime (10^{-6} M), where the electron donor–acceptor conjugates are present in their monomeric form, intramolecular ET processes were

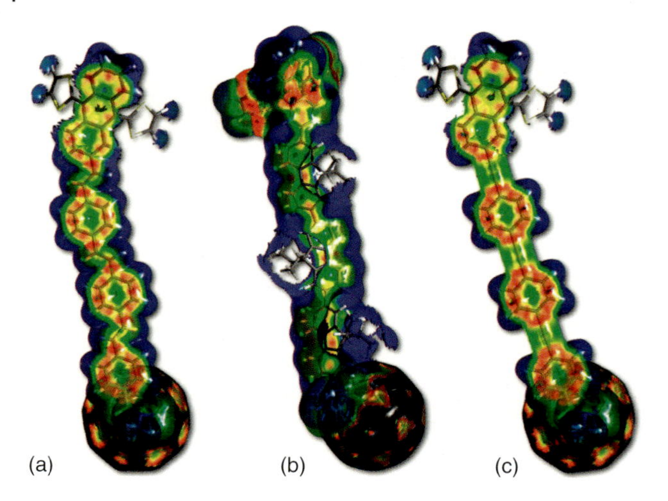

Figure 6.8 Local-electron-affinity maps of **1a** (a), **4** (b), and **2c** (c). Maxima are shown in red, minima in blue.

indicated by steady-state and time-resolved spectroscopic techniques. Kinetic analyses of the formation of the radical ion pair state, which is formed through bond in the monomer and through space in the dimer, trimer, and tetramer, revealed that the latter are metastable on the femto-/picosecond timescale. The charge recombination dynamics indicate lifetimes on the order of tens of nanoseconds for **3b** (26 ns), **3c** (41 ns), and **3d** (49 ns), for which the reactivity is driven by through-space interactions. **3a**, for which through-bond interactions lead to lifetimes of 128 ns, behaved differently.

At higher concentrations ($\geq 10^{-4}$ M) atomic force microscope (AFM) studies revealed that individual C_{60}–oMPE–exTTF molecules bind to give $(C_{60}\text{–oMPE–exTTF})_x$ aggregates due to π–π-stacking interactions between exTTF and C_{60}. The presence of these interactions was confirmed in complementary work on exTTF tweezer conjugates and pristine C_{60} molecules [110], which showed that the interactions between the aromatic system of exTTF and C_{60} cause pincer-like 1:1 complexes to form. Furthermore, the donor–acceptor supramolecular π–π-complexes exhibit intracomplex photoinduced CT processes between the electron donating exTTF moiety and the electron accepting fullerene. Very short donor–acceptor distances result in very short (picosecond range) lifetimes of the charge-separated states. These results all suggest that complexes are formed between individual C_{60}–oMPE–exTTF molecules via π–π-interactions. The interactions between the two redox-active moieties, exTTF and C_{60}, in the high-concentration regime lead to a substantially different photophysical response of the resulting intracomplex hybrids from that found in the low-concentration regime. In particular, ground-state CT interactions are present that result from a shift of electron density from exTTF to C_{60} because of the short distance between donor and acceptor. As a consequence, excitation into the CT bands in the high-concentration regime leads to the immediate formation of the radical ion pair state. In particular, the exTTF radical cation and the C_{60} radical anion features evolve nearly instantaneously and confirm the rapid formation of the intracomplex radical ion pair state. The lifetime of this intracomplex radical ion pair state is remarkably long. It exceeds the timescale of the femtosecond experimental setup used for the investigation. Experiments on the nanosecond timescale, where the same radical ion pair features are generated, allow the lifetimes to be determined as $4.0 \pm 0.7\,\mu s$ – **3b**: 3.2 μs; **3c**: 4.7 μs; **3d**: 4.6 μs.

Noteworthy, the analysis of the frontier orbital schemes and their energies as optimized at the B3PW91/6-31G(d,p) level, corroborates that the highest occupied orbitals are localized on the electron donating exTTF, whereas the lowest unoccupied orbitals are part of C_{60}. Important is that the bridge orbitals are energetically well separated from the frontier orbitals, resulting in virtually no overlap between the electron donor, the bridge and the electron acceptor. In line with this picture, the HOMO-1 to LUMO+2 orbitals of all C_{60}–oMPE–exTTF systems reveal almost equal orbital energies independently of the length of the bridge. Furthermore, semiempirical configuration interaction (CI) calculations with the AM1* Hamiltonian, which includes d-orbitals on sulfur, suggested two distinct CT states that are dominated by the HOMO → LUMO and HOMO → LUMO+1 excitations in all C_{60}–oMPE–exTTF systems.

6.3.1.3 C_{60}–oFL–exTTF

Recently oFL molecular wires have been integrated into dumbbell architectures (i.e., C_{60}–oFL_n–C_{60}) [111]. These oFL wires transfer energy upon photoexcitation if an acceptor, such as a fullerene, is present. In light of the excellent electron donating properties of exTTF and the favorable optical properties of oFL molecular wires, C_{60}–oFL–exTTF (**4**) donor–acceptor conjugates were synthesized as promising systems for photoinduced ET reactions (Figure 6.6) [112]. One unusual feature of oFLs is that their oxidation potential remains constant with increasing number of oligomeric units. As a consequence, strong electronic coupling between the bridge and the redox-active moieties must be present. One may therefore expect a rather weak distance dependence in the CT behavior of C_{60}–oFL_n–exTTF DBA triads. The donor–acceptor conjugates **4a, b** exhibit efficient CT processes upon photoexcitation over distances of more than 24 Å. The charge transport mechanisms are comparable to those established for the corresponding C_{60}–oPPV–exTTF and C_{60}–oPPE–exTTF systems. The ability of the oFLs to conduct charges lies between those of oPPVs and oPPEs. The dependence of the rate constants ($8.9 \times 10^9\,s^{-1}$ and $4.0 \times 10^9\,s^{-1}$ for charge separation and $7.2 \times 10^5\,s^{-1}$ and $4.4 \times 10^5\,s^{-1}$ for charge recombination) on the donor–acceptor distance gives an attenuation factor β of 0.09 $Å^{-1}$. On the basis of orbital energies obtained by DFT methods (B3LYP/6-31G(d)) good electronic communication between exTTF, oFL, and C_{60} was corroborated. The energies of the building blocks (exTTF, *N*-methylfulleropyrolidine, the pristine oFL oligomers, and C_{60}-oFL) reveal that, for instance, the HOMO/LUMO energies of exTTF in vacuo (−4.7 and −1.2 eV for HOMO and LUMO, respectively) match the energies of the oFL building blocks (around −5.0 and −1.0 eV) perfectly. This, in turn, favors electron injection from the donor to the bridge.

Attaching the fullerene moiety to the oFL chains lowers their LUMO energy to that of pristine *N*-methylfulleropyrolidine. Importantly, this occurs independently of the length of the oFL-bridge. Nonetheless, the HOMO energy remains unchanged in comparison to pristine oFLs, reflecting the invariance of their oxidation potential. In addition, semiempirical MO-based (AM1*) local electron-affinity calculations confirm the ET pathway from the donor over the bridge to the fullerene acceptor in C_{60}–oFL–exTTF (Figure 6.8). A continuous distribution of electron-acceptor capability throughout the molecule and a channel of high local electron affinity through the bridge that maximizes at C_{60} visualize the CT features of these systems.

Notably, equal quantum chemical investigations performed with the hypothetically designed trimer suggest that these CT features will be retained in oligomers of considerable length. It is therefore of interest to probe longer oFL conjugates and examine their CT properties, especially in light of different CT mechanisms. The fact that the oxidation potential of oFLs remains unchanged upon changing the number of fluorene units makes such studies particularly appealing. ET mediated by oFL-bridges should be independent of the donor–acceptor distance in a whole series of donor–oFL–acceptor conjugates.

6.3.2 C_{60}–WIRE–ZnP/H_2P

Exchanging exTTF by free-base (H_2P) or zinc porphyrins (ZnP), has significant influence on the molecular-wire behavior. The electron-rich nature of the porphyrins, for instance, extends the π-conjugation by transferring electron density to the adjacent parts of the bridges. This, in turn, impacts the electron-injection process and therefore facilitates the CT processes. Thus, the ZnP/H_2P donor affects the charge separation rates and the β-values compared to those of the corresponding exTTF conjugates. The extended π-conjugation depends on the connection pattern between the porphyrins and the oligomers, for instance, β- versus *para*-substitution. In the following the influence of porphyrins as donors on the molecular wire behavior of oPPV and oPPE oligomers will be considered.

6.3.2.1 C_{60}–oPPV–ZnP/H_2P

Photo- and electroactive triads in which π-conjugated oPPV oligomers of different lengths (trimer and pentamer, **5a, b**) are connected to a zinc tetraphenylporphyrin donor and C_{60}, were designed, synthesized, and tested as ET model systems (Figure 6.9) [113]. Detailed time-resolved transient absorption spectroscopic studies, concentrating mainly on long-range charge separation (in THF $-3.2 \pm 0.6 \times 10^9\ s^{-1}$), and charge recombination events (in THF $-9.3 \pm 0.15 \times 10^5\ s^{-1}$) revealed attenuation factors β of 0.03 ± 0.005 Å^{-1}. The larger values than found for the corresponding exTTF systems indicate the less effective *para*-conjugation between the porphyrin donor and the π-conjugated oligomer unit of the bridge and the donor–bridge energy gap. This loss of conjugation can be explained by the dihedral angles between the ZnP phenyl ring and the first ring of the oligomer unit ($\sim 30°$) and the deviation from planarity along the oligomer. Once more, energy matching

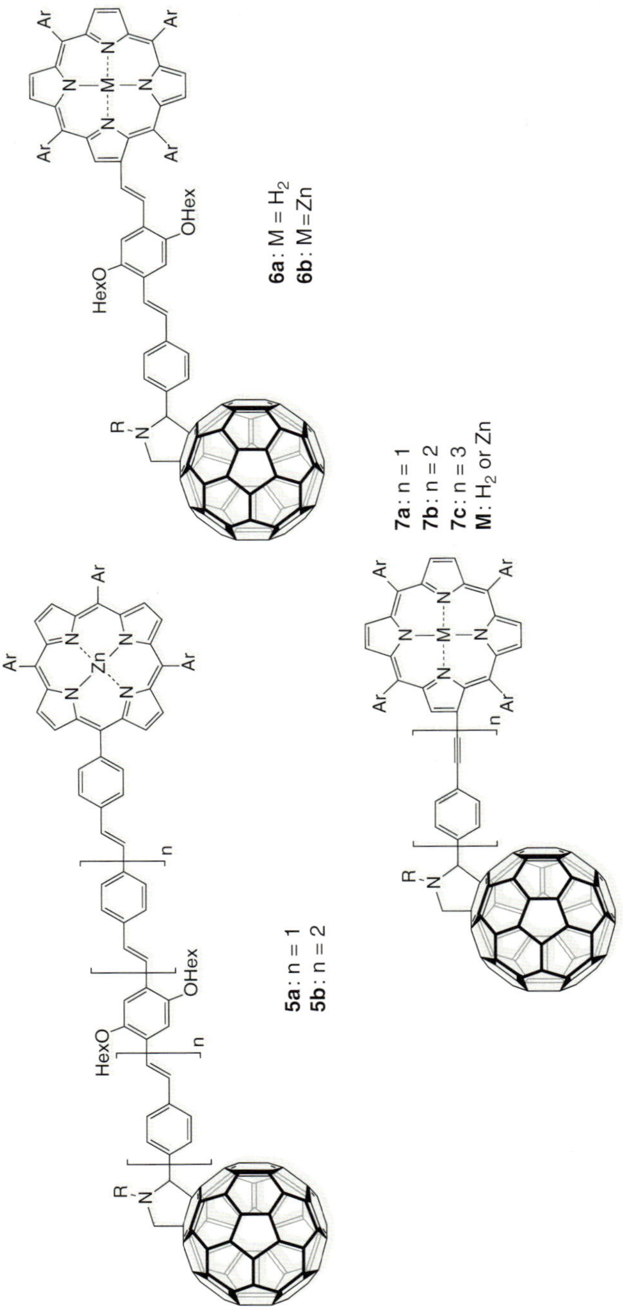

Figure 6.9 Examples of C_{60}–WIRE–ZnP/H_2P DBA conjugates.

between the orbital levels of the donor/acceptor and oPPVs emerged as a key parameter for molecular-wire-like behavior, since it favors rapid and efficient electron or hole injection into the oPPV wires. The energetic matching of the HOMOs of the fullerene with those of the oPPV bridges, which implies a hole-transfer process in these C_{60}–oPPV–(*para*)ZnP systems, facilitates charge injection into the bridge. In conjunction with the *para*-conjugation pattern of the oPPVs leads to donor/acceptor coupling constants of $\sim 2.0\,cm^{-1}$, even at electron donor acceptor separations of 40 Å, and assists ET reactions, whose rates exhibit only very weak distance dependencies. This is true despite the torsional freedom of the donor–bridge and bridge–acceptor contacts. To analyze the charge recombination mechanism, the radical ion pair lifetimes were probed between 268 and 365 K. The resulting Arrhenius plots were separated into two distinct sections: the low (<300 K) and the high-temperature regimes (>300 K). The weak temperature dependence in the 268–320 K range suggests that a stepwise charge recombination can be ruled out, leaving electron tunneling via superexchange as the operative mode. This picture is consistent with the thermodynamic barrier necessary to overcome in forming $C_{60}^{\bullet -}$–oPPV–(*para*)$ZnP^{\bullet +}$. At higher temperatures above 300 K, the situation changes, and the charge recombination is accelerated. The observed strong temperature dependence suggests a thermally activated charge recombination. The activation barriers (E_a), derived from the slopes of the Arrhenius plots (0.2 eV), correspond to the difference between the HOMO energy levels of C_{60} and oPPV.

As the β-values obtained for the C_{60}–oPPV–(*para*)ZnP systems are very low ($\sim 0.03\,\text{Å}^{-1}$), determining those for the corresponding β-substituted C_{60}–oPPV–(β)ZnP/H_2P triads was also of interest. In particular, assessing the *para*-conjugation along the oPPV dimer as a π-conjugated system of precise length, and using the β-substituted freebase (H_2P) and Zn-porphyrins (ZnP) as electron donors makes it possible to compare the efficiencies of the formation of charge-separated states with other related π-conjugated oligomers. Therefore, one analogous C_{60}–oPPV–(β)ZnP/H_2P system (dimers **6a**, **b**, Figure 6.9) was synthesized and investigated [114]. Photophysical studies revealed that efficient ET processes upon photoexcitation afford the corresponding radical ion pairs of C_{60}–oPPV–(β)ZnP/H_2P. Thus, the results confirm that β-substitution of the porphyrin moiety also favors the electronic coupling and the electronic communication between the donor and acceptor units. On the other hand, the influence of the metal ion in the porphyrin donor seems to be negligible. Work is currently in progress to prepare related systems with varying lengths of the oPPV-bridge to determine the attenuation factor in these β-substituted systems.

6.3.2.2 C_{60}–oPPE–ZnP/H_2P

To determine the influence of the donor moiety on the attenuation factor of the specific molecular-wire-bridging structure similarly to the investigations described above, β-substituted C_{60}–oPPE–(β)ZnP/H_2P (**7**) DBA arrays were synthesized (Figure 6.9) [115]. To enhance the electronic interaction through the extended π-system, the molecular bridges were again directly linked to the β-pyrrole position of the porphyrin ring. In this series of DBA systems, the *meso*-phenyl ring of the macrocycle is again not used to link the porphyrin and fullerene moieties. Importantly, the carbon–carbon triple bond directly attached to the pyrrole ring leads to π-conjugation that extends over the whole oPPE residue. Thus, the distance between the porphyrin moiety and the fullerene is completely conjugated and only interrupted by the sp^3 carbon atom of the pyrolidine ring, where the bridge is attached to the fullerene (see Figure 6.9). Hence, these structures are ideally suited for transporting charges effectively through the bridge to the electron accepting fullerene. In addition, the conformation of the bridge, namely, the relative positions of the phenyl rings, impacts the π-conjugated path. A planar configuration, for instance, preserves π-conjugation along the bridge. On the other hand, orthogonally arranged phenyl rings inevitably interrupt the extension of the π-conjugated system. For the validation of the orbital overlap between the phenyl subunits in the bridge and the donor and acceptor moieties it is necessary to evaluate the orbital energy levels (or alternatively the ionization potentials and electron affinities) of the modular components (i.e., donor, bridge, and acceptor). The matching of the orbital energies, especially the HOMO and LUMO energies, determines the orbital overlap between the components and, in turn, the mechanism of the CT processes.

Nevertheless, the photophysical response obtained contrasted quite strongly that obtained for the C_{60}–oPPE–(β)ZnP/H_2P architectures described above. In particular, a charge separated state is obtained in polar media but the lifetimes tend to increase with increasing distance between the two photoactive units. These lifetimes range from 300 to 700 ns. It is safe to assume a through bond mechanism for donor–acceptor separations of up to 23 Å, where the bridge plays a crucial role. On the other hand, in the corresponding C_{60}–oPPE–(β)ZnP/H_2P monomer (**7a**) and dimer (**7b**) transfer via a superexchange mechanism seems to be the most probable operative mode. Extremely fast charge separation dynamics in (**7a**) and relatively slow charge recombination dynamics in (**7c**), on the other hand, are due to the short and large donor–acceptor separation distances in **7a** and **7c**, respectively.

Replacing the exTTF with ZnP in C_{60}–oPPE–(β)ZnP/H_2P leads, however, only to minor changes in the overall reaction pattern. Thorough analysis of the electronic structure using DFT calculation methods at the B3PW91/6-31G(d) and the B3PW91/6-31++G(3df,2pd) levels revealed a strong localization of the HOMOs and LUMOs on the electron donating ZnP/H_2P and electron accepting C_{60}, respectively. In comparison with the exTTF derivatives, ZnP and H_2P exhibit stronger electron donating character, which is also reflected by lower attenuation factors, β, with a value of 0.11 Å^{-1} for C_{60}–oPPE–(β)ZnP/H_2P compared with 0.20 Å^{-1} for C_{60}–oPPE–exTTF. Additional evidence for the improved ET features came from AM1* local electron-affinity calculations. These reveal a more continuous distribution of local electron affinity. Furthermore, strong electronic coupling between ZnP or H_2P and C_{60} due to the β-substitution pattern, which does not vary with the length of oPPE, leads to nearly distance-independent ET processes. A schematic representation of the possible reaction pathways is given in the energy diagram of Figure 6.10.

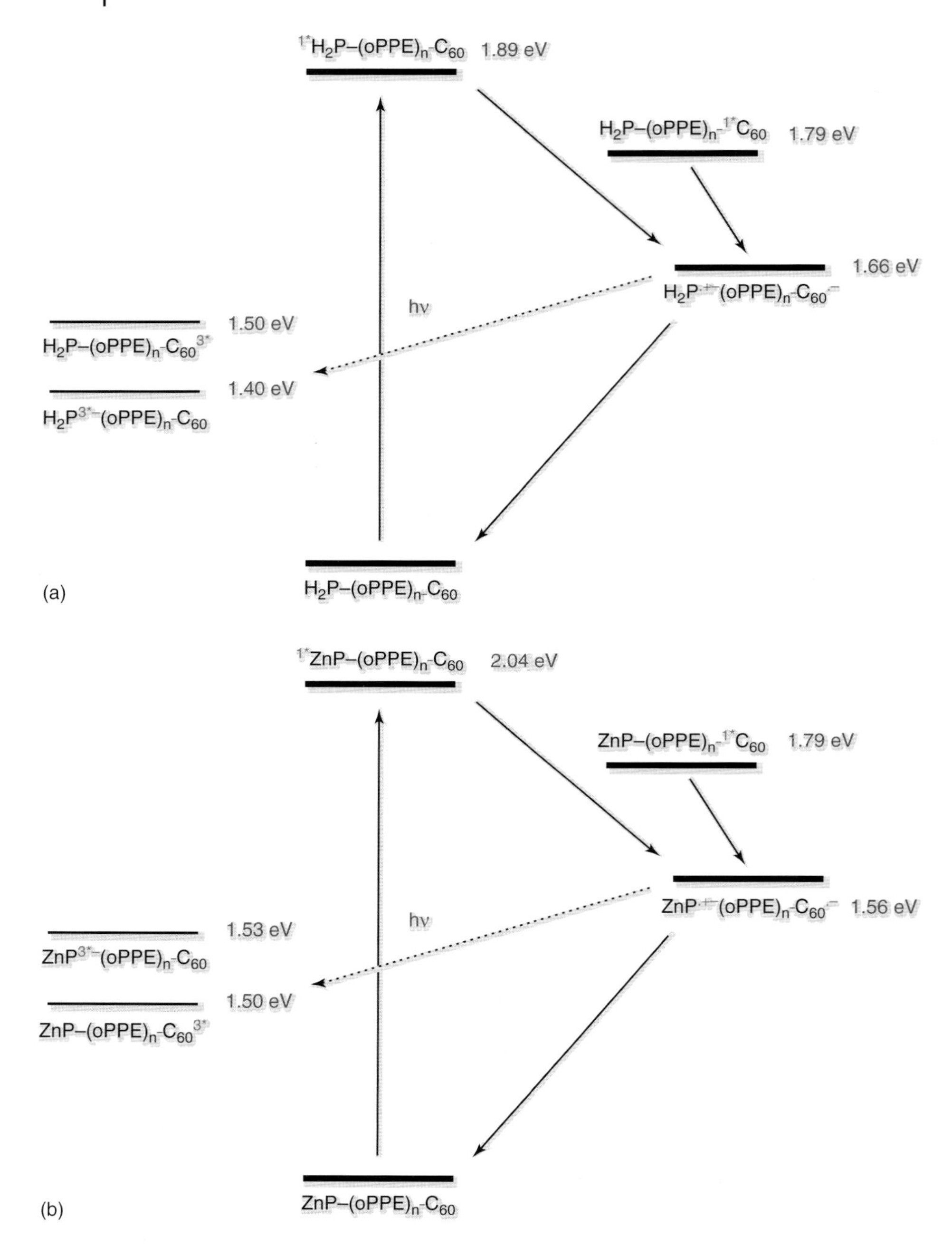

Figure 6.10 Schematic illustration of the reaction pathways in photoexcited H_2P**7b** (a) and ZnP**7b** (b) with the state energies as calculated according to the method by Weller [116] in THF. The different deactivation pathways are indicated by arrows.

6.3.3
C_{60}–WIRE–Fc

Using π-conjugated oligomers to link fullerenes to electron donating ferrocenes (Fcs) leads to further examples of DBA systems. Photoinduced ET reactions from Fc to C_{60}, involving C_{60} singlet and triplet excited states, are energetically feasible, and the ferrocenyl centers can be oxidized reversely. The lowest excited singlet state of Fc (2.46 eV) [117] is higher than the lowest excited state of monofunctionalized fullerenes (1.76–1.79 and 1.5 eV for the excited singlet and triplet states, respectively) [118, 119], ruling out any undesired singlet or triplet energy transfer from the excited fullerene to the Fc moiety. In this case, the spin density of the Fc radical cations is delocalized over the donor and the linker. This, in turn, affects the charge separation and charge recombination dynamics in a favorable manner.

6.3.3.1 C_{60}–ZnP–Fc

An interesting alternative involves testing the CT efficiencies of polyporphyrin bridges that are linked via butadiynes (Figure 6.11). Thus, a series of C_{60}–ZnP–Fc (**8**) DBA systems (i.e., monomer, dimer, and tetramer) was investigated [120]. Photoexcitation of the polyporphyrin results in formation of long-range spatially separated radical ion pair states. Enthrallingly, the formation occurs via a sequence of ET steps. Charge separation and charge recombination rates were elucidated using transient absorption and fluorescence spectroscopy. The charge recombination rates are remarkably fast ($15 - 1.3 \times 10^8\ s^{-1}$). This is a consequence of bridge-mediated electronic coupling. Plotting these rates versus donor–acceptor distances fails to provide a straightforward relationship. Nevertheless, it is plausible to separate the data points into two groups. The slope of a line that connects the first two data points (shorter bridges) corresponds to an attenuation factor β of $0.18\ \text{Å}^{-1}$. This value lies at the lower end of those typically found for conjugated bridge structures and indicates weak distance dependence. On the other hand, connecting the second and third data points (longer bridges) results in a damping factor with practically no distance dependence ($\beta = 0.003\ \text{Å}^{-1}$). We therefore conclude that the long-range charge recombination process occurs via electron tunneling rather than hopping or triplet recombination. Obviously, the distance dependence of the ET processes in these systems should not be considered as a single parameter for the CT efficiency and the exact mechanism still remains to be elucidated. However, the observation that the porphyrin tetramer **8c** mediates long-range CT over a distance of 65 Å is remarkable. These findings truly represent a landmark for the application of such structures as molecular wires.

6.3.3.2 C_{60}–oT–Fc

Another structural example of efficient charge mediating bridges that covalently link Fc donors and C_{60} acceptors are oligothiophenes (oTs). Because of their high degree of conjugation, oTs are attractive candidates for molecular bridges in a large number of molecular electronics applications.

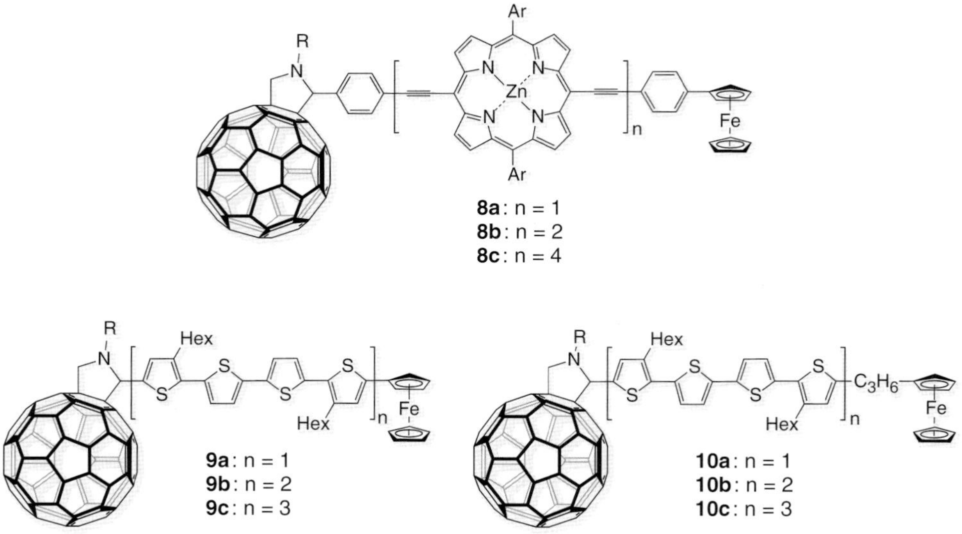
Ar
R
N
N
N
Zn
N
N
Ar
n
Fe
8a: n = 1
8b: n = 2
8c: n = 4
R
N
Hex
S
S
S
S
Hex
n
Fe
9a: n = 1
9b: n = 2
9c: n = 3
R
N
Hex
S
S
S
S
Hex
n
C3H6
Fe
10a: n = 1
10b: n = 2
10c: n = 3

Figure 6.11 Examples of C_{60}–WIRE–Fc DBA conjugates.

The CT processes in C_{60}–oT–Fc (**9**) and C_{60}–oT–tm–Fc (**10**) DBA systems were therefore investigated. In the latter, a trimethylene (tm) fragment is inserted between the Fc and the oT moieties (Figure 6.11). As a consequence, the conjugation between the two chromophores is perturbed. Time-resolved fluorescence and transient absorption spectroscopy of the C_{60}–oT–Fc conjugates disclosed that the excited-state deactivation in nonpolar toluene occurs exclusively via energy transfer. This process evolves from ^{1*}oT formation upon excitation and results in the generation of $^{1*}C_{60}$. In polar benzonitrile, on the other hand, charge separation is favored thermodynamically. The radical cation is generated instantaneously. Interestingly, the positive charge is homogeneously delocalized over the Fc and oT moieties giving $C_{60}^{\bullet-}$–(oT–Fc)$^{\bullet+}$. Upon systematic variation of the spacer length from 4 (**9a**) to 12 (**9c**) thiophene units, the lifetimes of $C_{60}^{\bullet-}$–(oT–Fc)$^{\bullet+}$ increased from 0.1 to 50 ns. In general, the varying oxidation potentials of the oT moieties seem to control the ET processes. For the tm-extended systems, C_{60}–oT–tm–Fc, a change in reactivity was observed in polar solvents. In spite of the instantaneous formation of a delocalized radical cation, a two-step mechanism was found as the operative mode. Charge separation first yields $C_{60}^{\bullet-}$–oT$^{+\bullet}$–tm–Fc. Then, in a sequential energy transfer scenario, in which C_{60}–oT^{1*}–tm–Fc deactivates via $^{1*}C_{60}$–oT–tm–Fc, the positive charge shifts from oT$^{+\bullet}$ to Fc forming the $C_{60}^{\bullet-}$–oT–tm–Fc$^{\bullet+}$ radical ion pair. The latter survived up to 330 ns when the number of the thiophene units was increased from 4 (**10a**) to 12 (**10c**). The longer lifetimes of $C_{60}^{\bullet-}$–oT–tm–Fc$^{\bullet+}$ compared with those of $C_{60}^{\bullet-}$–(oT–Fc)$^{\bullet+}$ are rationalized by the presence of the tm chain, which, in turn, disrupts the conjugation between the Fc and the oT moieties. β was evaluated for C_{60}–oT–tm–Fc using the corresponding donor–acceptor distances obtained from the optimized structures at the B3LYP/3-21G(d) DFT level.

In benzonitrile, for example, a value of 0.10 Å^{-1} was obtained [121]. β is significantly higher than for the corresponding C_{60}–oT–H_2P systems, which have a value of 0.03 Å^{-1}. There, the thiophene oligomers are directly linked to the *meso*-position of H_2P affording electron donor acceptor distances of up to 55.7 Å [122]. This, once again, highlights the critical role of the linkage between the donor and the bridge.

6.4 Conclusions

The above examples of molecule-assisted transport of charge and energy mediated by organic wire-like structures suggest that fully π-conjugated aromatic molecular wires are the best candidates for use in new electronic devices that constitute a new generation of functional materials for molecular electronics. In the systems tested, π-conjugation emerged as the sine qua non for effective long-range transport of charge carriers. Especially in view of distance-independent electron transport phenomena, the continuous distribution of the local electron affinity and the matching of the energy levels of the modular components are of utmost importance. However, simple changes made during the synthesis of molecular wires yield products with vastly different photophysical properties. Even minor alterations, such as exchanging the donor moiety, strongly affect the conduction behavior of the wires. Recently introduced elegant and versatile protocols for chemically functionalizing π-conjugated oligomers of precise length and constitution will be especially important. In fact, some of the aforementioned challenges may be easily handled by the controlled covalent functionalization of the molecular wire's termini. This includes, for example, attaching covalently linked molecular "handles" that ultimately serve as sources and drains for charge carriers. Recent work has shown how the connection between molecular building blocks and electrodes greatly affects the current–voltage characteristics. Therefore, control over handling the linkage is of primary importance. Besides potential applications in the field of molecular electronics, the design and development of light harvesting, photoconversion, and as a long term aim catalytic modules should not be overseen.

Still, many questions remain to be answered, the most important of which is how to integrate the molecular wires into future electronic and optoelectronic devices. Such integration efforts raise many problems, which likely will require several iterations in molecular wire research. For instance, challenges related to the controlability of the rate processes by synthetic efforts, positioning, and excitation. Certainly, these challenges will continue to make molecular wires posing tantalizing objects for the science and technology communities of the future. It is, however, currently not unreasonable to predict that the first molecular wire(s) that appear in commercial molecular electronic devices will bear no resemblance with those we have discussed.

References

1. Jortner, J. and Ratner, M. (1997) *Molecular Electronics*, Blackwell, London.
2. Guldi, D.M., Illescas, B.M., Atienza, C.M., Wielopolski, M., and Martín, N. (2009) *Chem. Soc. Rev.*, **38**, 1587–1597.
3. Craeger, S., Yu, C.J., Bamad, C., O'Connor, S., MacLean, T., Lam, E., Chong, Y., Olsen, G.T., Luo, J., Gozin, M., and Kayyem, J.F. (1999) *J. Am. Chem. Soc.*, **121**, 1059–1064.
4. Sachs, S.B., Dudek, S.P., Hsung, R.P., Sita, L.R., Smalley, J.F., Newton, M.D., Feldberg, S.W., and Chidsey, C.E.D. (1997) *J. Am. Chem. Soc.*, **10**, 563–564.
5. Sykes, H.D., Smalley, J.F., Dudek, S.P., Cook, A.R., Newton, M.D., Chidsey, C.E.D., and Felberg, S.W. (2001) *Science*, **291**, 1519–1523.
6. Donhauser, Z.J., Mantooth, B.A., Kelly, K.F., Bumm, L.A., Monnell, J.D., Stapleton, J.J., Price, D.W. Jr., Rawlett, A.M., Allara, D.L., Tour, J.M., and Weiss, P.S. (2001) *Science*, **292**, 2303–2307.
7. Chen, J., Reed, M.A., Rawlett, A.M., and Tour, J.M. (1999) *Science*, **286**, 1550–1552.
8. Tour, J.M., Rawlett, A.M., Kozaki, M., Yao, Y., Jagessar, R.C., Dirk, S.M., Price, D.W., Reed, M.A., Zhou, C.-W., Chen, J., Wang, W., and Campbell, I. (2001) *Chem. Eur. J.*, **7**, 5118–5134.
9. Haag, R., Rampi, M.A., Holmlin, R.E., and Whitesides, G.M. (1999) *J. Am. Chem. Soc.*, **121**, 7895–7906.
10. Holmlin, E.H., Ismagilov, R.F., Haag, R., Mujica, V., Ratner, M.A., Rampi, M.A., and Whitesides, G.M. (2001) *Angew. Chem., Int. Ed.*, **40**, 2316–2320.
11. Scandola, F., Chiorboli, C., Indelli, M.T., and Rampi, M.A. (2001) in *Electron Transfer in Chemistry*, vol. III, Chapter 2. (ed. V. Balzani), John Wiley & Sons, Inc., Weinheim.
12. De Cola, L. and Belser, P. (2001) in *Electron Transfer in Chemistry*, vol. V, Chapter 3 (ed. V. Balzani), John Wiley & Sons, Inc., Weinheim.
13. Davis, W.B., Svec, W.A., Ratner, M.A., and Wasielewski, M.R. (1996) *Nature*, **396**, 60–63.
14. Emberly, E.G. and Kirczenow, G. (1998) *Phys. Rev. B*, **58**, 10911–10920.
15. Nitzan, A. and Ratner, M.A. (2003) *Science*, **300**, 1384–1389.
16. Wasielewski, M.R. (1992) *Chem. Rev.*, **92**, 435–461.
17. Schatz, G.C. and Ratner, M.A. (2002) *Quantum Mechanics in Chemistry*, 2nd edn, Dover Publications, Mineola, New York.
18. Linderberg, J. and Ohrn, Y. (2004) *Propagators in Quantum Chemistry*, John Wiley & Sons, Inc., Hoboken, NJ.
19. Pople, J.A. and Beveridge, D.L. (1970) *Approximate Molecular Orbital Theory*, McGraw-Hill, New York.
20. Jortner, J. (1976) *J. Chem. Phys.*, **64**, 4860–4867.
21. Marcus, R.A. and Sutin, N. (1985) *Biochim. Biophys. Acta*, **811**, 265–322.
22. Kramers, K.H. (1934) *Physica*, **1**, 182–192.
23. Anderson, P.W. (1950) *Phys. Rev. Lett.*, **79**, 350–356.
24. Anderson, P.W. (1959) *Phys. Rev. Lett.*, **115**, 2.
25. Nitzan, A. (2001) *Annu. Rev. Phys. Chem.*, **52**, 681–750.
26. Berlin, Y.A., Burin, A.L., and Ratner, M.A. (2002) *Chem. Phys.*, **275**, 61–74.
27. Grozema, F.C., Berlin, Y.A., and Siebbeles, L.D.A. (2000) *J. Am. Chem. Soc.*, **122**, 10903–10909.
28. Finklea, H.O. and Hanshewm, D.D. (1992) *J. Am. Chem. Soc.*, **114**, 3173–3181.
29. Slowiski, K., Chamberlain, R.V., Miller, C.J., and Majda, M. (1997) *J. Am. Chem. Soc.*, **119**, 11910–11919.
30. Chidsey, C.E.D. (1991) *Science*, **251**, 919–922.
31. Leland, B.A., Joran, A.D., Felker, P.M., Hopfield, J.J., Zewail, A.H., and Dervan, P.B. (1985) *J. Phys. Chem. A*, **89**, 5571–5573.
32. Oevering, H., Paddon-Row, M.N., Heppener, M., Oliver, A.M., Cotsaris, E., Verhoeven, J.W., and Hush, N.S. (1987) *J. Am. Chem. Soc.*, **109**, 3258–3269.

33. Closs, G.L. and Miller, J.R. (1988) *Science*, **240**, 440–447.
34. Klan, P. and Wagner, P.J. (1998) *J. Am. Chem. Soc.*, **120**, 2198–2199.
35. Osuka, A., Maruyama, K., Mataga, N., Asahi, T., Yamazaki, I., and Tamai, N. (1990) *J. Am. Chem. Soc.*, **112**, 4958–4959.
36. Helms, A., Heiler, D., and McClendon, G. (1992) *J. Am. Chem. Soc.*, **114**, 6227–6238.
37. Weiss, E.A., Ahrens, M.J., Sinks, L.E., Gusev, A.V., Ratner, M.A., and Wasielewski, M.R. (2004) *J. Am. Chem. Soc.*, **126**, 5577–5584.
38. Osuka, A., Satoshi, N., Maruyama, K., Mataga, N., Asahi, T., Yamazaki, I., Nishimura, Y., Onho, T., and Nozaki, K. (1993) *J. Am. Chem. Soc.*, **115**, 4577–4589.
39. Barigelletti, F., Flamigni, L., Balzani, V., Collin, J.-P., Sauvage, J.-P., Sour, A., Constable, E.C., and Cargill Thompson, A.M.W. (1994) *J. Am. Chem. Soc.*, **116**, 7692–7699.
40. Martin, N., Giacalone, F., Segura, J.L., and Guldi, D.M. (2004) *Synth. Met.*, **147**, 57–61.
41. Atienza, C., Martín, N., Wielopolski, M., Haworth, N., Clark, T., and Guldi, D.M. (2006) *Chem. Commun.*, **30**, 3202–3204.
42. Benniston, A.C., Goulle, V., Harriman, A., Lehn, J.-M., and Marczinke, B. (1994) *J. Phys. Chem. A*, **98**, 7798–7804.
43. Osuka, A., Tanabe, N., Kawabata, S., and Speiser, I.S. (1996) *Chem. Rev.*, **96**, 1953–1976.
44. Marczinke, B. (1994) *J. Phys. Chem. A*, **98**, 7798–7804.
45. Osuka, A., Tanabe, N., Kawabata, S., Grosshenny, I.V., Harriman, A., and Ziessel, R. (1995) *Angew. Chem., Int. Ed.*, **34**, 1100–1102.
46. Grosshenny, I.V., Harriman, A., and Ziessel, R. (1995) *Angew. Chem., Int. Ed.*, **34**, 2705–2708.
47. Newton, M.D. (1991) *Chem. Rev.*, **91**, 767–792.
48. Marcus, R.A. (1987) *Chem. Phys. Lett.*, **133**, 471–477.
49. Ogrodnik, A. and Michel-Beyerle, M.E. (1989) *Z. Naturforsch., A: Phys. Sci.*, **44a**, 763–764.
50. Kilsa, K., Kajanus, J., Macpherson, A.N., Martensson, J., and Albinsson, B. (2001) *J. Am. Chem. Soc.*, **123**, 3069–3080.
51. Lukas, A.S., Bushard, P.J., and Wasielewski, M.R. (2002) *J. Phys. Chem. A*, **106**, 2074–2082.
52. Marcus, R.A. (1965) *J. Chem. Phys.*, **43**, 679–701.
53. McConnell, H.M. (1961) *J. Chem. Phys.*, **35**, 508–515.
54. Closs, G.L., Piotrowiak, P., McInnis, J.M., and Fleming, G.R. (1988) *J. Am. Chem. Soc.*, **110**, 2652–2653.
55. Roest, M.R., Oliver, A.M., Paddon-Row, M.N., and Verhoeven, J.W. (1997) *J. Phys. Chem. A*, **101**, 4867–4871.
56. Paddon-Row, M.N., Oliver, A.M., Warman, J.M., Smit, K.J., Haas, M.P., Oevering, H., and Verhoeven, J.W. (1988) *J. Phys. Chem. A*, **92**, 6958–6962.
57. Jortner, J., Bixon, M., Langenbacher, T., and Michel-Beyerle, M.E. (1998) *Proc. Natl. Acad. Sci. USA*, **95**, 759–765.
58. Bixon, M., Giese, B., Langenbacher, T., Michel-Beyerle, M.E., and Jortner, J. (1999) *Proc. Natl. Acad. Sci. USA*, **96**, 11713–11716.
59. Davis, W.B., Wasielewski, M.R., Mujica, V., and Nitzan, A. (1997) *J. Phys. Chem. A*, **101**, 6158–6164.
60. Kharkats, Y.I. and Ulstrup, J. (1991) *Chem. Phys. Lett.*, **182**, 81–87.
61. Sourtis, S.S. and Mukamel, S. (1995) *Chem. Phys.*, **197**, 367–388.
62. Felts, A.K., Pollard, W.T., and Friesner, R.A. (1995) *J. Phys. Chem. A*, **99**, 2929–2940.
63. Fan, F., Yang, J., Cai, L., Price, D.W. Jr., Dirk, S.M., Kosynkin, D.V., Yao, Y., Rawlett, A.M., Tour, J.M., and Bard, A.J. (2002) *J. Am. Chem. Soc.*, **124**, 5550–5560.
64. Zangmeister, C.D., Robey, S.W., van Zee, R.D., Yao, Y., and Tour, J.M. (2004) *J. Phys. Chem. B*, **108**, 16187–16193.
65. Zhu, X.Y. (2004) *J. Phys. Chem. B*, **108**, 8778–8793.

66. Bredas, J.L., Calbert, J.P., Filho, D.A., and Cornil, J. (2002) *Proc. Natl. Acad. Sci. USA*, **99**, 5804–5809.
67. Bredas, J.L., Beljonne, D., Coropceanu, V., and Cornil, J. (2004) *Chem. Rev.*, **104**, 4971–5004.
68. Van Vooren, A., Lemaur, V., Ye, A., Beljonne, D., and Cornil, J. (2007) *Chem. Phys. Chem.*, **8**, 1240–1249.
69. Nitzan, A. (2001) *J. Phys. Chem. A*, **105**, 2677–2679.
70. Kumar, K., Kurnikov, I.V., Beratan, D.N., and Waldeck Zimmer, D.H. (1998) *J. Phys. Chem. A*, **102**, 5529–5541.
71. Roesch, N. and Voityuk, A.A. (2004) *Top. Curr. Chem.*, **237**, 37–72.
72. Marcus, R.A. (1964) *Annu. Rev. Phys. Chem.*, **15**, 155–196.
73. Marcus, R.A. (1993) *Angew. Chem. Int. Ed.*, **105**, 1161–1172.
74. Cave, R.J. and Newton, M.D. (1996) *Chem. Phys. Lett.*, **249**, 15–19.
75. Cave, R.J. and Newton, M.D. (1997) *J. Chem. Phys.*, **106**, 9213–9226.
76. Voityuk, A.A. and Rösch, N.J. (2002) *Chem. Phys.*, **117**, 5607–5616.
77. Parcell, K.F. and Blaive, B. (1988) in *Photoinduced Electron Transfer*, Part A (eds M.A. Fox and M. Chanon), Elsevier, Amsterdam.
78. Hu, X. and Schulten, K. (1997) *Phys. Today*, **50**, 28–34.
79. Rice, M.J. and Gartstein, Y.N. (1996) *Phys. Rev. B*, **53**, 10764–10770.
80. Wu, M.W. and Conwell, E.M. (1998) *Chem. Phys.*, **227**, 11–17.
81. Paddon-Row, M.N. (1994) *Acc. Chem. Res.*, **27**, 18–25.
82. Guldi, D.M. (2002) *Chem. Soc. Rev.*, **31**, 22–36.
83. Bixon, M. and Jortner, J. (1999) *Adv. Chem. Phys.*, **106**, 35–202.
84. Müller, G.M., Lupton, J.M., Feldmann, J., Lemmer, U., Scharber, M.C., Sariciftci, N.S., Brabec, C.J., and Scherf, U. (2005) *Phys. Rev. B*, **72**, 195208–195218.
85. Kavarnos, G.J. and Turro, N.J. (1986) *Chem. Rev.*, **86**, 401–449.
86. Kuznetsov, A.M. and Ulstrup, J. (1999) *Electron Transfer in Chemistry and Biology: An Introduction to the Theory*, John Wiley & Sons, Ltd, New York.
87. Reed, M.A., Zhou, C., Muller, C.J., Burgin, T.P., and Tour, J.M. (1997) *Science*, **278**, 252–254.
88. Datta, S., Tam, W., Hong, S., Riefenberger, R., Henderson, J.I., and Kubiak, C.P. (1997) *Phys. Rev. B*, **79**, 2530–2533.
89. Cygan, M.T., Dunbar, T.D., Arnold, J.J., Bumm, L.A., Shedlock, N.F., Burgin, T.P.I., Allara, D.L., Tour, J.M., and Weiss, P.S. (1998) *J. Am. Chem. Soc.*, **120**, 2721–2732.
90. Seminario, J.M., Zacaias, A.G., and Tour, J.M. (1998) *J. Am. Chem. Soc.*, **120**, 3970–3974.
91. Wagner, R.W. and Lindsey, J.S. (1994) *J. Am. Chem. Soc.*, **116**, 9759–9760.
92. Hsaio, J.-S., Krueger, B.P., Wagner, R.W., Johnson, T.E., Delaney, J.K., Mauzerall, D.C., Fleming, G.R., Lindsey, J.S., Bocian, D.F., and Donohoe, R.J. (1996) *J. Am. Chem. Soc.*, **118**, 11181–11193.
93. Li, F., Yang, S.Y., Ciringh, Y., Seth, J., Martin, C.H., Singh, D.L., Kim, D., Birge, R.R., Bocian, D.F., Holten, D., and Lindsey, J.S. (1998) *J. Am. Chem. Soc.*, **120**, 10001–10017.
94. Winkler, J.R. and Gray, H.B. (1992) *Chem. Rev.*, **92**, 369–379.
95. Ulman, A. (1996) *Chem. Rev.*, **96**, 1533–1554.
96. Dubois, L.H. and Nuzzo, R.G. (1992) *Annu. Rev. Phys. Chem.*, **43**, 437–463.
97. Mann, B. and Kuhn, H.J. (1971) *J. Appl. Phys.*, **42**, 4398–4405.
98. Wong, E.W., Collier, C.P., Belohradsky, M., Raymo, F.M., Stoddart, J.F., and Heath, J.R. (2000) *J. Am. Chem. Soc.*, **122**, 5831–5840.
99. Collier, C.P., Wong, E.W., Belohradsky, M., Raymo, F.M., Stoddart, J.F., Keukes, P.J., Williams, R.S., and Heath, J.R. (1999) *Science*, **285**, 391–394.
100. Reed, M.A. and Lee, T. (2003) *Molecular Nanoelectronics*, American Scientific Publishers, Stevenson Ranch, CA.
101. Foley, E.T., Yoder, N.L., Guisinger, N.P., and Hersam, M.C. (2004) *Rev. Sci. Instrum.*, **75**, 5280–5287.
102. Xiao, X.Y., Xu, B.Q., and Tao, N.J. (2004) *Nano Lett.*, **4**, 267–271.

103. Slowinski, K. and Majda, M. (2000) *J. Electroanal. Chem.*, **491**, 139–147.
104. Imahori, H. and Sakata, Y. (1999) *Eur. J. Org. Chem.*, **10**, 2445–2457.
105. Becher, J., Zhan-Ting, L., Blanchard, P., Svenstrup, N., Lau, J., Brondsted Nielsen, M., and Leriche, P. (1997) *Pure Appl. Chem.*, **69**, 465–470.
106. Yamada, J. and Sugimoto, T. (2004) *TTF Chemistry*, Springer, Berlin, Heidelberg and Kodansha Scientic, Ltd, Tokyo.
107. (a) Giacalone, F., Segura, J.L., Martín, N., and Guldi, D.M. (2004) *J. Am. Chem. Soc.*, **126**, 5340–5341; (b) Giacalone, F., Segura, J.L., Martín, N., Ramey, J., and Guldi, D.M. (2005) *Chem. Eur. J.*, **11**, 4819–4834.
108. Wielopolski, M., Atienza, C., Clark, T., Guldi, D.M., and Martín, N. (2008) *Chem. Eur. J.*, **14**, 6379–6390.
109. Molina-Ontoria, A., Fernández, G., Wielopolski, M., Atienza, C., Sánchez, L., Gouloumis, A., Clark, T., Martín, N., and Guldi, D.M. (2009) *J. Am. Chem. Soc.*, **131**, 12218–12229.
110. Shankara Gayathri, S., Wielopolski, M., Pérez, E.M., Fernández, G., Sánchez, L., Viruela, R., Ortí, E., Guldi, D.M., and Martín, N. (2009) *Angew. Chem. Int. Ed.*, **48**, 815–819.
111. van der Pol, C., Bryce, M.R., Wielopolski, M., Atienza-Castellanos, C., Guldi, D.M., Filippone, S., and Martín, N. (2007) *J. Org. Chem.*, **72**, 6662–6671.
112. Atienza-Castellanos, C., Wielopolski, M., Guldi, D.M., Pol, Cvd., Bryce, M.R., Filippone, S., and Martín, N. (2007) *Chem. Commun.*, **48**, 5164–5166.
113. de la Torre, G., Giacalone, F., Segura, J.L., Martín, N., and Guldi, D.M. (2005) *Chem. Eur. J.*, **11**, 1267–1280.
114. Santos, J., Illescas, B.M., Wielopolski, M., Silva, A.M.G., Tome, A.C., Guldi, D.M., and Martín, N. (2008) *Tetrahedron*, **64**, 11404–11408.
115. Lembo, A., Tagliatesta, P., Guldi, D.M., Wielopolski, M., and Nuccetelli, M. (2009) *J. Phys. Chem. A*, **113**, 1779–1793.
116. Weller, A.Z. (1982) *Phys. Chem.*, **133**, 93–98.
117. Sohn, Y.S., Hendrickson, D.N., and Gray, H.B. (1971) *J. Am. Chem. Soc.*, **93**, 3603–3612.
118. (a) Williams, R.M., Zwier, J.M., and Verhoeven, J. (1995) *J. Am. Chem. Soc.*, **117**, 4093–4099; (b) Anderson, J.L., An, Y.-Z., Rubin, Y., and Foote, C.S. (1994) *J. Am. Chem. Soc.*, **116**, 9763–9764.
119. Bensasson, R.V., Bienvenue, E., Janot, J.-M., Leach, S., Seta, P., Schuster, D.I., Wilson, S.R., and Zhao, H. (1995) *Chem. Phys. Lett.*, **245**, 566–570.
120. Winters, M.U., Dahlstedt, E., Blades, H.E., Wilson, C.J., Frampton, M.J., Anderson, H.L., and Albinsson, B. (2007) *J. Am. Chem. Soc.*, **129**, 4291–4297.
121. (a) Nakamura, T., Kanato, H., Araki, Y., Ito, O., Takimiya, K., Otsubo, T., and Aso, Y. (2006) *J. Phys. Chem. A*, **110**, 3471–3479; (b) Kanato, H., Takimiya, K., Otsubo, T., Aso, Y., Nakamura, T., Araki, Y., and Ito, O. (2004) *J. Org. Chem.*, **69**, 7183–7189.
122. (a) Ikemoto, J., Takimiya, K., Aso, Y., Otsubo, T., Fujitsuka, M., and Ito, O. (2002) *Org. Lett.*, **4**, 309–311; (b) Nakamura, T., Fujitsuka, M., Araki, Y., Ito, O., Ikemoto, J., Takimiya, K., Aso, Y., and Otsubo, T. (2004) *J. Phys. Chem. B*, **108**, 10700–10710.

Part III
Charge Transport through Wires in Solution

Charge and Exciton Transport through Molecular Wires. Edited by L.D.A. Siebbeles and F.C. Grozema

ISBN: 978-3-527-32501-6

7
Electron and Exciton Transport to Appended Traps

John R. Miller, Andrew R. Cook, Kirk S. Schanze, and Paiboon Sreearunothai

7.1
Introduction

The idea that long, conjugated molecules could act as "molecular wires" received a big boost from microwave conductivity (μC) experiments by the Delft group [1–6]. Those experiments attached charges to conjugated polymers by pulse radiolysis and observed that conductivity increased as carriers attached to conjugated polymers. The μC experiments changed the perception that charge mobilities might be inherently slow in conjugated polymers, an impression that came from experience in films, finding that mobilities along isolated chains could be large, with important consequences for organic electronics and photovoltaics [7, 8]. This chapter describes the complementary method of transport to appended traps (ATs). The basic idea is illustrated in Figure 7.1; Figure 7.2 shows energy level diagrams for the capture processes.

An AT captures a charge or exciton created in a wire in two steps (i) transport along the wire, k_{trans} and (ii) reaction of the charge or exciton with the trap, k_{rxn}. Transport is not generally well characterized by a single rate, but may be nearly so (see below).

7.1.1
Appended Trap and Microwave Conductivity Comparison

The μC method has important advantages. The μC method is applicable to almost any chain; while AT measurements require attachment of the traps, usually by chemical synthesis requiring expertise and effort. μC can measure moderate to very high mobilities, all with instruments having time resolution of several nanoseconds; high mobilities appear as larger amplitudes. AT measurements of high mobilities require measurement with fast time resolution and may therefore require high solubility to obtain fast attachment of the charges. The effects of this limitation will be seen in the discussion of available data presented below.

The primary complementary advantage of the AT method is that charges must traverse substantial lengths along the chain to reach the trap. The method therefore

Charge and Exciton Transport through Molecular Wires. Edited by L.D.A. Siebbeles and F.C. Grozema

ISBN: 978-3-527-32501-6

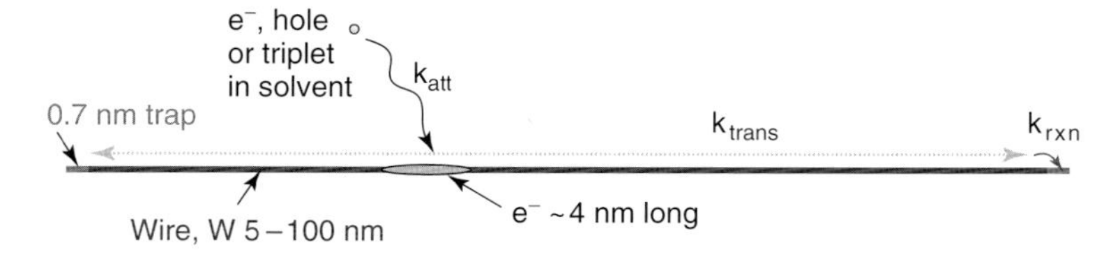

Figure 7.1 Schematic of the appended trap technique to determine transport. An electron, hole, or triplet excited state created in the solvent diffuses and attaches with rate k_{att} to the polymer chain ("wire"). In the wire it is depicted as a polaron. Singlet excitons (and sometimes triplets) can be directly created in the wire. The charge or exciton then transports along the chain with rate k_{trans} until captured with rate k_{rxn} at a trap chemically bonded to the end of the chain. Alternatively traps may be pendantly attached along the chain, or may be included within the chain by copolymerization. The figure also depicts the charge in the chain as spread over a substantial length, measured to be almost 4 nm for electrons in polyfluorene [9]. Fluorescence spectra [10] indicate a similar length for the singlet exciton. Much experience, including Refs. [9, 10] and reference cited therein, indicate that while spread, charges, and excitons are partly localized as polarons.

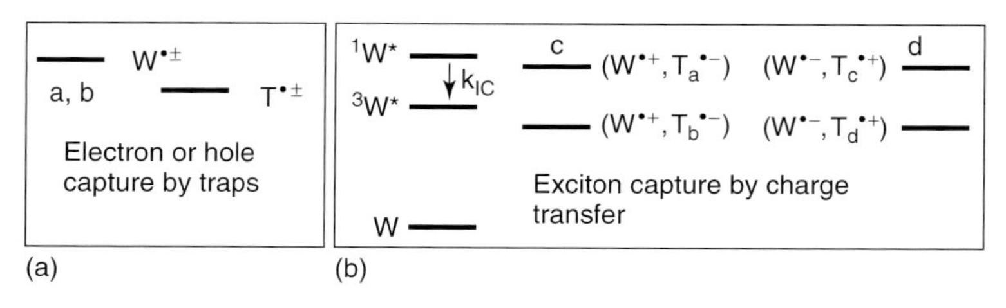

Figure 7.2 Energetics for capture of polarons in wires by transfer of electrons (or holes) to the trap. In (A) **a, b** an electron, **a** (or **b** a hole) in the wire (W) transfers to the trap (T). The trap must be more easily reduced (or oxidized) than W. In (B) a singlet ($^1W^*$) or triplet ($^3W^*$) exciton in the wire can (**c**) transfer electrons to the trap or (**d**) accept electrons from the trap. Four different trap molecules are shown: T_a and T_c can react only with singlets ($^1W^*$); T_b and T_d can react with singlets or triplets. $^1W^*$ and $^3W^*$ could alternatively transfer energy to traps with sufficiently low-lying excited states (not shown).

determines in a simple way if charges actually move long distances. Because μC is a high-frequency AC method it cannot easily distinguish whether charges can in fact traverse the entire chain. If charges were barricaded in a high mobility region between defects, they might still show high μC if the region is not too small. Extremely high mobilities with end effects observed in a ladder polymer is a stunning exception [5], and it may be possible to extract information from phases of μC signals, but for most cases the AT method has the greatest potential to determine transport over very long distances. A related advantage is that AT measurements can produce time-domain information. If transport were dispersive, for example, if transport were fast in some chains and slow in others, or if mobility were time dependent, these behaviors can, in principle, be seen in AT data. At this writing the techniques are still being developed; these insights have yet to be substantially realized.

Microwave conductivity is useful only in media that do not themselves strongly absorb microwaves. For fluids, this limits the μC method to nonpolar media, so it cannot tackle the interesting question: is transport of charges, thought to be polarons, slowed by energy dissipation in rotations of polar molecules around the wire? The AT method has not yet, but can approach this question.

Perhaps the most cogent advantage is that the AT method can also determine exciton transport, sometimes utilizing the same trap-containing molecules used to measure charge transport, and usually under similar or identical conditions. The potential to compare transport of charges and excitons along the same conjugated chains motivates much of our work on this subject.

This chapter will briefly describe experimental techniques, allude briefly to experimental information on the natures of charges in conjugated molecular chains, and then discuss findings to date.

7.1.2
Transport Mechanisms

Transport along the wire moves the polaron in a series of steps as depicted in Figure 7.3. Each step takes the polaron from its present location to a nearby one. At

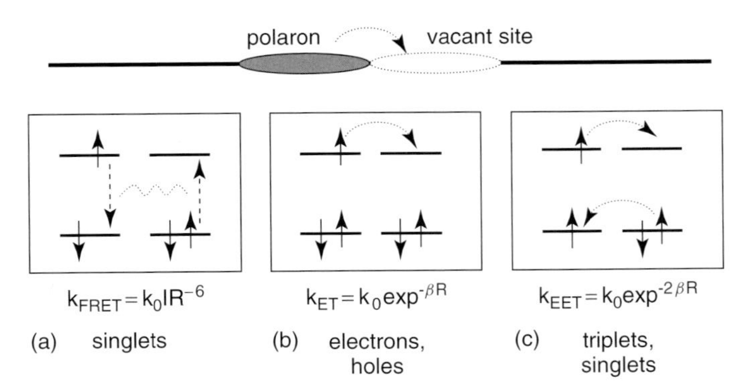

Figure 7.3 Transport of a polaron along a "wire" as a series of electron or exciton transfer steps. The electronic matrix element, V arises principally from three mechanisms. (a) The first, the dipole coupling of Forster resonant energy transfer (FRET) [11, 12] is important only for transport of singlet excitons. Transfer of electrons (b) or holes, occurs via two-center, one-electron exchange matrix elements. The Dexter mechanism (c) proceeds via two-center, two-electron exchange matrix elements, which can be envisaged as an electron moving to the right and another to the left (or transfer of a hole to the right). Dexter two-electron exchange is the only effective mechanism for triplets and contributes to transfer of singlets. Electronic matrix elements from all three mechanisms decrease rapidly with distance, R. FRET with a sixth power dependence on R is the most long range. Single electron transfer falls exponentially with distance, while two-electron exchange falls much faster with approximately twice a large an exponent. The parameter β is often $\sim 1\text{Å}^{-1}$ through saturated material, but may be smaller through conjugated material.

the present location of the polaron the positions of atoms (nuclear configurations) in the polymer and surrounding medium have partly or completely relaxed in response to the presence of the charge or exciton. The new location is a nearly identical, but unrelaxed, site. The rate, k, for each step is described by the Fermi golden rule expression, which is the product of an electronic matrix element, $V(r)$ and a Franck–Condon-weighted density (FCWD) of states. The FCWD reflects the relaxation and may often be characterized by reorganization energies. For exciton transport it can be expressed as a convolution of emission and absorption spectra:

$$k = \frac{2\pi}{\hbar} \left| V(r) \right|^2 \text{FCWD} \tag{7.1}$$

The electronic matrix element arises from three principal mechanisms, each of which is applicable to one or two types of polaron motion. Simplified pictorial descriptions are given in Figure 7.3.

7.2
Experimental Methods to Investigate Transport to Appended Traps

7.2.1
Injection of Electrons and Holes

Carrier injection into wires is most readily accomplished by pulse radiolysis that ionizes the solvent to create electrons or holes, which then diffuse to react with the polymer chain. Recent development of an instrument to measure transient absorption after single accelerator pulses extends the time resolution to 15 ps with minimal damage to the molecules under study [13]. At ~1 ns and longer conventional detection is possible. The charge attachment process may be complicated, but recent experiments have elucidated the kinetics for reaction with long-chain molecules [14]. In addition to the attachment process it is necessary to account for geminate and homogeneous recombination of electrons (or holes) with counterions, and similar recombination processes of anions (or cations) of the polymers (Figure 7.4).

7.2.2
Exciton Creation in Wires

Singlet excited states (excitons) in "wires" are readily created by photoexcitation. Detection can be as simple as quantitative determination of steady-state fluorescence (or phosphorescence) in molecules without (I_0) and with (I) ATs. From simple competition kinetics (Stern–Volmer) one obtains the quenching rate constant

$$k_q = k_f(I_0/I - 1) \tag{7.2}$$

in terms of the fluorescence decay rate without traps, k_f. Equation (7.2) is effective if the rate constant k_q for transport to and reaction with the trap groups is described as a first-order (single exponential) process. The application to capture of triplets

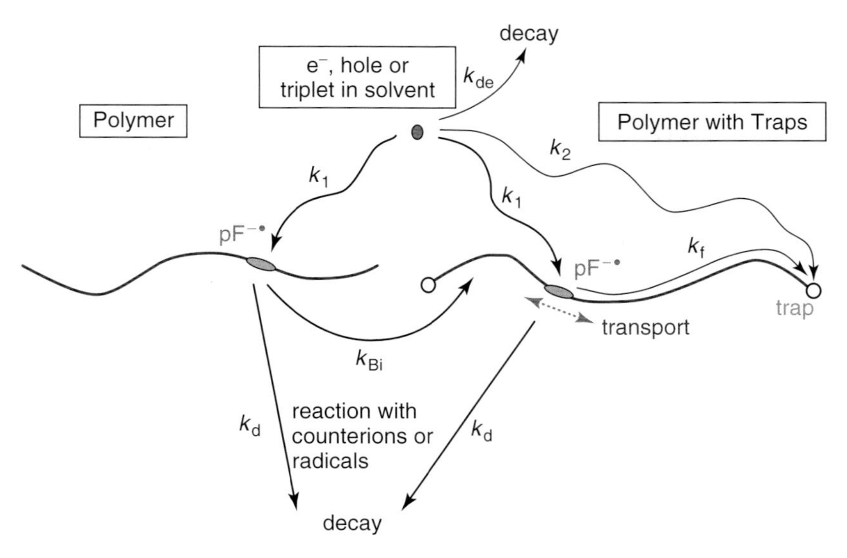

Figure 7.4 Reaction scheme for determination of transport of charges to appended traps. Solvated electrons, holes, or triplets created in the solvent attach to the polymer chains with rate constant k_1 to create polarons in the chain; some attach directly to the traps. Transport and reaction with the traps occurs with rate k_f, comprised of k_{trans} and k_{rxn} of Figure 7.1. In case of some chains that have no traps, polarons on those chains (shown left) reach traps via bimolecular encounters with chains having traps with rate k_{Bi}. Decay (k_d) of the polarons by reactions with geminate and homogeneous counterions and impurities are determined in measurements on polymers without traps. Decays (k_{de}) of solvated electrons (or holes or triplets) are determined in experiment without polymers. (Adapted from Ref. [15]).

in phosphorescence quenching requires that quenching of singlet precursors be taken into account.

Singlet and triplet transport to traps is most effectively investigated by time-resolved methods. Measurement of transient $S_1 \rightarrow S_n$ or $T_1 \rightarrow T_n$ absorption after laser pulses can give femtosecond time resolution; alternatively fluorescence or phosphorescence decay rates can be measured, including the possibility of fast upconversion methods for fluorescence. Creation of triplets by photoexcitation of the wires is effective when intersystem crossing (k_{IC} in Figure 7.2) is exceptionally fast, as in Pt-containing wires [16]. Other heavy atoms can be incorporated into the chains, but even so fast transport of singlet excitons to traps may compete effectively with intersystem crossing. Except where singlet transport is very slow, the external heavy atom effect, for example, in halogenated solvents, is typically not useful. Triplets can alternatively be created in the solvent; however, time resolution is then limited by diffusion to the wire. Triplet creation in the solvent can occur by photoexcitation and sensitization, but pulse radiolysis [17–19] discussed above is usually more convenient and effective.

7.3
Results on Transport to Traps

Available observations span a range of behaviors. We will first discuss results on electron and hole transport. With concrete examples in mind we discuss a description of polaron transport as diffusion that may be relevant to both charges and excitons. Next is a description of the results on transport of triplet excitons, then singlet excitons. The subject of singlet excitons is larger, but will be treated briefly here focusing on results from our labs, intended to be complementary to the substantial discussion in Chapter 9 by Andrew and Swager.

7.3.1
Electron and Hole Transport to Traps

This section describes observations of electron and hole transport along polymer chains. In each case the data is from pulse radiolysis experiments in solution.

7.3.1.1 A Nonconjugated Chains

A early example of transport utilized the nonconjugated polymer poly(4-vinylbiphenyl-*co*-1-vinylpyrene), containing 0.34–1.58% pyrene units [20]. Attachment of electrons in methyl-THF yielded radical anions of the biphenyl units, which made up the majority of the chain. From the decay of the biphenyl anions and growth of pyrene anions, the authors concluded that hopping between adjacent biphenyl anions occurred with a rate of $5.2 \times 10^9\ s^{-1}$ [20]. Holes attached to the same polymer reached pyrene more slowly by a factor of 10.

7.3.1.2 Charge Transport in π-Conjugated Chains

Table 7.1 shows structures of conjugated polymers having AT groups and average time to for the charge in the chain to transport to and react with the trap group.

Transport has been found to be reasonably fast, at least compared to the capabilities of measurement: four of the six reported results were too fast to resolve. An exception is that while electron transport is fast in the sigma-conjugated polysilane **1**, holes observed in the polysilane chains did not noticeably decay in 200 ns, possibly because the hole is more stable on the chain than on the porphyrin trap group [21]. Similarly holes in molecules **2** and **3** are expected to be captured on the arylamine and terthiophene trap groups, but electrons are not, and electrons in the polyfluorene chains of **4** and **5** are expected to be captured on the naphthylimide and anthraquinone traps, but holes are not.

Electrons injected into molecules polymers **4** and **5** were observed as $pF^{-\bullet}$ in the chains, most of which reacted with the traps. Figure 7.5 displays the observed kinetics for these two molecules. It is apparent that timescales for decay due to transport to the traps overlaps with the growth of $pF^{-\bullet}$ as electrons are attached to the polymer molecules. Determination of rates is made possible partly by having capped and uncapped polymers of similar length to separately determine electron attachment kinetics.

Table 7.1 Structures of conjugated polymers used to investigate electron and hole transport to appended trap groups[a].

Structure		n	τ (ns)	References
1	⊖	~120	<10 ns	[21]
2	R = 2EtHex ⊕	~40	<500 ns	[22]
	R = Hex	8	<10 ns	[15]
3 (R = 2EtHex)	⊕	13	<10 ns	[23]
4	⊖	35	5.3 ns	[15]
5	⊖	12	1.2 ns	[15]

[a] The table gives estimates of the average lengths of the chains, n, in repeat units and average time to for the charge in the chain to transport to and react with a trap group. A⊖ or ⊕ sign signifies transport of electrons or holes. All of these materials are polymers having a range of lengths.

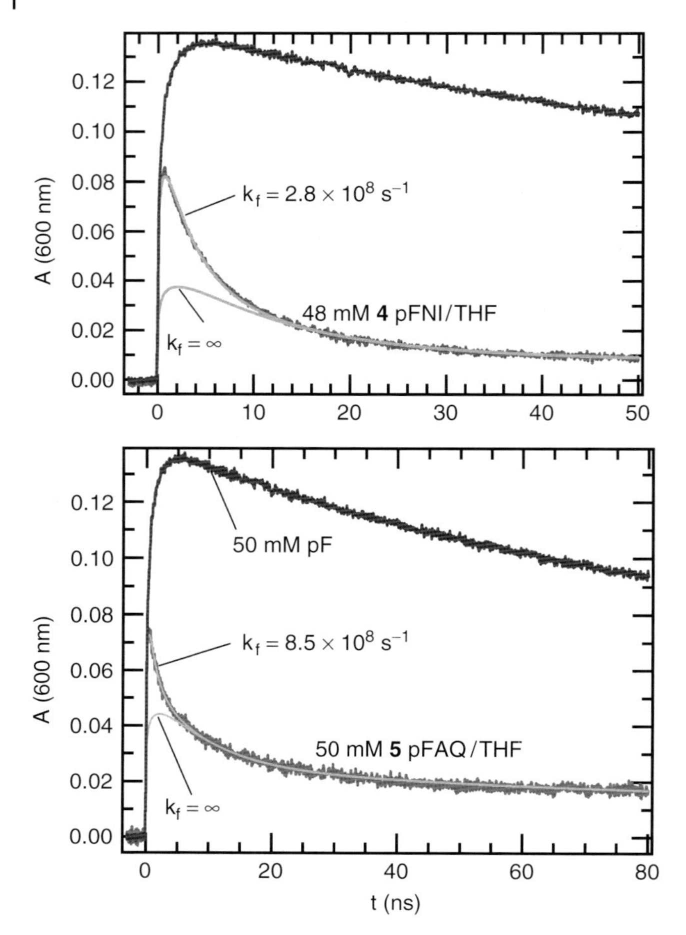

Figure 7.5 Transient absorption at the 600 nm absorption maximum of $pF^{-\bullet}$ for polymers **4** (pFNI) and **5** (pFAQ) along with the corresponding pF without trap groups in THF. Concentrations in repeat units are indicated. Kinetic fits giving intramolecular rate constants, k_f, for capture of $pF^{-\bullet}$ in the chains are shown, as are fits with k_f set to infinity. The considerably slower bimolecular electron transfer is accounted for. (Reproduced with permission from Ref. [15].)

For molecule **4** the length of the pF chain is 290 Å, so transport apparently occurs over substantial distances along the chains. We now discuss a model that can describe one-dimensional diffusion of these electrons to the traps.

7.3.2
One-Dimensional Diffusion of Polarons along Polymer Chains

If the transport of polarons, electrons, holes, or excitons consists of several small, uncorrelated steps, then the surviving fraction $P(t)$ of electrons as a function of time should be described by Eq. (7.3) [24, 25]:

$$P(t) = \frac{\sum_{n=0}^{\infty} \frac{1}{(2n+1)^2} e^{-\left(\frac{(2n+1)^2 \pi^2 Dt}{\alpha L^2}\right)}}{\sum_{n=0}^{\infty} \frac{1}{(2n+1)^2}} \quad (7.3)$$

Equation (7.3) is a well-known solution for one-dimensional diffusion of particles with the diffusion coefficient D, which can be related to the charge mobility value (μ) via the Einstein relation ($D = \mu k_B T/e$). If both of the chain ends are traps, the value of α is 1, whereas if only one of the ends is a trap, α is 4. The electrons (or holes or excitons) are assumed to be distributed uniformly along a chain of length L at an initial time. This uniform distribution is appropriate for creation of excitons by photoexcitation, which should produce excitons with equal probability at any location along the wire. It is also a good approximation for charges or excitons created by diffusional attachment, because the diffusing electrons, holes, or triplets can attach with almost equal probability anywhere along the chain. Usually such attachment reactions are diffusion controlled, in which case the uniform initial distribution is a reasonable approximation. A more exact treatment would also consider the bias toward attachment at and near the ends of the chains [26].

For **4** and **5**, the only cases in which transport was actually observed, the kinetics in Figure 7.5 were adequately described by single rates. Application of Eq. (7.3) to the data in Figure 7.5 gave mobilities of 9.1×10^{-3} and 3.2×10^{-3} cm^2/Vs for polymers **4** and **5**. Because the only difference in the polymer chain structures in **4** and **5** is the chain length (both polymers are polyfluorenes), the hopping times and mobilities should be the same Compared to **5**, in **4** the polymer chain is approximately three times longer and the transport to the traps takes three times as much time. This suggests that transport is linear in length. Diffusion (e.g., Eq. (7.3)), on the other hand would predict a nine times slower rate for a three times longer length. This is the reason for the discrepancy observed in mobilites. A second discrepancy between these observations and Eq. (7.3) is the single-exponential fits to the transport. Because some chains have the intended two traps (one at each end), but some have only one trap, two rates, different by a factor of 4, would be predicted.

The source of these discrepancies is not known at this time. It is possible that our knowledge of the lengths and the measurements of the rates are imperfect and that better measurements in the future will be in better accord with the predictions of diffusion. Alternatively, assumptions in the theory may be inadequate. Better measurements will certainly be possible in the future, as will measurements on other types of polarons.

7.3.3 Observed Transport of Triplet Excitons along Conjugated Chains

Silverman [16] found efficient capture of triplets by the 2,5-thienylene trap units in photoexcited **6** even when the fraction, y, of the traps was only 5%.

6

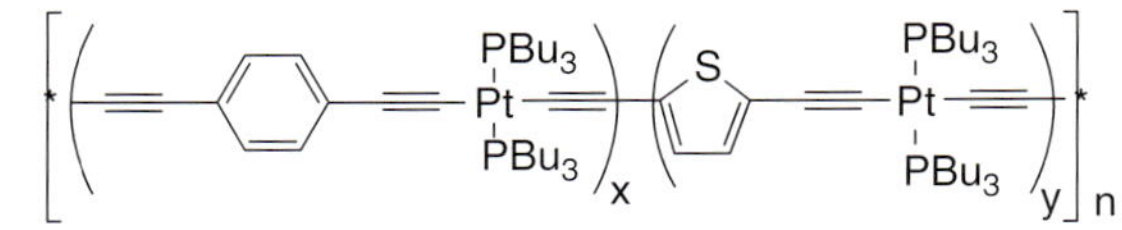

It was not possible to distinguish clearly whether triplets transported effectively or whether fast transport to the traps occurred for singlets prior to intersystem crossing. This result illustrates the difficulty of investigating triplet transport by direct photoexcitation noted in Section 7.2.2. In photoexcited **6** observation of phosphorescence indicated that some Pt–phenylene triplets were captured by the traps on the microsecond timescale.

A generally effective strategy is to create triplets in the solvent and then bring them to the chain molecules. While this method has time resolution limited by the need for triplets to diffuse to the polymer, it creates triplets that do not have as precursors singlets in the wires. In such experiments, triplets created in toluene solvent by pulse radiolysis attached to phenylene ethynylene chains in polymer **3** (but with $n = 10$) were found to be delocalized over 1.8 nm (1.3 repeat units, 2.6 phenylethnyls), shorter than the delocalization lengths of electrons, holes, or singlets. Still the triplets transported along the 13.6 nm long PPE chains and were captured at the terthiophene end traps in <5 ns [19]. The limited time resolution was due to the time for triplets to diffuse to and attach to the chains, which is, in turn, limited by solubility considerations.

7.3.4
Observed Transport of Singlet Excitons along Conjugated Chains

Transport of singlet excitons is crucial to the operation of organic photovoltaic cells [7], which, unlike inorganic solar cells, are designed to mix (interdigitate) donors and acceptors within a few nanometers of each other to cope with the short singlet exciton diffusion lengths in conjugated polymer films. Long-range exciton transport could have a major impact, so a question of great importance is – could singlet excitons transport rapidly within single chains as the Delft group found for charges? Table 7.2 pictures three polymers giving varied results for transport of singlet excitons to traps appended to the chains.

In a pioneering investigation, Swager [27] found transport to be fast, estimated here as $\sim$71 ps, in **7**. However other landmark results by transient absorption [29] and fluorescence [12] from the indenofluorene chains of **8** led to the conclusion that singlet exciton transport is intrinsically slow [29] along isolated single chains. We note however that the estimated 500 ps time [12] for transport compares favorably with charge transport processes noted above. Interchain quenching in films of **8** was found to be much faster. For transport in single chains it was noted that conformation might play an important role and that more rigid ladder polymers might give faster transport. Becker and Lupton investigated **8** with single molecule spectroscopy [30]. Their results confirmed the earlier findings of slow exciton transport in **8**, but discerned that transport varied among individual molecules and was much faster in some molecules having favorable conformations.

Table 7.2 Transport of singlet excitons to traps attached to polymer chains.

Structure	n	τ (ns)	References
7 (OR, RO; n = 21–23)	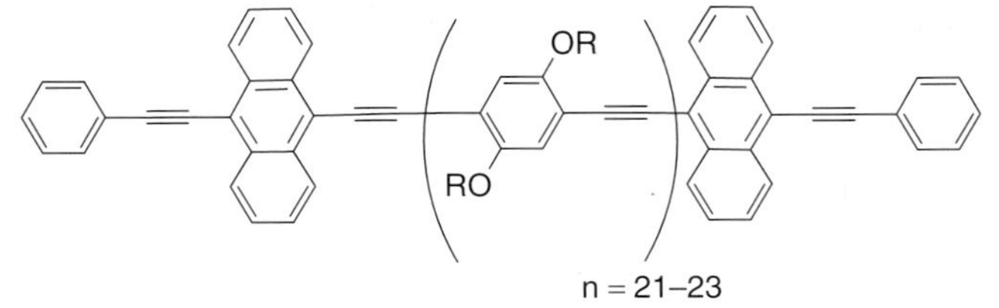 21–23	~71 ps[a]	[27,28]
8 (R R, R R; O, N; n)	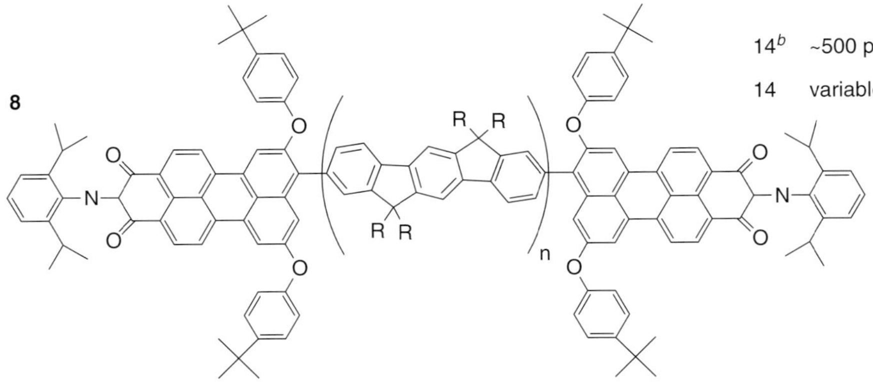 14[b]	~500 ps	[12,29]
	14	variable[c]	[30]
9 (R R; R = C_8H_{17}; N, S, N; n; m)	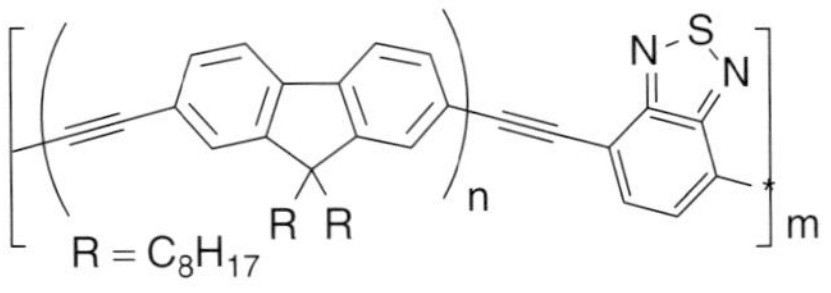 ~7	~1 ns[c]	[31]

[a]The τ ~71 ps time to transport to the traps in 7 is a Stern–Volmer estimate from the 95% quenching found, and the lifetime of ~1.5 ns in the polymer without traps [27].
[b]Due to substantial polydispersity the length is somewhat uncertain. Substantially longer lengths are present, leading to a range of rates.
[c]Different conformations give very different rates [30] (see text). For 9, a fluorescence lifetime of ~1 ns in the absence of traps was assumed.

The picture that exciton transport depends on molecular conformation and chemical structure is corroborated by measurements of quenching of fluorene ethynylene fluorescence by copolymerized 2,1,3-benzothiadiazole (BT) units. Shi [31] produced varied compositions, finding 55% of the fluorene ethynylene fluorescence was quenched by BT in **9** with 15% BT. Because this corresponds to only seven repeat units, this result implies even slower exciton transport. Observations of slower transport may have their root in the freely rotating ethynylene–fluorene bonds.

Contrary observations are “amplified quenching” and “superquenching” of fluorescence [32–34] of polyelectrolytes by quenchers like methyl viologen that bind at very low concentration due, usually, to Coulomb forces. The sensitive (nanomolar) quenching is attractive for creating sensors [28, 35]. It implies that excitons created in the polymer readily find the quencher over long chain lengths. However, detailed studies of quenching in these polyelectrolyte systems suggests that long-range exciton diffusion is facilitated by 3D transport within polymer aggregates that are formed by self-assembly of the polymer chains in solution [36, 37].

7.3.5 The Impact of Polydispersity: Diffusion to Traps in a Polymer Having a Distribution of Lengths

A problem encountered in applying Eq. (7.3) to real polymer chains is that, due to synthesis rarely produces uniform lengths. Information about distribution of chain lengths can be characterized (by calibration against standard polymers such as polystyrene) in terms of polymer polydispersity index (PDI) and its average length. The standard deviation (σ) is linked to PDI via $\sigma^2 = M_n^2(\text{PDI} - 1)$, where M_n is the number average molecular weight of the polymer. Real distributions of lengths take on a variety of shapes. To illustrate the effect of polydispersity we employ the simplification of assuming a Gaussian distribution function ($f(L)$). The surviving fraction of electrons (or excitons) in such case is then just the integration of Eq. (7.3) over the distribution function $f(L)$. Figure 7.6 shows the surviving fraction $P(t)$ of polarons from a numerical integration. In the Gaussian distributions, population decays faster than those in the uniform distribution at first due to electron population in short chains, and slower decay at longer time due to electrons left in the long chains. In general, if the distribution is wide

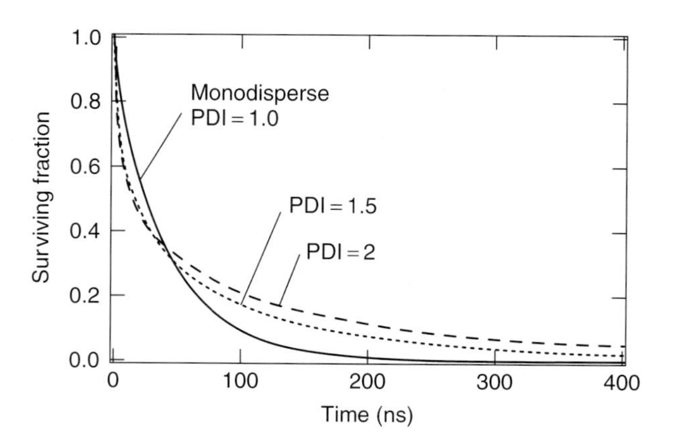

Figure 7.6 Comparison of calculated surviving populations on wires having a uniform single chain length, 66.4 nm, or Gaussian distributions of chain lengths with the same average length, and polydispersities, PDI = 1.5 and 2.0. Chains have perfectly absorbing traps at both ends. The diffusion coefficient is 1×10^{-4} cm^2/s.

(large polydispersity), the surviving fraction is going decay fast at first, crossing the population curve for that of uniform distribution, and a slower decay at long time.

7.4
Comparisons and Perspectives

Assessments of transport using attached traps have produced valuable information about excitons. For determinations of charge transport experimental results cited in Section 7.3 indicate that transport can be fast. An exciting result is that transport does occur over long distances. But many observations find only a lower limit. It is clear that progress is needed in experimental techniques to bring this method to its full potential. Specifically, synthetic methods need to be developed to the point where it is possible to synthesize long conjugated polymer chains with low polydispersity and that are end functionalized with high fidelity. In addition, the time resolution and wavelength range accessible for ultrafast pulse radiolysis could be further improved. For the two cases where experiments could actually observe the transport, important details could not be resolved leaving open questions. From these observations mobilities could be obtained for electron transport in polyfluorene chains in THF solution. Those mobilities, while respectable, are smaller by about two decades than mobilities of $\sim 0.5\,cm^2/V$ observed by μC for polyfluorene in decalin solution [4]. Among the questions remaining to be answered are whether the lower mobilities are typical of long distance transport. In that case the μC mobilities should be viewed as local to favorable regions of polymers that contain difficult-to-traverse regions that impede long distance motion of the charges. Alternatively the difference may reflect the slowing of polaron motion due to dissipation in the more polar THF medium. This question should be tested in the future.

With little interaction with dipoles in the medium, exciton transport might be expected to be faster than for charges. While described as slow for indenofluorene polymers, exciton transport competes favorably with charge transport. It is "slow" in the sense that it does not enable long-distance (>20 nm) transport within the ~ 0.5 ns lifetime of the singlet excitons. But no fundamental reason that exciton transport should be slow. We suggest that excitons may travel more that 100 nm along wires within their lifetimes. The faster transport in **7** [27], still faster results in liquid crystals (see Chapter 9), and improved transport in favorable conformations [30] all point to this possibility, and reinforce the speculation [12] that more structurally rigid polymers could give very fast transport. These questions merit further examination.

References

1. Hoofman, R.J.O.M., De Haas, M.P., Siebbeles, L.D.A., and Warman, J.M. (1998) Highly mobile electrons and holes on isolated chains of the semi-conducting polymer poly(phenylene vinylene). *Nature (London)*, **392**, 54–56.

2. Grozema, F.C., Siebbeles, L.D.A., Warman, J.M., Seki, S., Tagawa, S., and Scherf, U. (2002) Hole conduction along molecular wires: sigma-bonded silicon versus. pi-bond-conjugated carbon. *Adv. Mater. (Weinheim, Germany)*, **14**, 228–231.
3. Candeias, L.P., Grozema, F.C., Padmanaban, G., Ramakrishnan, S., Siebbeles, L.D.A., and Warman, J.M. (2003) Positive charge carriers on isolated chains of MEH-PPV with broken conjugation: optical absorption and mobility. *J. Phys. Chem. B*, **107**, 1554–1558.
4. Grozema, F.C. and Warman, J.M. (2005) Highly mobile electrons and holes on polyfluorene chains formed by charge scavenging in pulse-irradiated trans-decalin solutions. *Radiat. Phys. Chem.*, **74**, 234–238.
5. Prins, P., Grozema, F.C., Schins, J.M., Patil, S., Scherf, U., and Siebbeles, L.D.A. (2006) High intrachain hole mobility on molecular wires of ladder-type poly(*p*-phenylenes). *Phys. Rev. Lett.*, 96.
6. Prins, P., Grozema, F.C., Galbrecht, F., Scherf, U., and Siebbeles, L.D.A. (2007) Charge transport along coiled conjugated polymer chains. *J. Phys. Chem. C*, **111**, 11104–11112.
7. Gunes, S., Neugebauer, H., and Sariciftci, N.S. (2007) Conjugated polymer-based organic solar cells. *Chem. Rev.*, **107**, 1324–1338.
8. Bao, Z., Dodabalapur, A., and Lovinger, A.J. (1996) Soluble and processable regioregular poly(3-hexylthiophene) for thin film field-effect transistor applications with high mobility. *Appl. Phys. Lett.*, **69**, 4108–4110.
9. Takeda, N., Asaoka, S., and Miller, J.R. (2006) Nature and energies of electrons and holes in a conjugated polymer, polyfluorene. *J. Am. Chem. Soc.*, **128**, 16073–16082.
10. Jo, J.H., Chi, C.Y., Hoger, S., Wegner, G., and Yoon, D.Y. (2004) Synthesis and characterization of monodisperse oligofluorenes. *Chem. A Eur. J.*, **10**, 2681–2688.
11. Forster, T. (1959) 10th Spiers memorial lecture transfer mechanisms of electronic excitation. *Discuss. Faraday Soc.*, **27**, 7–17.
12. Hennebicq, E., Pourtois, G., Scholes, G.D., Herz, L.M., Russell, D.M., Silva, C., Setayesh, S., Grimsdale, A.C., Mullen, K., Bredas, J.L., and Beljonne, D. (2005) Exciton migration in rigid-rod conjugated polymers: an improved Forster model. *J. Am. Chem. Soc.*, **127**, 4744–4762.
13. Cook, A.R. and Shen, Y. (2009) Optical fiber-based single-shot picosecond transient absorption spectroscopy. *Rev. Sci. Instrum.*, **80**.
14. Sreearunothai, P., Asaoka, S., Cook, A.R., and Miller, J.R. (2009) Length and time dependent rates in diffusion-controlled reactions with conjugated polymers. *J. Phys. Chem. A*, **113**, 2786–2795.
15. Asaoka, S.T., Takeda, N., Iyoda, T., Cook, A., and Miller, J. (2008) Electron and hole transport to trap groups at the ends of conjugated polyfluorenes. *J. Am. Chem. Soc.*, **130**, 11912–11920.
16. Silverman, E.E., Cardolaccia, T., Zhao, X.M., Kim, K.Y., Haskins-Glusac, K., and Schanze, K.S. (2005) The triplet state in Pt-acetylide oligomers, polymers and copolymers. *Coord. Chem. Rev.*, **249**, 1491–1500.
17. Bensasson, R. and Land, E.J. (1971) Triplet–triplet extinction coefficients via energy transfer. *J. Chem. Soc., Faraday Trans.*, **67**, 1904–1915.
18. Lavalett, D., Bensasson, R., Amand, B., and Land, E.J. (1971) Experimental determination of triplet–triplet extinction coefficients and oscillator strengths. *Chem. Phys. Lett.*, **10**, 331–333.
19. Funston, A.M., Silverman, E.E., Schanze, K.S., and Miller, J.R. (2006) Spectroscopy and transport of the triplet exciton in a terthiophene end-capped poly(phenylene ethynylene). *J. Phys. Chem. B*, **110**, 17736–17742.
20. Yoshida, H., Ogasawara, M., and Tanaka, M. (1992) Pulse-radiolysis study on electron migration along polymer-chain ion radicals on poly(4-vinylbiphenyl-*co*-1-vinylpyrene). *Radiat. Phys. Chem.*, **39**, 35–40.
21. Matsui, Y., Nishida, K., Seki, S., Yoshida, Y., Tagawa, S., Yamada, K., Imahori, H., and Sakata, Y. (2002) Direct observation of intramolecular

electron transfer from excess electrons in a sigma-conjugated main chain to a porphyrin side chain in polysilanes having a tetraphenylporphyrin side chain by the pulse radiolysis technique. *Organometallics*, **21**, 5144–5147.

22. Burrows, H.D., de Melo, J.S., Forster, M., Guntner, R., Scherf, U., Monkman, A.P., and Navaratnam, S. (2004) Hole formation and transfer in poly[9,9-di(ethylhexyl)fluorene] and an amine end-capped derivative in solution. *Chem. Phys. Lett.*, **385**, 105–110.

23. Funston, A.M., Silverman, E.E., Miller, J.R., and Schanze, K.S. (2004) Charge transfer through terthiophene end-capped poly(arylene ethynylene)s. *J. Phys. Chem. B*, **108**, 1544–1555.

24. Zachmanoglou, E.C. and Thoe, D.W. (1987) *Introduction to Partial Differential Equations with Applications*, Dover, New York.

25. Carslaw, H.S. and Jaeger, J.C. (1986) *Conduction of Heat in Solids*, 2nd edn, Oxford University Press, New York.

26. Tsao, H.K., Lu, S.Y., and Tseng, C.Y. (2001) Rate of diffusion-limited reactions in a cluster of spherical sinks. *J. Chem. Phys.*, **115**, 3827–3833.

27. Swager, T.M., Gil, C.J., and Wrighton, M.S. (1995) Fluorescence studies of poly(P-Phenyleneethynylene)s – the effect of anthracene substitution. *J. Phys. Chem.*, **99**, 4886–4893.

28. Swager, T.M. (1998) The molecular wire approach to sensory signal amplification. *Acc. Chem. Res.*, **31**, 201–207.

29. Beljonne, D., Pourtois, G., Silva, C., Hennebicq, E., Herz, L.M., Friend, R.H., Scholes, G.D., Setayesh, S., Mullen, K., and Bredas, J.L. (2002) Interchain vs. intrachain energy transfer in acceptor-capped conjugated polymers. *Proc. Natl. Acad. Sci. USA*, **99**, 10982–10987.

30. Becker, K. and Lupton, J.M. (2006) Efficient light harvesting in dye-end-capped conjugated polymers probed by single molecule spectroscopy. *J. Am. Chem. Soc.*, **128**, 6468–6479.

31. Shi, C.J., Wu, Y., Zeng, W.J., Xie, Y.Q., Yang, K.X., and Cao, Y. (2005) Saturated red light-emitting copolymers of poly(aryleneethynylene)s with narrow-band-gap (NBG) units: synthesis and luminescent properties. *Macromol. Chem. Phys.*, **206**, 1114–1125.

32. Lu, L.D., Helgeson, R., Jones, R.M., McBranch, D., and Whitten, D. (2002) Superquenching in cyanine pendant poly(L-lysine) dyes: dependence on molecular weight, solvent, and aggregation. *J. Am. Chem. Soc.*, **124**, 483–488.

33. Fan, C.H., Wang, S., Hong, J.W., Bazan, G.C., Plaxco, K.W., and Heeger, A.J. (2003) Beyond superquenching: hyper-efficient energy transfer from conjugated polymers to gold nanoparticles. *Proc. Natl. Acad. Sci. USA*, **100**, 6297–6301.

34. Rininsland, F., Xia, W.S., Wittenburg, S., Shi, X.B., Stankewicz, C., Achyuthan, K., McBranch, D., and Whitten, D. (2004) Metal ion-mediated polymer superquenching for highly sensitive detection of kinase and phosphatase activities. *Proc. Natl. Acad. Sci. USA*, **101**, 15295–15300.

35. McQuade, D.T., Pullen, A.E., and Swager, T.M. (2000) Conjugated polymer-based chemical sensors. *Chem. Rev.*, **100**, 2537–2574.

36. Tan, C., Atas, E., Muller, J.G., Pinto, M.R., Kleiman, V.D., and Schanze, K.S. (2004) Amplified quenching of a conjugated polyelectrolyte by cyanine dyes. *J. Am. Chem. Soc.*, **126**, 13685–13694.

37. Jang, S., Cheng, Y.C., Reichman, D.R., and Eaves, J.D. (2008) Theory of coherent resonance energy transfer. *J. Chem. Phys.*, **129**, 101104-1–101104-4.

8
Electron Lattice Dynamics as a Method to Study Charge Transport in Conjugated Polymers

Sven Stafström and Magnus Hultell

8.1
Introduction

In this chapter we present a method, electron lattice dynamics (ELD), which describes charge transport in conjugated polymeric and molecular-based systems. ELD is focused on the details of how the charge carriers interact with the nuclear degrees of freedom of the polymer or molecular backbone during the dynamic transport process. This interaction, which can also be termed electron–phonon coupling, is an intrinsic property of molecular and polymer electronic systems based on unsaturated carbon atoms [1] and affects the equilibrium nuclear positions as well as the electron and nuclear dynamics. In general, the electron–phonon coupling is stronger the smaller and more restricted the dimensionality of the system is. Extended carbon bases systems such as carbon nanotubes or graphene show almost no electron–phonon coupling, whereas similar but confined structures such as C_{60} and likewise extended but quasi one-dimensional conjugated polymer systems show a geometrical response to changes in the electron density matrix, in particular to changes in the off-diagonal bond-order elements.

The methodology presented here is designed to be applied to the studies of dynamical processes in polymeric and molecular systems, in particular to processes that involve charge transport. The strong electron–phonon coupling prevents the use of a rigid lattice in these studies. It is often also insufficient to treat the dynamics within the Born–Oppenheimer adiabatic approximation [2]. In the Born–Oppenheimer approximation the calculation of dynamical processes is divided into two stages, first the electronic problem is solved keeping the nuclei fixed. In the second stage, the nuclear dynamics on a given potential energy surface is treated. This treatment is based on the assumption that the spacing of the electronic eigenvalues is large compared to the energetics of the nuclear motion. However, when this condition is violated, events of exchange of energy between the nuclear kinetic energy and electronic excitations can occur. Typical processes for which this happen are the formation and breaking of covalent bonds, inelastic molecular collisions, intermolecular electron transfer by means of tunneling or hopping and decay of excited electronic states. The treatment of such processes

Charge and Exciton Transport through Molecular Wires. Edited by L.D.A. Siebbeles and F.C. Grozema

ISBN: 978-3-527-32501-6

has to be able to account for nonadiabatic effects, i.e., transitions between different quantum states (or different potential energy surfaces) of the electronic system. Our ELD approach is based on earlier developed methodologies [3–8]. There are also similar initiatives that are developing in parallel with our approach [9, 10]. In this context we would also like to mention the more rigorous approach developed by Öhrn *et al.*, which is referred to as electron-nuclear dynamics (END) [11].

The ELD method involves the simultaneous solution of the time-dependent Schrödinger equation (TDSE) and the Ehrenfest equation of motion for the (classical) nuclear degrees of freedom [12]:

$$i\hbar|\dot{\Psi}(t)\rangle = \hat{H}|\Psi(t)\rangle \tag{8.1}$$

$$M_n\ddot{\mathbf{r}}_n = \dot{\mathbf{p}}_n = -\nabla_{\mathbf{r}_n} V_{\text{nuc}} - \langle\Psi(t)|\nabla_{\mathbf{r}_n}\hat{H}|\Psi(t)\rangle \tag{8.2}$$

Here $|\Psi(t)\rangle$ is the time-dependent wavefunction, $\hat{H}$ is the Hamiltonian, and $\mathbf{r}_n$ and $\mathbf{p}_n$ are the position and momentum of the nth nucleus, respectively. The classical nuclear potential energy is denoted by V_{nuc}.

Before the ELD methodology is presented in detail we will discuss the basic and most important aspects of transport in conjugated polymeric systems and also make comparisons with transport in molecular-based materials such as molecular crystals [13] and molecularly doped polymers [14]. The focus on polymeric materials is motivated by their great potential in terms of applications as well as from the point of view of complexity of charge transport processes. This complexity originates from the internal structure of polymeric systems, which leads to charge transport in terms of both intra- and interchain processes. This complexity can be directly addressed in the ELD approach, which handles both processes equally well.

As a model system for our presentation we use poly(*para*-phenylene vinylene) (PPV). PPV is one of the most well-studied polymers and a polymer which is used extensively in applications including organic light emitting diodes (OLED) [15], organic photovoltaic cells [16], and organic field effect transistor (OFET) devices [17]. In this context it should be noted that depending on the type of application discussed the circumstances for charge transport are quite different. However, when focussing on the most fundamental aspects of transport these devices have much in common. In all cases the charge carrier is believed to be polarons, i.e., a unit consisting of the charge carrier (electron or hole) together with its induced lattice deformation [18–21]. These polarons are transported through the material under the influence of an external electric field. The material itself contains disorder which limits the transport via processes such as trapping and backscattering. In fact, the overall charge transport properties do not reflect the intrinsic properties of individual polymer chains as much as the interchain ordering and electronic overlap. Nevertheless, the intrinsic properties of the polymeric system form the basis of the material properties and are, as such, important for the understanding of any physical property of the material. In the next section we therefore present the basic properties of PPV, before we, in the following sections, turn to a discussion about disorder and its effect on charge transport.

8.1.1
Ideal Conjugated Polymers

The intrinsic properties of conjugated polymers are derived from the equilibrium ground state structure obtained via energy minimization with respect to the charge density and the nuclear coordinates. The ideal ground state geometrical structure, which results from this energy minimization, includes the positions of the covalently bonded atoms as well as the perfectly ordered structure of the crystal which results from the nonbonded electrostatic and van der Waals interactions. Note that not only the ground state electronic properties are obtained from such an energy minimization, but also the excited state properties of the combined system of electronic and nuclear degrees of freedom. In the case of PPV (see Figure 8.1), the ground state geometry has the following properties: carbon–carbon bond lengths in the range from 1.35 to 1.45 Å, bond angles in the range from 115° to 126° and torsional angles of the order of 5° [22]. The solid-state structure is obtained from X-ray diffraction measurements [23] and show that PPV has a herringbone crystal structure with a monoclinic unit cell containing two chains. The crystal structure is shown in Figure 8.1, where $a = 8.07$ Å, $b = 6.05$ Å, and $c = 6.54$ Å, the monoclinic angle between b and c is 123°, the setting angle ϕ is 52°, and the unit cell has P21/a symmetry.

The most prominent phonon modes include the ring C–C stretch frequencies that group around 1274 to 1546 cm^{-1}, while the vinylene C–C stretch frequency appears at 1635 cm^{-1} [24]. The torsional modes, which include phenyl ring rotational motion, have frequencies from around 320 cm^{-1} and below [25, 26].

The electronic properties of the ideal crystalline PPV system exhibits delocalized, Bloch-type electronic states and a band structure with an electronic band gap of about 2.4 eV [27]. Also of great importance for the transport properties of PPV is the width of the highest occupied electronic band perpendicular to the (k-space) direction along the polymer chains. This band width is of the order of 0.5 eV [22].

The electron–phonon coupling introduced above is an intrinsic property of the polymer system. If an electron is added or removed from the ideal single-chain

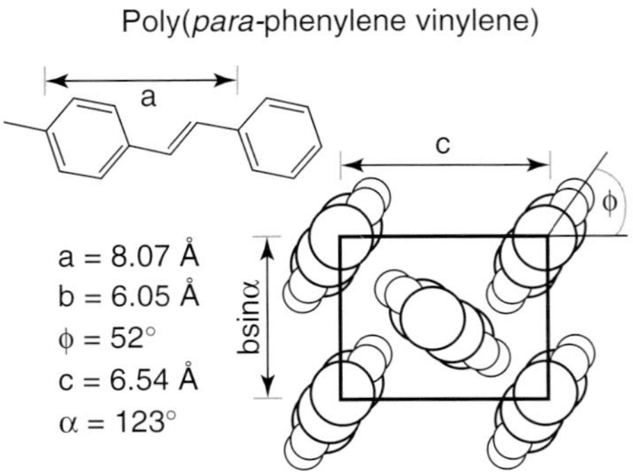

Figure 8.1 Geometrical structure of poly(*para*-phenylene vinylene) (PPV).

system, the coupling between intrachain phonons and the electronic system results in the formation of a polaron, which is localized over approximately 4–5 phenylene–vinylene units. The energy gained by forming this type of defect instead of hosting the electron (or hole) in the system in its neutral ground state geometry is referred to as the polaron formation energy. In the case of PPV this energy is ~0.3 eV [28]. In a similar manner, an excitation over the energy gap between the highest occupied molecular orbital (HOMO) and the lowest unoccupied molecular orbital (LUMO) also couples to phonons, which results in a localization of the excited state into a polaron exciton [29, 30].

The intrinsic properties of the conjugated systems are indeed important for the ability of organic systems to transport charge. In particular, in the case of both polymer and molecular-based materials, the electronic overlap between polymer chains or between molecular units is directly related to the electronic bandwidth and therefore also to the conductivity of the system. The charge carrier, i.e., the polaron, is also strongly affected by the interchain interactions. A direct relation between the stability of the polaron and the strength of this interactions is, to the best of our knowledge, not available in analytical terms. It is clear, however, that in the case of strong intrachain interactions in well-ordered polymeric systems the polaron is no longer localized to a single polymer chain but extends in 2D [31, 32] or 3D depending on the isotropy of the system [33, 34]. As a result of this delocalization, polaron binding energy is destabilized and if it was possible to tune the ratio between the electronic 2D or 3D bandwidth and the single-chain polaron binding energy it is clear that if this ratio is much larger than unity the polaron becomes unstable and charge transport can be described within the band structure picture. The numbers given above indicate a ratio between one and two for the ideal 3D PPV system and the stability of the polaron can be questioned [33]. However, this situation is highly unlikely to exist in reality since, as a result of the weak interchain or intermolecular interactions that exist in van der Waals bonded systems, these systems are very sensitive to disorder or fluctuations caused by temperature effects. Thus, for most molecular- or polymer-based materials there is no band structure and the stability of the polaron vs. three-dimensional delocalization is therefore less relevant.

Charge transport in highly ordered polymeric system can be described by an adiabatic process in which the self-localized polaron (1-, 2-, or 3D) can move through the system on the potential energy surface corresponding to the electronic ground state. This potential energy surface exhibits barriers for displacing the polaron from one central carbon atom to a neighboring carbon atom, which prevents transport at zero temperature and zero external electric field. However, in the ideal system these barriers are small and compared to other relevant energies at ambient temperatures and typical operating external electric field strengths polaron motion occurs either via a temperature-activated diffusion process or a drift process in which an external electric field acts as the driving force. This is actually the case for both intrachain transport in perfectly ordered conjugated polymeric systems [32] and in ideal pentacene molecular crystals [35].

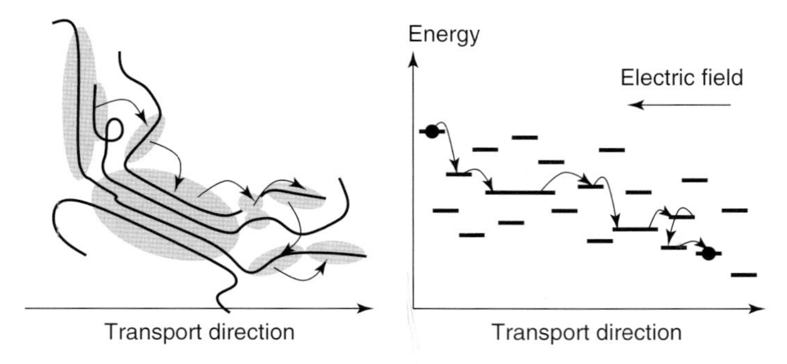

Figure 8.2 A schematic presentation of the electronic sites in conjugated polymers as distributed in space (a) and in energy along a cross-section (b). The polymer backbone and the conjugated segments are presented in (a) by the curved lines and the dot filled oval, respectively.

8.1.2 Disordered Conjugated Polymers

The degree of disorder, or rather the possibility to synthesize system with as low disorder as possible, is perhaps the most important aspect of realizing organic systems for high-performance electronic applications. In practice, however, disorder is inevitable due to the relatively weak van der Waals interchain interactions which allows for entropy to have a strong effect on the free energy. It should be pointed out that the disorder results from both static defects, i.e., impurities, misalignment of polymer chains or molecules, interruptions of the conjugation length due to chain twists, etc., and dynamic irregularities that involve atomic as well as molecular motions at finite temperatures. These effects are both intra- and interchain and manifest themselves in decreasing conjugation lengths and weakening of the interchain interactions, etc. (see Figure 8.2(a)). Such effects are also coupled to the localization of the electronic wavefunctions and are therefore decrementing the ability of the polymeric system to transport charge (see Figure 8.2(b)).

In the limit of strong disorder where all states will be localized (see Figure 8.3) and consequently, the charge carriers have to move nonadiabatically through the system via a sequential hopping process. Even though this situation can be described by the TDSE, it is common to consider the weak electronic coupling between such localized states as a perturbation. From the basic assumption of time-dependent perturbation theory, i.e., that the perturbed states can be expanded in terms of the unperturbed localized states with expansion coefficients that vary in time, the rate at which transitions occur from one unperturbed state (a) to another state (b) as a result of the (small) time-dependent perturbation, V_{ab}, which describes the coupling between these states is the "golden rule" expression:

$$P_{ab} = \frac{2\pi}{\hbar} | < \varphi_a | V_{ab} | \varphi_b > |^2 \delta(E_b - E_a) \tag{8.3}$$

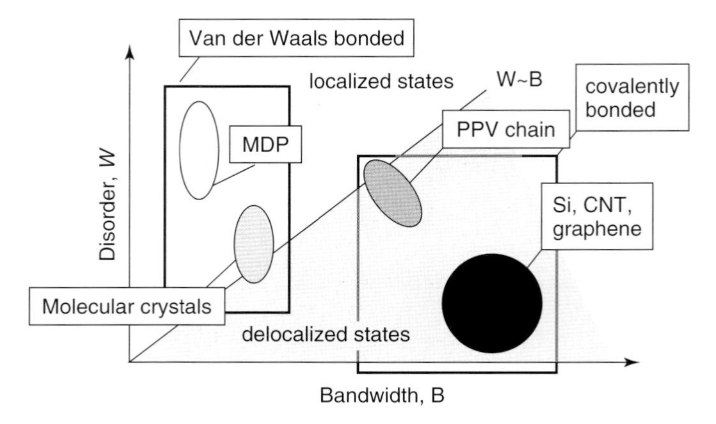

Figure 8.3 A schematic presentation of how the disorder and electron transfer integral affect the localization properties of some materials of interest for electronic applications.

In contrast to Eq. (8.1) above, the states involved in the hopping process are those obtained from the solution of the *time-independent* Schrödinger equation, i.e.,

$$\hat{H}|\varphi_a\rangle = E_a|\varphi_a\rangle \tag{8.4}$$

The rate equation above is the basis of a number of nonadiabatic transport theories such as the Miller and Abrahams formalism [36], Holstein small polaron model [18, 37], and the Marcus theory [38], where the latter two also include the electron–phonon coupling. This coupling is taken into account as a linear change of the site energy with one or more normal mode coordinates. In the rate expression above, the electron–phonon coupling gives rise to a factor which describes the overlap between the vibrational wavefunctions of the initial and final states, weighted by the density of electronic states (see, e.g., Brédas *et al.* [39]).

The applicability of the different approaches discussed above is directly related to the electronic structure which originates from the properties of the ideal system but strongly modified due to static and dynamic disorder. In molecularly based systems such a molecularly doped polymers, V_{ab} can be varied by preparing systems with different strength in the electronic overlap between nearest-neighbor molecules. This allows for studies of the transition from band-like (adiabatic) transport to hopping (nonadiabatic) transport [40]. A similar possibility exists in polymeric systems in which side chains and regioregularities can have large effects on the transport properties [41].

8.1.3 Mobility

The ability of a system to transport charge is quantified in terms of the charge carrier mobility. In the case of electronic devices that operate under the influence of an external electric field, the mobility, μ, is related to the drift speed of the charge carrier, v_d, as follows:

$$v_d = \mu E_0 \tag{8.5}$$

where E_0 is the amplitude of the electric field. The mobility can also be defined for zero bias, i.e., diffusive motion. In this case the mobility depends explicitly on the temperature, T in the following way:

$$\mu = \frac{e}{k_B T} D \tag{8.6}$$

where D is the diffusion constant.

From a theoretical point of view, calculation of mobility is rather straightforward and can be performed using both the ELD approach and the perturbation approach discussed above. The main problem that any theoretical approach is facing is, however, to model the disorder present in the system since detailed information regarding the deviation from the crystalline structure is impossible to get. Extensive studies have been performed based on uncorrelated disorder with a Gaussian distribution accounting for various types of imperfections in the lattice. This model was first suggested by Bässler and is referred to as the Gaussian disorder model [14]. Models describing this transport usually do not include explicitly the electron–phonon coupling and are based on the Miller and Abrahams [36] rate equation.

Experimental studies of the charge carrier mobility provides very useful information regarding the transport properties of electronic materials. A detailed review of results of such studies and their interpretation lies outside the scope of this chapter (see instead recent reviews by Coropceanu *et al.* [42] and by Grozema and Siebbeles [43]). However, we note that even though the mobility is specific for a particular sample and includes the properties of the ideal system in combination with extrinsic disorder effects, it also depends on parameters such as charge carrier concentration, electric field strength, and temperature. This dependence is important since it gives information about the physical processes that are involved in the transport. In particular, the temperature dependence can separate different types of transport processes: band transport, tunneling (or superexchange), and activated adiabatic or nonadiabatic transport. For band-like transport, increasing temperature causes additional scattering and consequently a lowering of the mobility whereas in localized systems the charge carrier has to pass energy barriers and the mobility can increase with temperature. In the latter case the additional scattering caused by lattice phonons can dominate over the temperature activated processes. This can happen in highly ordered pentacene [44] and rubrene [45, 46] molecular crystals which show a maximum mobility at a temperature that usually lies in the regime from 200 K up to room temperature. There are also cases, primarily found in highly ordered molecular crystals [13], that show band-like behavior even at low temperatures. In contrast to these more or less crystalline molecular materials, other molecular-based systems, e.g., moleculary doped polymers, have a structure in which the molecules are separated by nanometer distances and the intermolecular interaction, V_{ab}, is very small. As a consequence of this structure the charge transport becomes thermally activated and the mobility increases exponentially with temperature at all temperatures [14].

Most conjugated polymers show temperature-activated transport, which is a result of the disorder that is built into the system upon synthesis. The conjugation length is also quite restricted, the polymer chains are electronically divided into segments that constitute the unit over which the electrons can delocalize. This can be a single short chain segment (or "spectroscopic unit") [47, 48] or a region characterized by well-aligned chains (see Figure 8.2). The size of these segments or regions is of the order of 20–100 Å. In practice this means that charge transport in conjugated polymers consists of both intra- and interchain transport. In addition, the interfaces between the regions of ordered polymer chains also act as barriers for charge transport. The height of these barriers and the type of transport across them depends on how well the π-systems of these separate systems overlap. This issue is to some extent addressed below, but we probably need to go to even larger systems in order to fully account for the type of interactions present in these regions.

The room temperature mobilities for polymers of interest for organic electronics applications are of the order 10^{-4}–$10^{-5}\,cm^2/Vs$ [49, 50]. Under special conditions, well-ordered materials can be produced with much higher mobilities. Among main-chain π-conjugated polymers, regioregular head-to-tail poly(3-hexylthiophene) (P3HT) has been known to give the best FET performances. A field-effect mobility of $0.1\,cm^2/Vs$ was achieved and in accordance with the discussion above, the origin of the high mobility can be ascribed to the presence of well-developed intermolecular $\pi-\pi$ stacking of thiophene rings [17, 51]. A similar dependence of the charge carrier mobility on the supramolecular order induced by alkyl side chains is supported by results on PPV [52] as well as symmetrically substituted PPV derivatives with a mobility reaching ($10^{-2}\,cm^2/Vs$) [53]. It should be noted that the mobility in PPV oligomers can reach the same values as for the more extended polymeric systems [54]. Apparently, the order and thereby the intermolecular interactions strength is favored in molecular materials, which is also evident from the high values of the mobility reported in crystals of thiophene oligomers [55]. Still, the intermolecular interaction is the limiting factor whereas the intrachain mobility is considerably larger [56].

With this brief introduction to charge transport in conjugated polymeric materials and molecular crystals we now proceed to discuss the details of the ELD method and to which extent this method can address the specific questions related to charge transport properties.

8.1.4
Electron-Lattice Dynamics I

ELD is a model that is designed to account for both the intrinsic charge transport mechanisms and the extrinsic disorder effects (both static and dynamic) which leads to electron localization. A number of studies of charge transport in conjugated polymers have been made based on this or similar models [5–8, 57].

There are several approximation levels that can be used to describe electron-lattice dynamics [58]. In the simplest form, the adiabatic approximation, the potential energy and the ionic forces are derived within the Born–Oppenheimer approximation based on the solution to the time-independent Schrödinger equation. In this approximation, the electrons are assumed to follow the nuclei in such a way that they are always in an eigenstate associated with the instantaneous positions of the nuclei, which are seen as merely providing an external potential for the electrons. This description of the system can then be used to define a set of fixed potential energy surfaces (PES) corresponding to the positions of the nuclei and to the degree of excitation of the electrons. In our case, however, this approach is not applicable since, as discussed above, we would like to include electron transfer by means of tunneling or an activated hopping processes into our model, i.e., nonadiabatic processes.

8.2 Methodology

In this section we present the methodology of the ELD approach. The basic concepts of the electronic system are introduced in Section 8.2.1, followed by a discussion in Section 8.2.2 concerning the main approximations in the ELD equations of motion, in particular the approximations related to the treatment of the electron–phonon interaction. Sections 8.2.3 and 8.2.4 present the electronic and lattice Hamiltonians, respectively, and finally Section 8.2.5 presents the explicit ELD equations.

8.2.1 Hamiltonian

Bonds in molecular or polymer systems based on sp^2 hybridized carbon atoms can be divided into two parts: σ-bonds and π-bonds; the σ-bond is formed from two sp^2 hybridized orbitals on two neighboring carbon atoms, whereas the π-bond results from overlap (conjugation) of the $2p_z$-orbitals over the entire system. Since the orbital overlap that gives rise to the σ-bond is much stronger than the π-overlap, there is a larger energy splitting between the (occupied) bonding and (unoccupied) antibonding σ-orbitals than for the corresponding π-orbitals. Consequently, electronic processes such as charge transport, which involve energies of the order of electron volts or less, do not include the change in time of the occupation of σ-orbitals. Studies of charge transport in π-conjugated system can therefore be made based on the approximation of separability between the π- and σ-systems:

$$E = \langle \Psi | \hat{H} | \Psi \rangle = \langle \Psi_\sigma | \hat{H}_\sigma | \Psi_\sigma \rangle + \langle \Psi_\pi | \hat{H}_\pi | \Psi_\pi \rangle \tag{8.7}$$

With the approximation that the occupation number of the σ-orbitals does not change during the dynamics that we are studying we can replace the quantum mechanical description of this subsystem with a classical potential energy surface, E_σ. Furthermore, we also treat the σ-orbitals as fully localized to each covalent

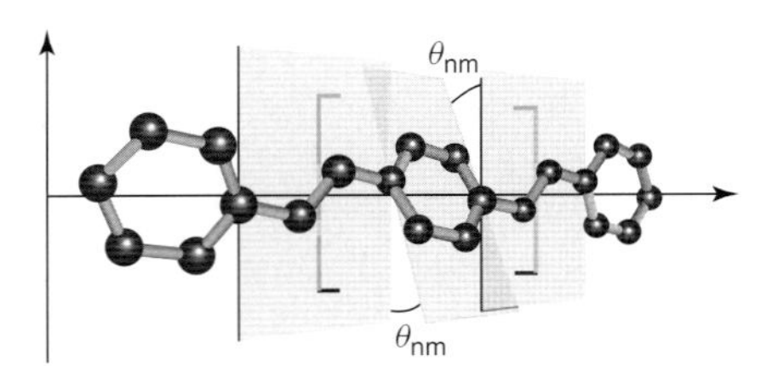

Figure 8.4 Torsions in PPV.

bond involving two neighboring atoms. The potential energy of the σ-system, V_σ can now be expressed in terms of local variables such as the bond lengths (R_{nm}), bond angles ($\vartheta_{n'nm}$), and dihedral (or torsion) angels (θ_{nm}).

$$\langle \psi_\sigma | \hat{H}_\sigma | \psi_\sigma \rangle = V_\sigma (R_{nm}, \vartheta_{n'nm}, \theta_{nm}) \tag{8.8}$$

The dihedral angel (θ_{nm}) is illustrated in Figure 8.4. Note that we use the bond index nm to denote this angle to emphasis the rotation around the bond, even though torsion angles in general have to be specified using four atoms. The explicit expression of V_σ is presented in Section 8.2.4. In contrast to the σ-system, the π-electrons have to be treated quantum mechanically. Typical features of the π-system are the delocalized nature of the electronic wavefunctions and the relatively small energy gap (in the range from one to three electron volts) between bonding and antibonding orbitals. Both these features are essential in order to achieve charge transport through such a system: the mobility is related to delocalization of the electronic wavefunctions and the small gap energies facilitate creation of charge carriers, either by means of charge injection or by thermal- or photoexcitation. The π-orbitals are characterized by their bonding, nonbonding or antibonding nature with respect to specific C–C bonds. A change in the occupation number of such an orbital naturally results in a change in the bond-order which gives rise to a force on the carbon atoms. This force corresponds to the electron–phonon coupling. The relation between the size and dimensionality of the system and the strength of this coupling is now evident since a spatially confined π-orbital causes a larger change in bond order.

8.2.2 Electron-Lattice Dynamics II

As stated in the introduction above, the equation of motion of the π-electrons is given by the time-dependent Schrödinger equation (TDSE),

$$i\hbar|\dot{\Psi}_\pi(t)\rangle = \hat{H}_\pi |\Psi_\pi(t)\rangle \tag{8.9}$$

Simultaneously to solving the TDSE we determine the ionic motion in the evolving charge density distribution by solving the lattice equation of motion within the potential field of the σ- and π-electrons. The equation is based on the Ehrenfest theorem

$$M_n \langle \ddot{\mathbf{r}}_n(t) \rangle = -\langle \nabla_{\mathbf{r}_n} V(\mathbf{r}_n) \rangle \tag{8.10}$$

which, in accordance with the $\pi-\sigma$ separation discussed above, results in the following expression:

$$M_n \ddot{\mathbf{r}}_n(t) = -\nabla_{\mathbf{r}_n} V_\sigma - \langle \Psi_\pi(t) | \nabla_{\mathbf{r}_n} \hat{H}_\pi(t) | \Psi_\pi(t) \rangle \tag{8.11}$$

Here, the classical σ-part is separated from the quantum mechanical π-part, for which the potential energy is expressed in terms of an expectation value of the operator equation. The potential describing the σ-part, V_σ, also includes the potential energy of the nuclei.

The Ehrenfest method consists of two separate approximations: the electrons influence the ions through a classical field generated by its time-dependent electron density (a mean-field approximation); the nuclei obey classical (Newtonian) mechanics. The effect of the mean-field approximation (averaging over electronic trajectories) is to break the microscopic correlations between the force experienced by the nucleus due to the electrons and the momentum of the nucleus [58]. As briefly discussed in the introduction above, this approximation is well justified in the case of charge transport. Studies of the dynamics in the photoexcited state (see Section 8.3.4) could in principle be more strongly affected by correlation effects but the relatively slow deexcitaion process and the small energy spacing between the electronic energy levels make this effect less pronounced as will be discussed further below.

The approximative treatment of the nuclei (ions) as classical particles also needs a motivation. There are three different energy quantities that are important in this context: the phonon energy ($\hbar\omega_\xi$, where ξ denotes a particular normal mode), the activation energy (E_{act}), and the energy related to a finite temperature ($k_B T$). In all processes that include lattice vibrations, a classical treatment assumes that the number of phonons involved is large, i.e., $\hbar\omega_\xi \ll k_B T$. In addition, in the case of thermally activated transport, the classical approach is justified if this activation can reach the nuclear coordinates of the crossing region, which requires that $E_{\text{act}} \ll k_B T$. If the temperature decreases such that $E_{\text{act}} \gg k_B T$, then the transition has to occur via nuclear tunneling, which requires a quantum mechanical consideration of the vibrational coordinates.

In the case of adiabatic transport, typically polaron transport in a system with delocalized states, the activation energies are very small since the polaron is rather extended and therefore insensitive to the exact position in the system. Typically, the energy barriers are of the order of a few meV in molecular-based systems and even less in intrachain polymer systems [12, 35]. The timescale of this motion is also fast compared to nuclear motions which also justifies the use of a classical description.

When the electronic states are localized due to disorder the situation is more complex. The transport is nonadiabatic in this case and involves transition from one localized state to another, a process which is normally associated with a larger activation energy since the polaron is destroyed on the donor and created at the acceptor sites. The activation energy (or reorganization energy) is in this case the sum of the energy to created and annihilate a polaron. However, if we have in mind that the polarons we are discussing are rather delocalized (large polarons [18], see also discussion above related to 2D and 3D polarons) and have a considerable overlap with the acceptor levels, the path for charge transfer is somewhere in between that of the adiabatic polaron drift and the small polaron hopping. Consequently, the barrier is not as large as one might expect simply by adding the polaron energies of small polarons. The driving force due to the external electric field also results in a lowering of the energy barrier for transport. In addition, in realistic samples there will always exist a manifold of acceptor levels and it is highly probable that at some point in time one of these levels corresponds to a transition with a very low activation energy (see Figure 8.2). All these effects point to low energy barriers for nonadiabatic polaron transport in polymers, a situation which can be accurately described by a classical description of the lattice vibrations. However, when moving into the "deep" nonadiabatic regime, with mobilities of the order of 10^{-6} cm^2/Vs or lower, as would be the case for strongly disordered systems, with small overlap between the donor and acceptor states, the ELD approach is less adequate and transport should be studied using the perturbation approach discussed above (see Eq. (8.3)).

8.2.3 Electronic Hamiltonian

For the Hamiltonian, $\hat{H}$, we use a tight-binding model developed from the Su-Schrieffer–Heeger (SSH) model [19] for the polymer chain and assume, in accordance with the discussion above, π–σ separability. The contribution from the π-electrons to the Hamiltonian (including the contribution from an externally applied electric field, $\hat{H}_E(t)$) can then be written in the form

$$\hat{H}_\pi(t) = -\sum_{\langle nm \rangle} \beta_{nm}(t)[\hat{c}_n^\dagger \hat{c}_m + \hat{c}_n \hat{c}_m^\dagger] + \hat{H}_E(t) \tag{8.12}$$

where $\langle nm \rangle$ denotes summation over covalently bonded atoms, β_{nm} the resonance integral between the 2p$_z$-orbitals on sites n and m, and $\hat{c}_n^\dagger$ ($\hat{c}_n$) the operator for creating (annihilating) an electron on site n. The resonance integrals are treated in the Mulliken approximation [59] with the analytical functions for the overlap integrals taken from Ref. [60]. Expanding these integral expressions to first order around the undimerized state, we derive the following equation for the dependence of β_{nm} on the bond length distortions $\Delta r_{nm}(t)$ and the torsion angle θ_{nm} between the 2p$_z$-orbitals on sites n and m along the bond axis:

$$\beta_{nm}(t) = \cos(\theta_{nm}(t))[\beta_0 - \alpha \Delta r_{nm}(t)] \tag{8.13}$$

The second factor in this expression is exactly identical to the SSH expression [19], where, β_0 is the reference resonance integral and α the electron–phonon coupling constant. By deriving this factor from the Mulliken approximation it follows that the values of β_0 and α are dependent on each other and can be obtained from the two parameterized functions

$$\beta_0 = A(15 + 15a\zeta + 6(a\zeta)^2 + (a\zeta)^3) \tag{8.14}$$

$$\alpha = Aa\zeta^2(3 + 3a\zeta + (a\zeta)^2) \tag{8.15}$$

where $A = k(e^{-a\zeta}/15)$, a is the undimerized bond length distance, and $\zeta = 3.07\,\text{Å}^{-1}$. In the case of PPV, for which $k = 11.04\,\text{eV}$ and $a = 1.4085\,\text{Å}$ [61], the numerical values of β_0 and α are 2.66 eV and 4.49 eV/Å, respectively. These values are close to those presented in the original SSH model [19]. It should be noted that in the literature we can find many different values of these parameters, in particular for α. Our experience is that it is possible to obtain qualitatively similar electronic structure results for many different combinations of parameters. However, if we also include properties that also include the σ systems, such as the frequencies of the normal modes, the parameter space becomes considerably more restricted (see discussion concerning the σ-energy parameters in Section 8.2.4 below). Furthermore, following our treatment of the effect of torsional dynamics on the charge transport properties of PPV [62], $\theta_{nm} \neq 0^\circ$ only at those bonds which interconnect the phenylene rings with neighboring vinylene segments. We will henceforth in Section 8.3 use the short-hand notation θ_j, β_j, and Δr_j, respectively, for all θ_{nm}, β_{nm}, and Δr_{nm} associated with C–C phenylene–vinylene bonds indexed by j.

The model also takes into consideration the contribution to the Hamiltonian from an externally applied electric field, $\mathbf{E}(t)$, (in the Coulomb gauge) such that

$$\hat{H}_E(t) = -e \sum_n \mathbf{r}_n \mathbf{E}(t)[\hat{c}_n^\dagger \hat{c}_n - 1] \tag{8.16}$$

where e being the absolute value of the electron charge. If not otherwise specified, $\mathbf{E}(t)$ is directed along the long molecular axis with a smooth adiabatic turn on followed by a constant value E_0.

8.2.4
Lattice Energy

The essential elements in implementing the ion dynamics are the choice of the functional forms of the different energy contributions to the system. In the general case of three-dimensional molecules a standard classical force field potential such as CHARMM (Chemistry at HARward Macromolecular Mechanics) [63, 64] can be used to calculate the energy contribution from the σ-electrons. We use a slightly simplified version of this potential in our studies [65]:

$$V_\sigma(t) = \frac{K_1}{2} \sum_{\langle nm \rangle} (r_{nm}(t) - a)^2 + \frac{K_2}{2} \sum_{\langle nml \rangle} (\vartheta_{nml}(t) - \vartheta_0)^2 + K_3 \sum_{\langle nm \rangle} [1 - \cos(\theta_{nm}(t) - \theta_0)] \tag{8.17}$$

Together with the kinetic energy of the ions the total lattice energy is

$$E_{\text{latt}}(t) = V_\sigma(t) + \frac{1}{2} \sum_{n=1}^{N} M_n \dot{\mathbf{r}}_n^2 \tag{8.18}$$

The summations in the equation above run over unique bonds, bond angles, and torsional angles and K_1, K_2, and K_3 are the force constants associated with deviations in bond lengths $\Delta r_{nm}(t) = r_{nm}(t) - a$, bond angles $\Delta\vartheta_{nml}(t) = \vartheta_{nml}(t) - \vartheta_0$, and torsion angles $\Delta\theta_{nm}(t) = \theta_{nm}(t) - \theta_0$ from the undimerized planar structure, i.e., $\{a, \vartheta_0, \theta_0\}$, respectively.

With PPV as our model system the values of K_1, K_2, and K_3 are chosen so as to provide a ground state geometry which is as close as possible to that obtained from first principles calculations and in addition provides realistic phonon mode frequencies: K_1 is set to 37.0 eV/Å^2 which in combination with the parameters of the electronic Hamiltoninan (Eq. (8.13)) provides bond lengths and C–C stretch frequencies that are close to those obtained from first principles calculations [22]. The bond angles potential has no direct coupling to the electronic systems and the value of K_2 is determined entirely from the ground state geometry of PPV: $K_2 = 20\,\text{eV/rad}^2$ [62]. In order to obtain reference values for the frequencies and magnitudes of out-of-plane phenylene ring torsion in PPV, we employed the Tinker software package [66] and performed a set of molecular dynamics simulations at room temperature on small PPV oligomers such as *trans*-stilben. For these simulations, we used an MM3 force field and a modified Pariser–Parr–Pople (PPP) method for the self-consistent field (SCF) molecular orbital calculations for the π-system. In particular, we find that the time period for the torsion of the phenylene rings with respect to the vinylene segment in *trans*-stilben is roughly 1.4 ps. With this as a reference, we determine the value of K_3 in Eq. (8.17), which regulates the time period for phenylene ring torsion in our model, to be $K_3 \simeq 0.1\,\text{eV}$ [62].

8.2.5
Electron-Lattice Dynamics III

Having defined the constituent parts of the π-electron Hamiltonian, the σ potential and kinetic energiy of the ions it follows that Eqs. (8.9) and (8.11) are coupled via the π-electronic wavefunction Ψ_π and therefore must be solved simultaneously. There are several approximation levels that can be used to describe electron-lattice dynamics [58]. In the simplest form, the adiabatic approximation, the potential energy and the ionic forces are derived within the Born–Oppenheimer approximation based on the solution to the time-independent Schrödinger equation. In our case this approach is not valid since, as discussed above, we would like to include electron transfer by means of tunneling or thermally activated hopping events in our model. Furthermore, the dynamics can be studied either within the mean-field approximation [4, 67] or with correlation effects taken into account. In our case, since we are interested in relatively large systems studied over several femtoseconds, we are effectively restricted to the former approximation level. This approach is the standard approach in studies of electron transport, in which the energy exchange between the electronic system and the lattice is not the dominating process [67]. The mean-field approach is, however, less good in capturing the details of processes such as Joule heating [58].

The time-dependent molecular orbitals of the π-system are denoted by $|\psi_\nu(t)\rangle$, where ν is the orbital index. These orbitals are now expressed in a basis set of atomic orbitals, ϕ_n, in the following way:

$$|\psi_\nu(t)\rangle = \sum_{n=1}^{N} C_{n\nu}(t)|\phi_n\rangle \tag{8.19}$$

where $C_{n\nu}(t)$ are the time-dependent expansion coefficients in the linear combination of atomic orbitals.

The mean-field approach discussed above corresponds to that the electrons influence the ions via the time-dependent electron density

$$\rho_{nm}(t) = \sum_{\nu=1}^{N} C^*_{n\nu}(t) f_\nu C_{m\nu}(t) \tag{8.20}$$

where $f_\nu \in [0, 1, 2]$ is the time-independent occupation number of the νth time-dependent molecular orbital. Using the generalized Hellmann–Feynman theorem for the ionic forces [4], Eq. (8.11) then resolves into

$$M_{n'}\ddot{\mathbf{r}}_{n'}(t) = -\nabla_{\mathbf{r}_{n'}} V_\sigma - \sum_{\substack{\nu=1\\ n,m=1}}^{N} f_\nu C^*_{n\nu}(t)\, \langle\phi_n|\nabla_{\mathbf{r}_{n'}} \hat{H}(t)|\phi_m\rangle\, C_{m\nu}(t) \tag{8.21}$$

where the expansion coefficients $C_{n\nu}(t)$ are obtained from the following equation derived from Eq. (8.9):

$$i\hbar\dot{C}_{n\nu}(t) = -e\mathbf{r}_n\mathbf{E}_0(t)C_{n\nu}(t) - \sum_{m\in\langle nm\rangle} \beta_{nm}(t)C_{m\nu}(t) \tag{8.22}$$

In the simulations of excited state dynamics, energy is inserted in the system in the form of electron excitations. In this situation, the excess energy is transferred from the electronic system to kinetic energy of the ions and we need to include additional degrees of freedom to accommodate the excess energy. This is done by introducing a fictitious coupling to the surrounding via a viscous force acting on the ions in the system: $-\gamma\dot{\mathbf{r}}_{n'}(t)$. This force is added to Eq. (8.21) in the simulations presented in Section 8.3.4. We also make use of the lattice energy to include temperature in the simulations by initiating the simulations with a nonzero kinetic energy. In principle, the same procedure as that used in ordinary molecular dynamics simulations can also be applied here, even though so far we have not emphasized this in the development of the method.

The coupled differential equations (8.21) and (8.22) are solved numerically using a Runge–Kutta method of order 8 with step-size control [68], which in practice means a time step of about 10 as. Furthermore, we use a "global time step" of 1 fs and take as the starting wavefunction the solution to the time-independent Schrödinger equation of the atomic configuration at $t = 0$ fs.

In order to study also the nonadiabaticity of the charge transport process, we expand the time-dependent molecular orbitals $|\psi_\nu(t)\rangle$ in a basis of instantaneous eigenstates of the π-electron Hamiltonian, $|\varphi_\mu\rangle$,

$$H_\pi|\varphi_\mu\rangle = \varepsilon_\mu|\varphi_\mu\rangle \tag{8.23}$$

In the equation above the explicit time dependence is suppressed even though the solutions at different instances of course differ from each other via the time evolution of the π-electron Hamiltonian (see Eq. (8.22)). The linear expansion is expressed as

$$|\psi_{\nu}(t)\rangle = \sum_{\mu=1}^{N} \alpha_{\nu\mu}(t)|\varphi_{\mu}\rangle \tag{8.24}$$

where $\alpha_{\nu\mu}(t) = \langle\varphi_{\mu}|\psi_{\nu}(t)\rangle$ is the overlap between the time-dependent molecular orbitals and the instantaneous eigenstates. This overlap is unity if the dynamics follows a single Born–Oppenheimer potential energy surface. However, if two instantaneous energy eigenvalues, ε_{μ} and ε'_{μ}, are close to each other, the time-dependent orbital can very well be a linear combination of these two electronic instantaneous orbitals (and no longer an eigenstate to H_{π}). To resolve such a behavior we introduce the time-dependent occupation number, $n_{\mu}(t)$ of the instantaneous eigenstates μ as [67]

$$n_{\mu}(t) = \sum_{\nu=1}^{N} f_{\nu}|\alpha_{\mu\nu}(t)|^{2} \tag{8.25}$$

Again, in the case of adiabatic dynamics, $\alpha_{\mu\nu}(t)$ is diagonal and the time-independent and time-dependent occupation numbers are identical. If a (nonadiabatic) hopping event occurs there is a rapid change in $n_{\mu}(t)$ of the donator orbital from unity to zero and a corresponding change in the acceptor occupation number from zero to unity. In the result section below we will point at these different events that occur during the dynamics.

8.3
Results

In this section we present results from simulations that we have performed using the ELD approach. As stated above, we have limited the presentation to PPV systems and include below the cases of interchain and intrachain polaron transport as well as recent results of the effect of torsional dynamics. In the last part of this section we discuss the dynamics following electronic excitations across the HOMO–LUMO bandgap. The simulations include both adiabatic and nonadiabatic dynamics, where the nonadiabatic effects are monitored by the time evolution of the occupation number given by Eq. (8.25).

8.3.1
Intrachain Polaron Dynamics

The ground state of a single PPV chain with one excess electron (or hole) contains a polaron lattice defect to which the excess charge is localized. The extension of the polaron is about 4–5 phenylene–vinylene units. When an electric field is

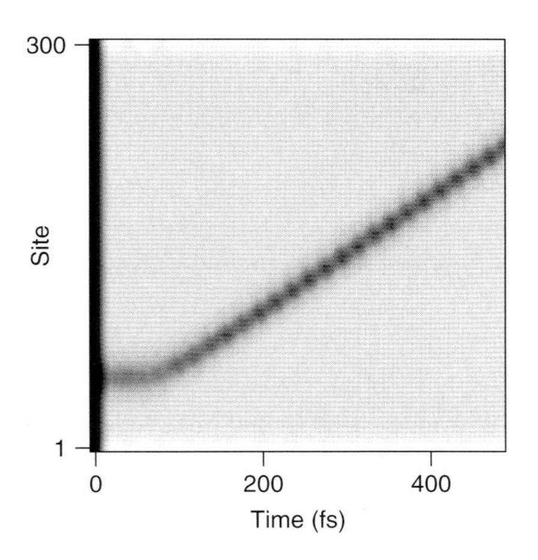

Figure 8.5 Time dependence of the charge density associated with a polaron moving in a single chain, $E_0 = 1.0 \times 10^5$ V/cm.

turned on the polaron experiences a slope in the potential energy surface and starts to move in the direction opposite to (along) the electric field. The dynamics of this process is depicted in Figure 8.5. In a single ideal polymer chain without defects or disorder, and if the field is switched on slowly (adiabatically) this motion occurs on a single potential energy surface. The surface has some barriers for motion since the dynamics involves displacement of the central part of the polaron from one C atom to the next. However, since the polaron is extended over several carbon atoms, these barriers are very small compared to both temperature fluctuations and the kinetic energy of the polaron, and in practice the polaron moves at a velocity which is determined by the field strength. From simulations at finite temperatures we observe very clearly how these phonons interact with the polaron and some of the phonon energy is transferred to kinetic energy of the polaron. When the polaron is moving under the influence of the external electric field, there is a constant dissipation of energy from the electronic system to the lattice in the form of a "bow-wave" that moves with the velocity of sound away from the polaron toward the ends of the polymer chain. If the electric field is switched off, however, this energy dissipation stops and the polaron moves up and down the polymer chain bouncing at the chain ends with very little energy dissipation.

The polaron drift process has some interesting features. In particular we have noted that [69] the velocity can exceed the sound velocity. This process is explained by the fact that the acoustic phonons that are associated with the polaron at subsound velocities, and that correspond to a local contraction of the polymer chain in the region of the polaron, can be left behind when the polaron moves along the chain. The supersonic polaron is therefore considerably lighter than the polaron at rest and also considerably more localized. In the case of trans-polyacetylene, we have calculated the polaron width to be reduced by a factor around 3 when it undergoes the transition into supersonic velocities which occurs at a field strength of about 1.3×10^4 V/cm.

8.3.2
Interchain Polaron Dynamics

We then turn to a system consisting of two chains placed besides each other and overlapping by 50 C sites. The first chain consists of 140 sites and contains the polaron at the start, the other chain is a perfectly dimerized 300 site chain. The orthogonal hopping integral, i.e., the hopping between a site on one chain and a nearest-neighbor site on an adjacent chain, was set to $t_{\perp} = 0.1$ eV. The diagonal hopping integral, i.e., the hopping between next nearest neighbors on adjacent chains, was set to $t_d = 0.05$ eV. Simulations were carried out for a number of different field strengths ranging from 10^4 to 6×10^5 V/cm. Three regions of markedly different behavior of the polaron dynamics are displayed in Figure 8.6, representative for field strengths in three different regions. In all cases the polaron starts to move on the short chain with constant speed as discussed above. When it reaches the end of chain 1, i.e., the region where the chains interact, the polaron (both charge and distortion) gets stuck for fields lower than $E_0 = 8 \times 10^4$ V/cm, as shown in Figure 8.6(a). For a slightly higher field strength, $E_0 = 9 \times 10^4$ V/cm, the polaron jumps over to the second chain, but it has too low energy to continue moving with constant velocity and just oscillates at the beginning of the chain (this behavior is not displayed in Figure 8.6). At a field strength of 1×10^5 V/cm the polaron remains at the first chain end for about 100 fs (see Figure 8.6(b)). It oscillates back and forth due to the potential of the electric field and the barrier energy barrier caused by the chain end. After this "waiting time" it has gained enough energy from the electric field to jump over to chain 2. It recollects and continues to move as a polaron along the second chain, see Figure 8.6(b) – first at a slightly reduced velocity as compared to the velocity on chain 1 but after some time it gains velocity and eventually behaves exactly as on the first chain. Thus our simulations show that interchain polaron transport is possible at sufficiently high field strengths. This behavior persists up to a field strength of about 3×10^5 V/cm but with shorter and shorter waiting times.

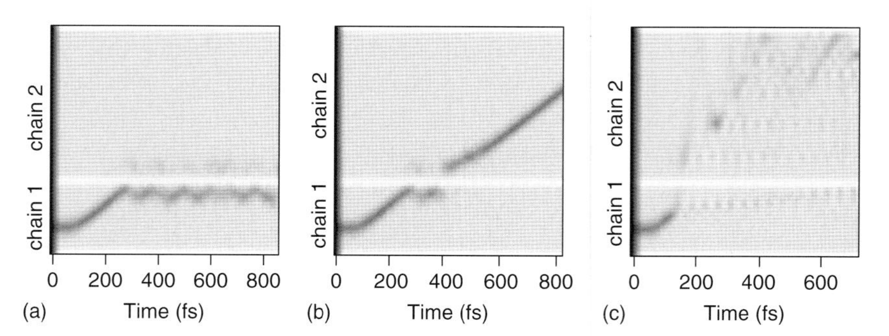

Figure 8.6 Time dependence of a polaron moving in a two-chain system with different electric field strengths E0. (a) Region I: $E_0 = 0.8 \times 10^5$ V/cm; (b) Region II: $E_0 = 1 \times 10^5$ V/cm; (c) Region III: $E_0 = 3 \times 10^5$ V/cm.

Now focus on the same system but for high fields, $E_0 \gtrsim 3 \times 10^5$ V/cm (see Figure 8.6(c)). Like in the previous case the polaron jumps over to chain 2, leaving a trace of phonons behind on chain 1. However, the charge appears to be delocalized on chain 2, having difficulty to recollect. It is possible to see that some localized lattice distortion is traveling along the chain and is scattered at the chain end but this can no longer be regarded as a polaron. Clearly, in this case the excess energy in the system creates too much disturbance, e.g., in the form of phonons in order for the polaron to be formed on the second chain. In a real physical system, e.g., a OLEDs or OFETs, excess energy will to some extent be transported away in the form of heat, which might lead to a stabilization of the polaron at slightly higher field strengths than 3×10^5 V/cm. For even higher fields, about 4×10^5 V/cm the polaron starts to dissociate already on the first chain as described for the single-chain system above.

In Figure 8.7, we present polaron dynamics in a three-chain system, each chain being 140 sites long. The first two chains are placed as described above, the third and second chains are spatially related to each other as the second chain to the first. The chain overlaps are again 50 sites. We study the system with the same intra- and interchain hopping as discussed above and for the electric field strength we choose to study the intermediate region, e.g., $E_0 = 2 \times 10^5$ V/cm, see Figure 8.7.

Even in this case we see that the polaron jumps nicely from chain 1 to chain 2 and from chain 2 to chain 3. The fact that it actually manages even the second jump indicates that the polaron on the second chain is not different from the initial one.

8.3.3
Torsional Dynamics

The next step in the simulations comprise the impact of torsion dynamics on the charge transport process along the PPV chain [62]. As pointed out by Papanek *et al.* [25], almost all vibrational modes below $\sim$322.62 cm^{-1} can be attributed to

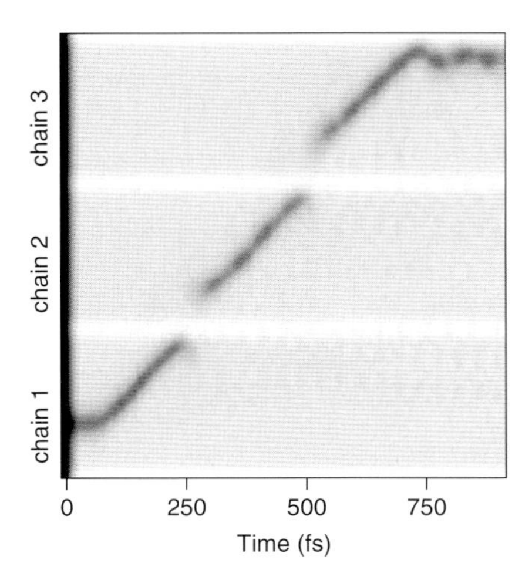

Figure 8.7 Time dependence of a polaron moving in a three-chain system, $E_0 = 2 \times 10^5$ V/cm.

ring librons. In particular, they found that both the phenyl rings and the vinylene segments can be regarded as almost rigid units in this low-frequency regime and that the vibrational modes involve the rotations about and bending of the C–C bonds connecting the phenylene and the vinylene units. In our simulations we keep both the phenyl ring and the vinylene units planar and only allow for torsions around the bond connecting these two units (see Figure 8.4 above). Such slow ring motions can occur at the same timescale as charge transport which makes studies of torsional dynamics in combination with charge transport particularly interesting.

Another important observation by Papanek *et al.* [25] is that the motion of the phenylene rings is decoherent. This feature can be introduced into our system by initiating the system in a coherent state, e.g., with $\theta_j(0) \neq 0^\circ$ and $\theta_0 = 0^\circ$, and wait for the system to reach the decoherent state. This process is, however, extremely time consuming and thus not suitable even for very small systems. Instead, we chose to initiate the system with all phenylene ring torsions fixed to the same value ($\{\theta_j(0)\} = 0^\circ$, 10°, and 20°) and let the torsion dynamics for each individual phenylene ring start randomly during a certain period of time (typically 900 fs). By means of this deterministic initiation procedure, the time evolution of phenylene ring torsion in the system, and thus also the resonance integral strength (see Eq. (8.13)), will be uniquely defined. It should be emphasized, however, that the charge transport processes observed in systems with different random sequences in the order in which the torsions start do not differ qualitatively from each other, wherefore the results presented and discussed below represent general features of this type of system.

For future references, it should also be pointed out that the kinetic energy in our system is slightly lower than in real PPV oligomers due to the restriction of keeping all torsion angles other than $\{\theta_j\}$ at 0° (or 180°). The fact that not all degrees of freedom are activated means that we cannot calculate the temperature directly from the average kinetic energy. Instead, we have made comparison with molecular dynamics simulations on PPV oligomers. These simulations show a standard deviation in the distribution of $\{\theta_j\}$ of 24° at room temperature [70], which is in good agreement with the large amplitude initial value of $\{\theta_j\}|_{t=0} = 20^\circ$ used in this work. The smaller amplitude value of $\{\theta_j\}|_{t=0} = 10^\circ$ also studied here thus corresponds consequently to low temperatures.

The chain is charged with one additional electron in the ground state polaronic configuration, but without an external electric field applied (unbiased system). In Figures 8.8(a)–(c) are shown the results for PPV oligomers with 31 rings and initial torsion angles, $\{\theta_j\}|_{t=0}$, at 0°, 10°, and 20°, respectively. The left panels display the time evolution of the net charge density per monomer, i.e., the charge density associated with the polaron, and in the right panels the corresponding dynamics of the occupation number (Eq. (8.25)) of the instantaneous eigenstates are shown.

In the case of $\{\theta_j\}|_{t=0} = 0^\circ$, the polaron is stable and resides at the center of the chain as expected for a finite-sized system. The presence of an acoustic phonon bouncing back and forth through the system does, however, introduce a periodic

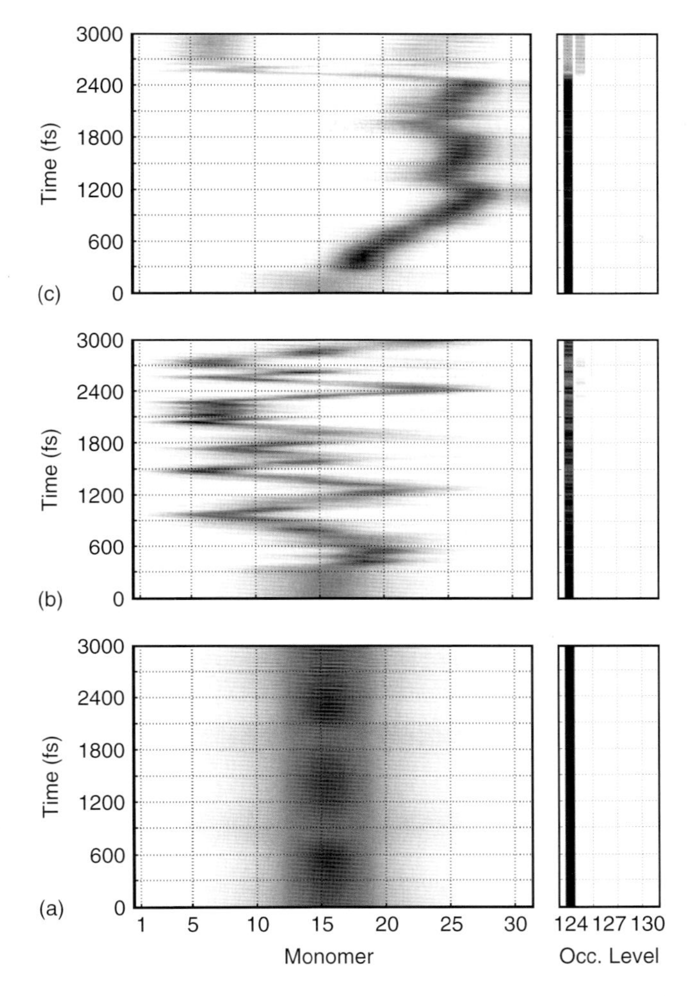

Figure 8.8 In the left panels of (a), (b), and (c) are shown the time evolution of the density of charge per monomer in PPV oligomers with 31 phenylene rings and identical onset procedures for ring torsion dynamics but with different initial torsion angles at $\{\theta_j\}|_{t=0} = 0^\circ$, 10°, and 20°, respectively. In the right panels of (a), (b), and (c) are shown the corresponding time evolution of the occupation number of the instantaneous eigenstates.

fluctuation in the density of charge. The width of the polaron agrees with that observed in an earlier work [32].

A dramatic change in the behavior of the system occurs when the torsion dynamics is initiated. In the left panels of Figures 8.8(b) and (c) for the systems with $\{\theta_j\}|_{t=0} = 10^\circ$, and 20°, we no longer observe a polaron resting at the middle of the PPV chain but rather a diffusive propagation of the charge carrier spurred into motion by the presence of ring torsion motion. We also find that the polaron motion is much more restricted in the system with $\{\theta_j\}|_{t=0} = 20^\circ$ than for $\{\theta_j\}|_{t=0} = 10^\circ$ and that it moves more slowly through the former system due to the greater impact of ring torsion on the reduction of the resonance integral strength, β_j (see Eq. (8.13)).

Another important observation is that the polaron is considerably more localized when $\theta_j(0) \neq 0^\circ$ compared to the completely planar system. For $\{\theta_j\}|_{t=0} = 20^\circ$ the polaron extends over 2–3 phenylene–vinylene units only. This is a consequence of the increase in total electron–phonon coupling which results from the cosine-term in Eq. (8.13) (in addition to the usual SSH coupling to bond-length changes [19]). We also note that the polaron can be destabilized by the motion of the phenylene rings. This is the case at $t \sim 2600$ fs in the system with $\{\theta_j\}|_{t=0} = 20^\circ$ and we observe in the left panel of Figure 8.8(c) a delocalization of charge to two different regions of the system. From the right panel of Figure 8.8(c) it is clear that this is a nonadiabatic event with the simultaneous occupation of two instantaneous eigenstates. Multiple level occupation of this kind is also observed in the occupation spectrum for the system with $\{\theta_j\}|_{t=0} = 10^\circ$, but correlates in this case to events when the charge carrier breaches potential energy barriers introduced through the dynamics of ring torsion.

To obtain a more detailed picture of the dynamics that governs diffusive charge transport processes, we reproduce in Figure 8.9(a) the time evolution of the density of charge per monomer displayed already in the left panel of Figure 8.8(c) for a PPV oligomer with $\{\theta_j\}|_{t=0} = 20^\circ$ (corresponding to room temperature) and in the right panel the corresponding dynamics of the resonance integrals, β_j, across each interconnecting bond between a phenylene ring and a vinylene group in that system. Also displayed Figure 8.9(b) is the superimposed trace of the center of the local density of charge. Keeping in mind that β_j does not change in time before the onset of ring torsion and that the darker regions in Figure 8.9(b) represent weak resonance integrals, it is obvious by following the trace of the polaron that the charge carrier is localized by the dynamics of the lattice to regions with consistently large values of the resonance integrals. According to Eq. (8.13) the modulation of β_j is governed by the dynamics of both θ_j and the variations in the interatomic distance across the associated bond. However, when comparing the time evolution of these three quantities it is found that the modulation of β_j due to bond length variations Δr_j is less pronounced than that for θ_j and of a much higher frequency than that observed in the right panel of Figure 8.9. The correlation between the dynamics of θ_j and β_j is, however, very strong. We therefore conclude that the intrachain diffusive charge transport process in PPV oligomers is controlled by the dynamics of ring torsion and that the bond length vibrations are of less importance for the mobility of the polaron. Note though that these vibrations are responsible for high-frequency shifts in the position of the center of the polaron between two neighboring monomers, which are also visible in Figures 8.8 and 8.9.

As discussed above, in the model we are using a value of K_3 (and therefore also the torsional frequency) determined on the basis of MD simulations. In the literature there are reports of different frequencies [25, 56] of the normal modes associated with the torsional degrees of freedom. In order to incorporate a wider range of possible torsional frequencies in our study we have performed simulations similar to those shown in Figure 8.9 but with different values of K_3 in the range

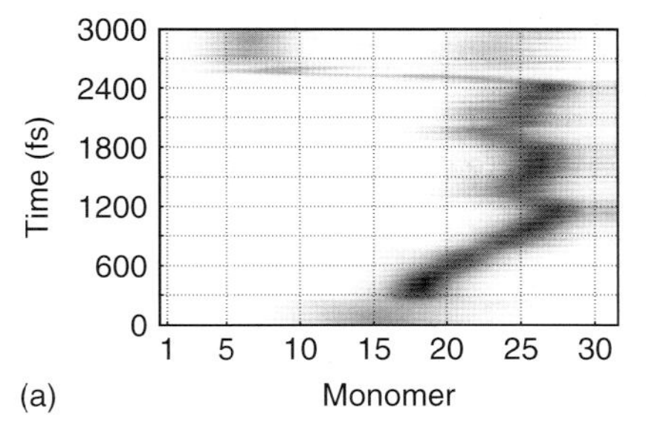

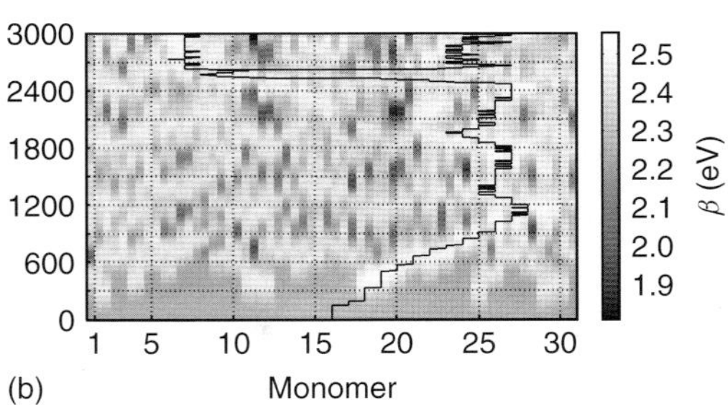

Figure 8.9 (a) It shows the time evolution of the density of charge per monomer in a PPV chain with 31 phenylene rings and initial torsion angles at $\{\theta_j\}|_{t=0} = 20^\circ$. The net charge density is depicted in grayscale. (b) It displays the resonance integral strength, β, across each bond between a phenylene ring and a vinylene group in this system. The grayscale to the right shows the value of β in electron volt. Note that the dark regions represent low values of β, which act as barriers for charge transport. The solid line (b) indicates the position of the center of the local density of charge in (a).

[0.01, 1.0] eV. The results from these simulations show that it is only the timescale of the system dynamics that changes with the value of K_3; the lower the torsional frequency, the longer time it takes for the charge to traverse the system, but the principal dynamical behavior remains the same.

Due to the close correspondence between β_j and θ_j, the right panel of Figure 8.9 also provides a fair representation of the ring torsion dynamics in the system. We note that while some ring torsion angles are suppressed others are enhanced such that the maximum torsion angles observed during the dynamics of the system actually reaches as high as 35° (given an initial torsion angle $\{\theta_j\}|_{t=0} = 20^\circ$), which, e.g., is the case for θ_{40} and θ_{41} of the 20th phenylene ring at $t \simeq 2.2$ ps. A similar behavior is also observed in the molecular dynamics simulations [70].

Following the polaronic trace superimposed on the dynamics of β_j in the right panel of Figure 8.9, we have identified two mechanisms that influence the diffusive motion of polaronic charge carriers in this type of systems. The first is the evolution of regions with low values of β_j that act as potential energy barriers for polaron propagation. The second is the evolution of regions along the chain where the resonance integrals are consistently strong and toward which the polaron is attracted. The interplay between these two mechanisms is such that the polaron will localize to the region of the chain with the highest values of β_j provided that the charge carrier is able to breach the intersecting potential energy barriers that arise due to ring torsion. Such regions can be regarded as traps for the charge carrier. One way to make jumps out of the trap more probable is to introduce an electric field. For this purpose we repeat the previously detailed simulations associated with the results presented in Figure 8.8 but with the external electric field supplied in accordance with Eq. (8.16).

The left panels of Figures 8.10(a) and (b) show the time evolution of the density of charge per monomer for two PPV oligomers with identical initial state configuration and onset conditions to the system in the left panel of Figure 8.9. The center panel shows the time evolution of the occupation number of the

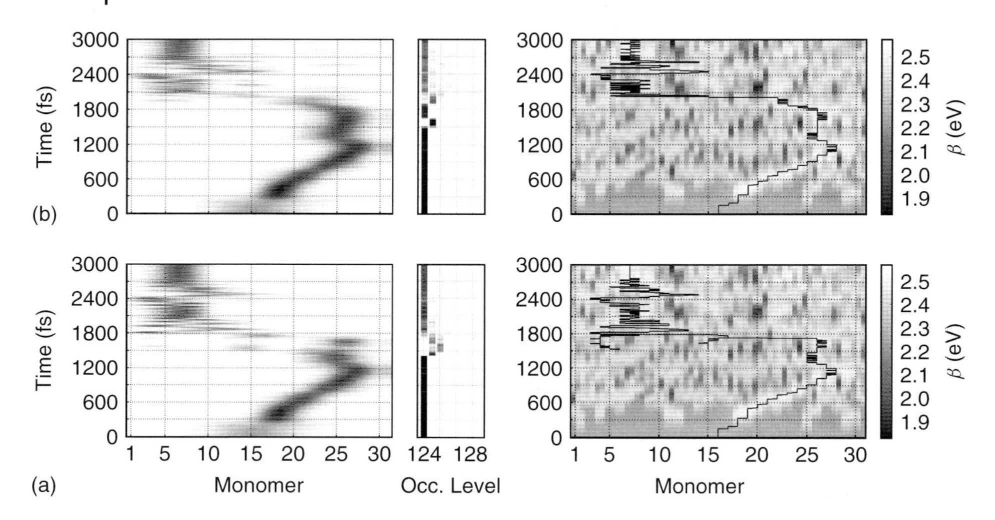

Figure 8.10 The left panels of (a) and (b) show the time evolution of the density of charge per monomer in two PPV chains with 31 phenylene rings and initial torsion angles at $\{\theta_j\}|_{t=0} = 20^\circ$, but with an external electric field ($E_0 = 5.0 \times 10^4$ V/cm^2) switch on smoothly at $t_s = 1400$ fs over a time period of 25 fs (a) and 200 fs (b), respectively. The solid horizontal lines indicate t_s and t_f. Note that in (a) these two times appear as one thicker line due to the short switch-on time. In the center panels are shown the corresponding time evolutions of the occupation number per molecular level. The LUMO of the neutral system is molecular level 124, followed by LUMO+1 (125), LUMO+2 (126), etc. The right panels show the time evolutions of the resonance integral strength across each bond between a phenylene ring and a vinylene group. The grayscale to the right shows the value of β in eV. Note that the dark regions represent low values of β, which act as barriers for charge transport. The solid lines in the right panels indicate the positions of the center of the local density of charge in the left panels.

instantaneous eigenstates associated with the two systems and the right panel the time evolution of β_j.

The external electric field is switched on adibatically at 1400 fs and reaches its maximum amplitude $E_0 = 5.0 \times 10^4$ V/cm at 1600 fs as indicated by the solid horizontal line. After this switch-on period the electric field strength is kept at $E_0 = 5.0 \times 10^4$, a field strength which represents typical values in molecular electronic devices [42].

With reference to the dynamics of β_j displayed in the right panels of Figure 8.10 this means that the field is introduced at a point in time when the charge is localized to a narrow region around the 26th monomer. The charge carrier approaches a region where the torsion of phenylene rings is so strong and the corresponding potential energy barriers consequently so high that not even the assistance of the electric field will enable the charge carrier to breach the barrier at a certain point in time. Eventually, though, the barrier height is reduced by the torsion dynamics of the rings and the charge carrier will be able to continue to propagate through the system. For the simulation displayed in Figure 8.10 this occurs at $t = 1900$ fs. However, the propagation is not an adiabatic polaron drift process. Instead we observe a temporary destabilization of the polaron and a corresponding change in the occupation from level 124 (which is the LUMO level of the neutral system) to level 125. Thus, there is a clear signature of a nonadiabatic charge transfer process in this case. The destabilization of the polaron is due to the fact that the acceptor level (level 125) initially is quite delocalized. We also notice that this state, when it becomes occupied, stabilizes due to the electron–phonon coupling and after around 200 fs (at $t = 2100$ fs) it crosses the donor level (level 124) and becomes the lowest occupied molecular orbital.

The fact that the charge carrier transverses the PPV chain despite the presence of barriers caused by ring torsions shows that the hopping contribution to the charge transport plays an important role. Intrachain charge transport in the presence of an external electric field is obviously not dramatically reduced by the disorder caused by the dynamics of the ring torsions.

Correlating the dynamics of the time-dependent occupation number displayed in the center panels of Figures 8.10 with the dynamics of the charge density (left panels) and the resonance integrals (right panels), we find that the actual transition across the barrier involves a situation where the charge density and the occupation number split between multiple regions and multiple instantaneous eigenstates, respectively. Such nonadiabatic transitions are likely to increase in numbers with increased amplitudes of ring torsion since the number of barriers with a height that require the assistance of the electric field for the charge carrier to be able to traverse the system then also increases. In other words, multiple occupancy of instantaneous eigenstates is more of a commonality in systems with higher magnitudes of ring torsion angles than in those systems where θ_j is lower, provided that the field is sufficiently strong for the charge carrier to actually breach the potential energy barrier induced by the dynamics of ring torsion.

8.3.4 Dynamics in the Photoexcited State

The simulations presented in this section describe the evolution of the PPV system following an excitation of the electronic system. The most relevant initial state geometry to use in this type of simulation is the equilibrium geometry of the ground state of the system. We will also discuss the effect of an initial state in which some lattice vibrations have been excited, which simulates the effect of a finite temperature. Note that the effects of the torsion dynamics have been omitted in these simulations, although, according to the previous discussion most certainly will affect the details dynamical behavior of the system. Simulations of this type will, however, be performed in the near future.

The first system to study here is a single PPV chain with 20 repeat units (158 carbon atoms) [30]. This is a typical conjugation lengths in a PPV compound [47]. The results from the simulation, in terms of the calculated changes in the bond lengths along the PPV chain as a function of time, are shown in Figure 8.11. The simulation starts (at $t = 0$) with a vertical electronic excitation from the HOMO to the LUMO. This excited state is initially delocalized over essentially the whole

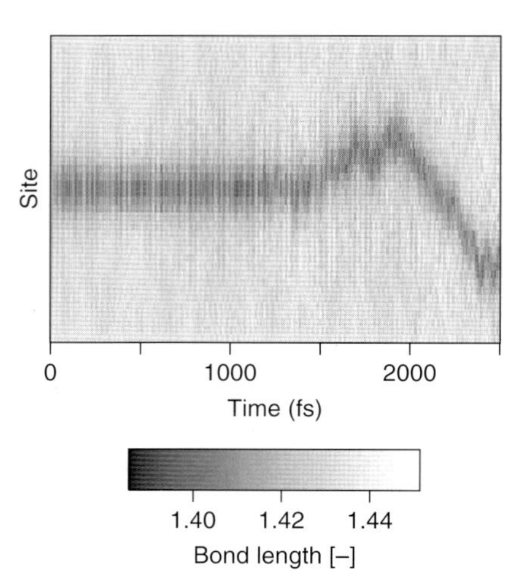

Figure 8.11 Time evolution of the bond length order parameter for a 20 PPV unit long chain excited from HOMO to LUMO.

system. After approximately 15 fs, the excited electron together with the hole in the HOMO level localizes to the center of the chain and form a polaron exciton with a width of approximately 4 phenylene–vinylene repeat units. This extremely fast process has not been resolved experimentally, but it was concluded by Kersting *et al.* [71] that the vibronic relaxations following an electron excitation is much faster than the sub-ps timescale which they could resolve. The dark region in the middle of the chain defines the region of the polaron exciton in which the chain is deformed and the electron–hole pair resides. The excess energy that is released from the polaron exciton when it undergoes the rapid structural relaxation is transferred to local acoustic lattice vibrations that move up and down the polymer chain with the velocity of sound.

During a first period of about 1.5 ps essentially no movement of the polaron exciton is observed. There is however a considerable internal dynamics, the position of the electron and hole wavefunctions change from side to side of the lattice defect. The atoms within the region of the polaron exciton are also vibrating but these vibrations are trapped by the exciton [72] and there is no collective movement of the defect. After approximately 1.5 ps, however, the exciton suddenly starts to move along the chain with a velocity of ~0.2 Å/fs. As suggested by Onsager [73], this motion follows a stochastical pattern, very similar to a Brownian type of motion and is caused by scattering of the polaron exciton by the phonons that exist outside the region of the exciton.

To further investigate the effect of these external phonons we let the chain contain some excess initial energy. Here, the excess energy is introduced in the form of random initial displacements of the individual carbon atoms (but could equally well have been introduced via an initial random kinetic energy associated with each atom). The total excess energy which is added to the system in this way is 1 eV or about 6 meV per carbon atom. This extra energy causes the exciton to move within a shorter period of time after the excitation, in this case in ~900 fs. The bond length order parameter displayed in Figure 8.12 shows a very clear

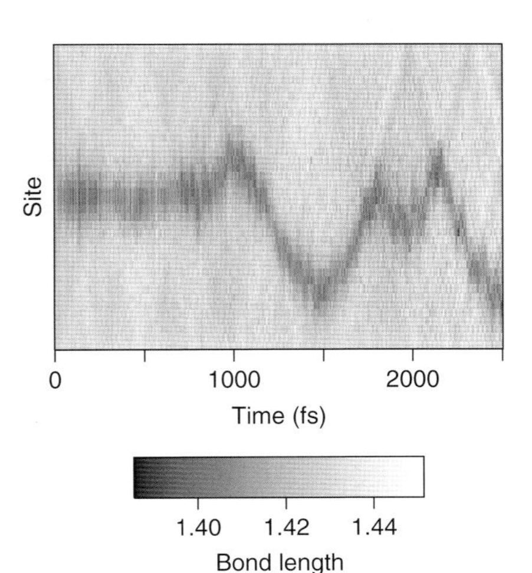

Figure 8.12 Time evolution of the bond length order parameter for a 20 PPV unit long chain excited from HOMO to LUMO with extra lattice energy.

picture of a random motion of the polaron exciton up and down the chain. It is also clear from this picture that the amplitudes of the lattice vibrations are much larger than that in the previous case. Some of the vibrations are quite localized and originate from the region of the creation of the exciton. In our representation of the order parameter, the brighter regions correspond to an expanded segment of the polymer chain. These regions bounce back and forth over the chain with a speed of approximately 0.24 Å/fs, which for parameter values used here corresponds to the sound velocity in the system.

If we take the next symmetric excitation, the HOMO-1 to LUMO+1 level, the exciton is more delocalized over the chain (see Figure 8.13). There are two regions along the chain to which the exciton initially is localized. After approximately 400 fs these two regions merge into one. This transition is associated with an electronic transition from LUMO+1 to LUMO of the excited electron and a corresponding transition of the hole from HOMO-1 to HOMO [74, 75]. If we go further and excite

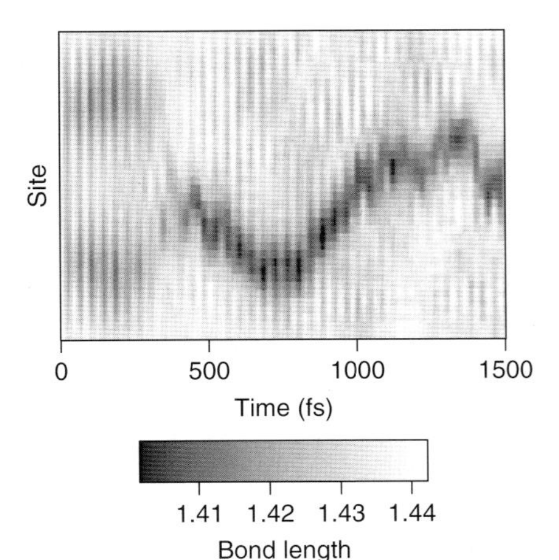

Figure 8.13 Time evolution of the bond length order parameter for a 20 PPV unit long chain excited from HOMO-1–LUMO+1 excitation.

from HOMO-2 to LUMO+2 we get three regions in which the exciton initially exists (not shown here). The excited state has a life-time of about 150 fs before it relaxes into the HOMO to LUMO excited state. The initial shape of the exciton reflects the symmetry of the wavefunctions, which affect the geometry according to Eq. (8.21). The shorter lifetime of the exciton is a result of a lower energy barrier for scattering out of this state. The HOMO-3–LUMO+3 excitation first relaxes into the HOMO-1–LUMO+1 excitation, which takes about 100 fs, and then reach the final excitonic state of the HOMO–LUMO excitation.

The polaron exciton generated by the HOMO-1 to LUMO+1 excitation starts to move directly after this final transition (see Figure 8.13). In this case, the energy released by the electron during the relaxation process generates enough phonons to scatter the polaron exciton and give it a nonzero velocity. Note that complete relaxation to the ground state is not allowed in our model due to symmetry restrictions. However, this dipole-allowed transition is very slow, of the order of microseconds [76], compared to the timescale of the relaxation processes studied here. Some transitions between higher lying states are also only dipole allowed which might affect the relaxation process slightly. However, we believe that our simulation gives a qualitatively correct description of the fast excitation dynamics of PPV.

Above we have discussed the relaxation process in terms of electronic relaxation and generation of lattice vibration, i.e., thermalization of hot excitations. The vibrations that couple most strongly to the electronic system are of course those that affect the bond lengths. The numerical values of these frequencies are, as discussed above, dependent on the parameters that are used in the Hamiltonian (see Sections 8.2.3 and 8.2.4). In our model we have not optimized these parameters to fit experimental data for these frequencies. Experimentally, these most predominant C–C stretching modes of PPV are found in bands around 1330 and 1628 cm^{-1} [77]. The difference between our results and the experimental data is partially due to the fact that we do not include the hydrogen atoms explicitly. Therefore, any contribution to the vibrational modes coming from degrees of freedom associated with these atoms will be absent in our model. The effect of these errors in vibrational frequencies on the results of our simulations is difficult to estimate. Since the frequencies are too large in our model the energy stored in the lattice is also larger than that in the real system. This indicates that the energy transfer from the electronic system, in the case of hot excitations, is too fast in our simulations.

In order to include the effect of interchain interactions into our simulations we have performed studies on a system consisting of two identical PPV chains, both with a length of 20 repeat units. The two chains have an overlap of five repeat units and are electronically coupled to each other via a nearest-neighbor interaction set to 0.1 eV.

The initial HOMO to LUMO excitation for the two chain system is distributed over both chains and localized to the region of the overlap. This localization reflects the spatial distribution of the HOMO and LUMO orbitals. After a very short time period, about 20 fs, the exciton is transferred to one chain but still localized to the overlap region. With two identical chains the localization to a single chain is arbitrary. Otherwise, with two chains of different lengths (20 and 16 repeat units in our simulation), the exciton appears on the longer chain. Similarly, when exciting from HOMO-1 to LUMO+1 we get an excitation localized to two different regions, one in the middle of each chain. After ~400 fs the excitation relaxes down to the HOMO–LUMO polaron exciton. In contrast to the case of a single chain, the energy of the lattice vibrations is not enough to scatter the exciton out of the potential well of the overlap in any of these two cases. We need an excitation from HOMO-2 to LUMO+2 to get enough excess lattice energy to create a mobile exciton.

8.4 Summary

In this chapter we have presented the ELD method together with a few examples that illustrate its features and the kind of information that is obtained from simulations based on this method. The main advantage of the ELD method is that it applies directly to regimes of interactions that relevant for many types of polymeric and molecular-based systems, i.e., reasonably well-ordered systems that provide either adiabatic paths for polaron transport or nonadiabatic paths with rather large overlap between the initial and final states. Such paths are either based on the tunneling across energy barriers or hopping over energy barriers that are small compared to the energy fluctuations caused by temperature effects. The nuclear dynamics is therefore treated classically and the equation of motion is obtained from the Ehrenfest methodology, which corresponds to a mean-field approximation in which the ions move in the potential field due to the average charge density.

Using modern computers we can perform simulations of systems containing up to approximately 500 carbon atoms and run the simulations during a period of time that is relevant for this systems, i.e., the time of flight for a polaron to traverse the system. Even though these are rather extended systems it is difficult to obtain data that are directly comparable to experimental observation on macroscopic systems. Thus, the main goal of future work is to scale up the calculations to even larger systems, in order to make the simulation results more useful by means of a direct experimental comparison. However, at the stage in the development of the method that is summarized here we have reached a number of important conclusions concerning polaron transport at the microscopic level. In particular we have pointed at the features of transport that are the main limiting factors for transport in highly ordered systems, namely the barriers caused by chain interruptions and ring torsions. In particular, we have identified the regions of large overlap between the neighboring units of the polymer chain as polaron traps. The possibility to cross the barriers that surround these traps is highly dependent on the field strength and it is likely that the polaron actually destabilized into an extended state during transport. Thus, charge transport can be described in a process in which the charge carriers from time to time are trapped as a polarons in regions of well-ordered chains. During the nonadiabatic transport between such regions the carrier is in a short-lived high-energy state that does not couple strongly to phonons. It should be noted that this picture of transport does not include situations for which the charge carrier has to hop across regions of very low electron densities. Such processes are very slow compared to the timescales studied here and therefore apply to highly disordered polymeric systems.

Finally, we would like to point out that ELD can be used for studies of several different dynamical processes in molecular or polymer-based systems. In addition to the polymeric systems discussed here we have studied both charge transport and internal conversion in pentacene molecular systems [35, 75] as well as charge separation in PPV-C_{60} [69]. We are now working on an extension of the model that include intermolecular electron–phonon coupling. In another project we are replacing the simple tight-binding Hamiltonian with a self-consistent field approach that will allow us to better describe the dynamics of, e.g., charge separation. We are also looking into the possibility to replace the classical description of the lattice with it quantum mechanical counterpart, which will make it possible to perform studies deeper into the nonadiabatic regime of electron transfer. Thus, there are many interesting studies yet to be performed using the ELD approach.

Acknowledgments

The authors would like to thank Åsa Johansson, Anders Johansson, and Linus Gisslén for their contributions to this work. Financial support from the Swedish Research Council and the Center of Organic Electronics (COE), Swedish Foundation of Strategic Research, is also gratefully acknowledged.

References

1. Salem, L. (1966) *The Molecular Orbital Theory of Conjugated Systems*, W. A. Benjamin, New York.
2. Born, M. and Oppenheimer, R. (1927) *Ann. Phys.*, **84**, 457.
3. Theilhaber, J. (1992) *Phys. Rev. B*, **46**(20), 12990.
4. Allen, R. E. (1994) *Phys. Rev. B*, **50**, 18629.
5. Mele, E. J. (1982) *Phys. Rev. B*, **26**(12), 6901.
6. Block, S. and Streitwolf, H. W. (1996) *J. Phys. Condens. Matter*, **8**, 889.
7. Kinoshita, M., Hirano, Y., Kuwabara, M. and Ono, Y. (1997) *J. Phys. Soc. Jpn.*, **66**, 703.
8. Chen, Q.-H., Ren, Y.-H. and Jiao, Zh.-K. (1998) *Eur. Phys. J. B*, **3**(3), 307.
9. Wu, C. Q., Qiu, Y., An, Z. and Nasu, K. (2003) *Phys. Rev. B*, **68**, 125416.
10. Liu, X., Gao, K., Fu, J., Li, Y., Wei, J. and Xie, S. (2006) *Phys. Rev. B*, **74**, 172301.
11. Deumens, E., Diz, A., Longo, R. and Öhrn, Y. (1994) *Rev. Mod. Phys.*, **66**(3), 917–983.
12. Johansson, A. A. and Stafström, S. (2001) *Phys. Rev. Lett.*, **86**, 3602.
13. Karl, N. (2001) *Organic Electronic Materials: Conjugated Polymers and Low Molecular Weight Organic Solids*, Springer, Berlin.
14. Bässler, H. (1993) *phys. stat. sol. (b)*, **175**, 15.
15. Burroughes, J. H., Bradley, D. D. C., Brown, A. R., Marks, R. N., MacKay, K., Friend, R. H., Burns, P. L. and Holmes, A. B. (1990) *Nature*, **347**, 539.

16. Sariciftci, N. S., Smilowitz, L., Heeger, A. J. and Wudl, F. (1992) *Science*, **258**, 1474.
17. Sirringhaus, H., Tessler, N. and Friend, R. H. (1998) *Science*, **280**(5370), 1741–1744.
18. Holstein, T. (1959) *Ann. Phys.*, **8**, 325.
19. Su, W. P., Schrieffer, J. R. and Heeger, A. J. (1979) *Phys. Rev. Lett.*, **42**, 1698.
20. Brazovskii, S. A. and Kirova, N. N. (1981) *Pis'ma Zh. Eksp. Teor. Fiz.*, **33**, 6.
21. Brédas, J. L., Chance, R. R. and Silbey, R. (1982) *Phys. Rev. B*, **26**(10), 5843.
22. Zheng, G., Clark, S. J., Brand, S. and Abram, R. A. (2004) *J. Condensed Matter*, **16**, 8609.
23. Chen, D., Winokur, M. J., Masse, M. A. and Karasz, Frank. E. (1990) *Structural phases of sodium-doped polyparaphenylene vinylene. Phys. Rev. B*, **41**, 6759.
24. Capaz, R. B. and Caldas, M. J. (2003) *Phys. Rev. B*, **67**, 205205.
25. Papanek, P., Fisher, J. E., Sauvajol, J. L., Dianoux, A. J., Mao, G., Winokur, M. J. and Karasz, F. E. (1994) *Phys. Rev. B*, **50**, 15668.
26. de Sousa, R. L., Alves, J. L. A. and Leite Alves, H. W. (2004) *Mater. Sci. Eng.: C*, **24**, 601.
27. Hrhold, H.-H. and Opfermann, J. (1970) *Makromol. Chem.*, **131**, 105.
28. Zuppiroli, L., Bieber, A., Michoud, D., Galli, G., Gygi, F., Bussac, M. N. and André, J.J. (2003) *Chem. Phys. Lett.*, **374**, 7.
29. Swanson, L. S., Lane, P. A., Shinar, J. and Wudl, F. (1991) *Phys. Rev. B*, **44**, 10617.
30. Gisslén, L., Johansson, A. A. and Stafström, S. (2004) *J. Chem. Phys.*, **121**, 1601.
31. Österbacka, R., An, C. P., Jiang, X. M. and Vardeny, Z. V. (2000) *Science*, **287**(5454), 839–842.
32. Johansson, A. A. and Stafström, S. (2003) *Phys. Rev. B*, **68**, 35206.
33. Gartstein, Y. N. and Zakhidov, A. A. (1986) *Solid State Commun.*, **60**(2), 105.
34. Mizes, H. A. and Conwell, E. M. (1993) *Phys. Rev. Lett.*, **70**(10), 1505.
35. Hultell, M. and Stafström, S. (2006) *Chem. Phys. Lett.*, **428**, 446.
36. Miller, A. and Abrahams, E. (1960) *Phys. Rev.*, **120**, 745.
37. Holstein, T. (1959) *Ann. Phys.*, **8**, 343.
38. Marcus, R. A. (1993) *Rev. Mod. Phys*, **65**, 559.
39. Brédas, J. L., Beljonne, D., Coropceanu, V. and Cornil, J. (2004) *Chem. Rev.*, **104**, 4971.
40. Schein, L. B., Glatz, D. and Scott, J. C. (1990) *Phys. Rev. Lett.*, **65**, 472–475.
41. Jiang, X., Harima, Y., Yamashita, K., Tada, Y., Ohshita, J. and Kunai, A. (2003) *Synth. Met.*, **135**, 351.
42. Coropceanu, V., Cornil, J., da Silva Filho, D. A., Olivier, Y., Silbey, R. and Brédas, J. L. (2007) *Chem. Rev.*, **107**, 926.
43. Grozema, F. C. and Siebbeles, A. L. D. (2008) *Int. Rev. Phys. Chem.*, **27**, 87.
44. Nelson, S. F., Lin, Y.-Y., Gundlach, D. J. and Jackson, T. N. (1998) *Appl. Phys. Lett.*, **72**, 1854.
45. Podzorov, V., Pudalov, V. M. and Gershenson, M. E. (2003) *Appl. Phys. Lett.*, **82**(11), 1739–1741.
46. Haas, S., Stassen, A. F., Schuck, G., Pernstich, K. P., Gundlach, D. J., Batlogg, B., Berens, U. and Kirner, H.-J. (2007) *Phys. Rev. B*, **76**(11), 115203.
47. Yaliraki, S. N. and Silbey, R. J. (1996) *J. Chem. Phys.*, **104**(4), 1245.
48. Beenken, Wichard J. D. and Pullerits, Tonu (2004) *J. Phys. Chem. B*, **108**(20), 6164–6169.
49. Assadi, A., Svensson, C., Willander, M. and Inganäs, O. (1988) *Appl. Phys. Lett.*, **53**(3), 195–197.
50. Blom, P. W. M., de Jong, M. J. M. and Vleggaar, J. J. M. (1996) *Appl. Phys. Lett.*, **68**(23), 3308–3310.
51. Sirringhaus, H., Brown, P. J., Friend, R. H., Nielsen, M. M., Bechgaard, K., Langeveld-Voss, B. M., Spiering, A. J. H., Janssen, R. A. J., Meijer, E. W., Herwig, P. and de Leeuw, D. M. (1999) *Nature*, **401**, 685.
52. Warman, J., Gelinick, G. H. and de Haas, M. P. (2002) *J. Phys. C*, **14**, 9935.
53. van Breemen, A. J. J. M., Herwin, C. H. T., Chlon, P. T., Sweelssen, J., Shoo, H. F. M., Benito, E. M., de leeuw, M., Wildeman, J. and Bloom, P. W. M. (2005) *Adv. Funct. Mater.*, **15**, 872.
54. Yasuda, T., Saito, M., Nakamura, H. and Tsutsui, T. (2006) *Jpn. J. Appl. Phys.*, **45**, L313.

55. Chabinyc, M. L. and Salleo, A. (2004) *Chem. Mater.*, **16**, 509–4521.

56. Prins, P., Grozema, F. C. and Siebbeles, L. D. A. (2006) *J. Phys. Chem. B*, **110**, 14659.

57. An, Z., Wu, C. Q. and Sun, X. (2004) *Phys. Rev. Lett.*, **93**(21), 216407.

58. Horsfield, A. P., Bowler, D. R., Fisher, A. J., Todorov, T. N. and Montgomery, M. J. (2004) *J. Phys. Condens. Matter*, **16**, 3609.

59. Mulliken, R. S. (1949) *J. Chem. Phys.*, **46**, 675.

60. Mulliken, R. S., Reike, C. A., Orloff, D. and Orloff, H. (1949) *J. Chem. Phys.*, **17**, 1248.

61. Hultell, M. and Stafström, S. (2007) *Phys. Rev. B*, **75**, 104304.

62. Hultell, M. and Stafström, S. (2009) *Phys. Rev. B*, **79**(1), 014302.

63. Brooks, B. R., Bruccoleri, R. E., Olafson, D. J., States, D. J., Swaminathan, S. and Karplus, M. (1983) *Charmm: A program for macromolecular energy, minimization, and dynamics calculations. J. Comp. Chem.*, **4**, 187–217.

64. MacKerel Jr., A. D., Brooks III, C. L., Nilsson, L., Roux, B., Won, Y. and Karplus, M. (1998) *CHARMM: The Energy Function and Its Parameterization with an Overview of the Program*, volume 1, of The Encyclopedia of Computational Chemistry, pp. 271–277. John Wiley & Sons, Chichester.

65. Ha, S. N., Giammona, A. and Field, M. (1988) *Carbohydr. Res.*, **180**, 207.

66. Ponder, J. W. TINKER 4.2: Software tools for Molecular Mechanics (*http://dasher.wustl.edu/tinker/*).

67. Streitwolf, H. W. (1998) *Phys. Rev. B*, **58**, 14356.

68. Brankin, R. W. and Gladwell, I. RK-SUITE_90: Software for ODE IVPs, (1994) (*www.netlib.org*).

69. Johansson, A. A. and Stafström, S. (2004) *Phys. Rev. B*, **69**, 235205.

70. Linares, M., Hultell, M. and Stafström, S. (2009) *Synth. Met.*,

71. Kersting, R., Lemmer, U., Deussen, M., Bakker, H. J., Mahrt, R. F., Kurz, H., Arkhipov, V. I., Bässler, H. and Göbel, E. O. (1994) *Phys. Rev. Lett.*, **73**(10), 1440–1443.

72. Heeger, A. G., Kivelson, S., Schrieffer, J. R. and Su, W. P. (1988) *Rev. Mod. Phys.*, **60**, 781.

73. Onsager, L. (1938) *Phys. Rev.*, **54**, 554.

74. Miyamoto, Y., Rubio, A. and Tomanek, D. (2006) *Phys. Rev. Lett.*, **97**, 126104.

75. Hultell, M. and Stafström, S. (2008) *J. Lumin.*, **128**, 2019.

76. You, W. M., Wang, C. L., Zhang, F. C. and Su, Z. B. (1993) *Phys. Rev. B*, **47**(8), 4765–4770.

77. Rakovic, D., Kostic, R., Gribov, L. A. and Davidova, I. E. (1990) *Phys. Rev. B*, **41**, 10744.

9
Charge Transport along Isolated Conjugated Molecular Wires Measured by Pulse Radiolysis Time-Resolved Microwave Conductivity

Ferdinand C. Grozema and Laurens D. A. Siebbeles

9.1
Introduction

Over the last 10 years, there has been a growing interest in molecular electronics because conventional chip manufacturing methods are rapidly approaching the fundamental size limits that are dictated by the wavelength of the light used for photolithography. Moore's law predicts that the dimensions of the interconnecting wires will decrease from about 50 nm presently used in computer chips to approximately 10 nm within the next 10 years [1]. Therefore it is of considerable fundamental and practical interest to study the properties of conjugated molecular wires since these are the smallest wires imaginable [2–5]. An advantage of using conjugated molecular wires is the possibility to control the chemical structure through organic synthetic methods [6, 7]. The electronic properties of conjugated molecular wires can be tuned to meet the desired properties for specific applications by modifications in the backbone or by attaching electronically active side chains. An example of such a modification is the inclusion of units in the backbone that can be switched optically from conjugated to nonconjugated conformations. This would result in a molecular wire that can be switched on and off by illumination with different wavelengths of light [8, 9].

There are two main ways in which charge transport/transfer through conjugated molecules is being studied [10]: direct current (DC) conductivity measurements on single molecules between electrodes [11, 12] and spectroscopic studies of charge transfer in donor–bridge–acceptor systems [13–15]. Both methods are discussed in detail in other chapters of this book.

Although measurements using these approaches have led to a wealth of information on charge transfer through molecules they also have their limitations. The conductance properties of single molecules in a DC measurement are often determined to a large extent by the contact between the electrodes and the molecule. In most single-molecule conductance experiments, the charge tunnels from one electrode to the other without actually becoming localized on the molecule. As a result, only limited information about the mobility of the charge on the wire is obtained and no absolute values for the mobility of charges on a molecular wire

Charge and Exciton Transport through Molecular Wires. Edited by L.D.A. Siebbeles and F.C. Grozema

ISBN: 978-3-527-32501-6

have been published. Also in donor–bridge–acceptor systems, the charge tunnels from donor to acceptor and the mobility on the bridging wire cannot be obtained from measurements of the tunneling rate.

In this chapter, we review an alternative AC conductivity technique to study charge transport along conjugated molecular wires. This so-called pulse-radiolysis time-resolved microwave conductivity (PR-TRMC) technique involves generation of charges on the conjugated chains by irradiation with high-energy electrons and determination of the mobility by use of an oscillating microwave field, without the need to apply electrodes [16–18]. This circumvents some of the inherent problems encountered in DC experiments and, importantly, the motion of the charge is probed while it is actually present on the conjugated chain. In the following sections, we discuss the way in which the experiments are performed and give an overview of results obtained for different conjugated polymer backbones.

9.2 Pulse-Radiolysis Time-Resolved Microwave Conductivity

As mentioned above, the PR-TRMC technique [16, 19–21] is a convenient way to study the mobility of charges along π-conjugated molecular wires. Since no electrodes are attached to the molecules, it is possible to determine the mobility of charges moving along molecular wires in solution. In the pulse radiolysis experiments, the sample of interest is irradiated with a short pulse of electrons with a kinetic energy of 3 MeV, which is produced with a Van de Graaff accelerator [22]. The use of ionizing high-energy electrons has advantages over generation of charge carriers by laser flash photolysis. High-energy electron irradiation is nonspecific in the sense that energy deposition in the sample is dependent only on the electron density of the medium, regardless of the color or morphology. Another advantage is the relatively large penetration depth of high-energy electrons. If a laser is used, excitations are created with a concentration depth profile that depends on the extinction coefficient of the material at the wavelength used. In the case of high-energy electrons, a close to uniform distribution of ionizations/excitations can be produced in bulk solids or liquid materials for depths of several millimeters [23].

The TRMC technique was developed in the early 1970s by Warman and De Haas for measuring changes in the conductivity of a dielectric medium upon generation of charges [24, 25]. Figure 9.1 schematically shows the principle of TRMC measurements. If a medium does not contain charges that are mobile, microwaves can travel through this medium without being attenuated and the transmitted microwave power is the same as the incoming microwave power. If mobile charges are generated in the medium by a pulse of high-energy electrons, they can carry out a forward and backward drift motion and follow the oscillating electric field of the microwaves. This motion leads to an absorption of microwave power and hence an attenuation of the microwaves that are transmitted by the medium.

The primary information obtained from TRMC measurements is the relative change in microwave power $\Delta P/P$, which is proportional to the change in

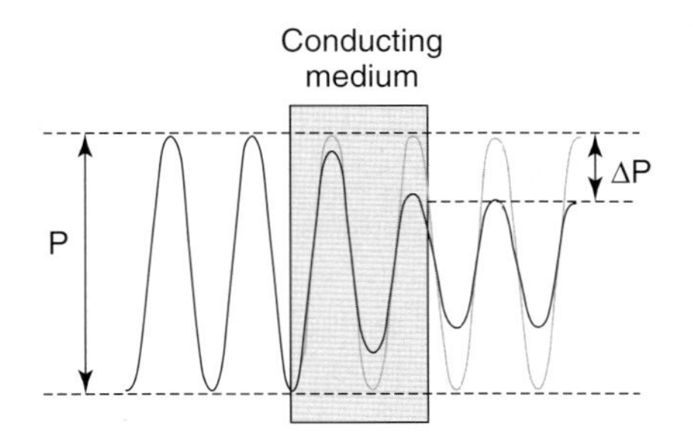

Figure 9.1 Schematic representation of the attenuation of microwaves when propagating through a conducting medium.

conductivity of the sample upon irradiation,

$$\frac{\Delta P}{P} = -A\Delta\sigma \tag{9.1}$$

where $\Delta\sigma$ is the change in conductivity and A is a sensitivity factor [25]. The magnitude of the sensitivity factor A is dependent on the geometry of the cell and on the frequency of the microwaves due to multiple reflections that occur at dielectric interfaces. The relation between A and the dimensions of the cell and the dielectric properties of the solution are known, thus the value of A can be calculated for an arbitrary solvent [25].

Pulsed irradiation of a sample leads to the formation of transient charged species. If these charges are mobile, their formation and decay can be monitored as a function of time using TRMC as discussed above. As an example, a TRMC transient is shown for pure benzene in Figure 9.2. The conductivity transients are normalized to the radiation dose (D amount of energy deposited per unit mass of sample expressed in Gy (= J/kg)), or equivalently, to the initial number of electron–hole pairs generated in the sample.

During the 2-ns 3-MeV electron pulse from the accelerator, the conductivity of the sample increases as mobile charges are created, reaching a maximum at the

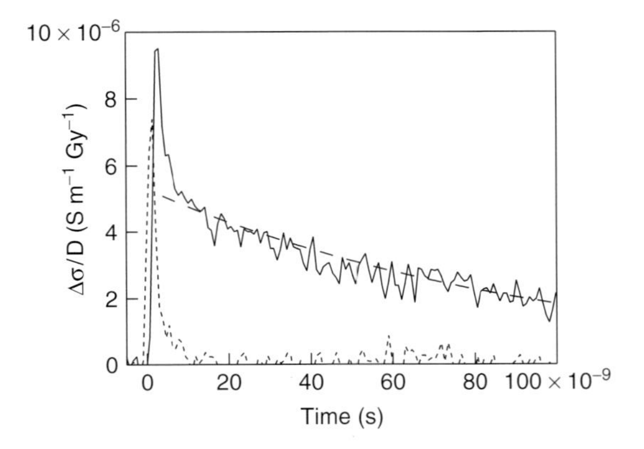

Figure 9.2 Radiation-induced conductivity in pure benzene (solid line) and oxygen-saturated benzene (dotted line) upon irradiation with a 2-ns 3-MeV electron pulse. The dashed line is an exponential fit to the transient for pure benzene.

end of the pulse. After the pulse, the conductivity signal decays due to trapping and recombination of the charged species [17]. The conductivity in pure benzene is mostly due to the motion of excess electrons that are generated upon irradiation. These excess electrons have a high mobility (0.13 cm^2/Vs) [26], whereas positively charged benzene ions have to move by molecular diffusion leading to a low mobility and hence a negligible contribution to the conductivity signal [27]. In the experiments discussed in this chapter, we are primarily interested in the motion of positive charges along conjugated polymer chains. Therefore, the experiments are performed in oxygen-saturated benzene solutions. The presence of oxygen leads to a rapid trapping of the excess electrons according to Eq. (9.2):

$$e^- + O_2 \rightarrow O_2{}^- \tag{9.2}$$

The trapping of electrons on O_2 leads to a very rapid decay of the conductivity after the irradiation pulse in oxygen-saturated benzene, as shown in Figure 9.2.

When a conjugated polymer is added to the oxygen-saturated benzene solution, the radiation-induced conductivity transients change rather dramatically. This is shown in Figure 9.3 for a solution containing one of the most extensively studied conjugated polymers, poly(2-methoxy-5-[2′-ethyl-hexyloxy]-phenylene vinylene) (MEH-PPV). As noted above, benzene radical cations have a very low mobility and do not contribute to the conductivity signal in pure benzene. However, they can diffuse through the solvent and eventually undergo charge transfer to the conjugated polymer chains via Eq. (9.3):

$$Bz^+ + PPV \rightarrow Bz + PPV^+ \tag{9.3}$$

Hole transfer from the benzene solvent to poly(phenylene-vinylene) (PPV) via Eq. (9.3) occurs due to the lower ionization potential of the polymer. The conductivity is seen to increase after the electron pulse on a timescale of hundreds to thousands

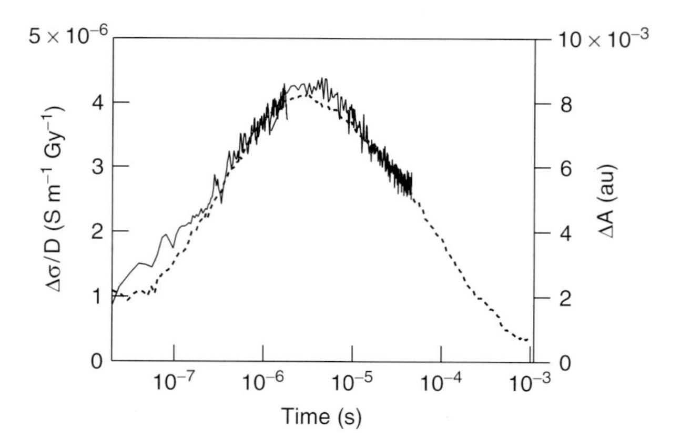

Figure 9.3 Radiation-induced conductivity (dashed lines, left-axis) and transient optical absorption measured at 1.3 eV (solid line, right-axis) for an oxygen-saturated solution of MEH-PPV in benzene as a function of time.

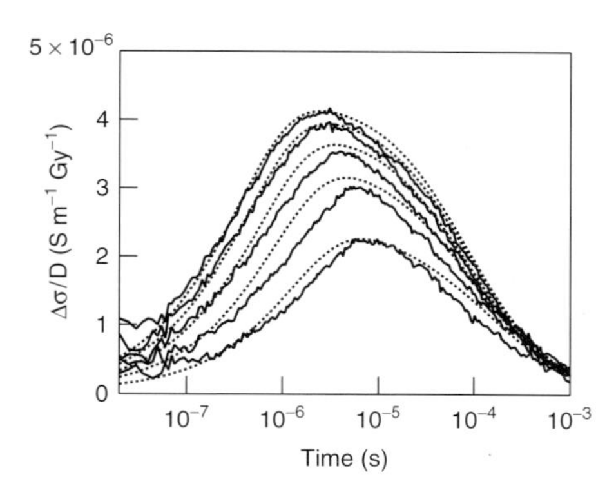

Figure 9.4 Radiation-induced conductivity in benzene solution of MEH-PPV for different concentration using a 10-ns pulse. From top to bottom: 1.25, 1.0, 0.75, 0.5, and 0.25 μM (polymer concentrations). The dotted lines are kinetic fits to the conductivity transient where all parameters except the polymer concentration were kept constant.

of nanoseconds. This after-pulse growth becomes faster with increasing polymer concentration (see Figure 9.4), in accordance with expectations based on Eq. (9.3) as the underlying source. These results provide immediate qualitative evidence that the mobility of positive charges on MEH-PPV chains is considerably higher than that of Bz^+ ions in benzene. Note that the large increase in mobility is due to the motion of charge along the polymer chain rather than diffusional motion of the polymer chains as a whole. As an independent confirmation that the conductivity transient is indeed due to the motion of positive charges along the polymer chains, we have also performed time-resolved optical absorption measurements. If the transient absorption is monitored at the characteristic energy at which radical cations in MEH-PPV absorb (1.3 eV) the exact same formation and recombination kinetics are observed, see Figure 9.3 [28].

The conductivity eventually decays on a timescale of tens to hundreds of microseconds. The timescale of the decay of the conductivity has been shown to become shorter as the pulse duration increases, that is, as the concentration of ion pairs formed in the pulse is increased. Accordingly, we attribute this decay to the second-order charge recombination of PPV^+ with $O_2{}^-$ according to

$$PPV^+ + O_2{}^- \rightarrow PPV + O_2 \tag{9.4}$$

Using Eq. (9.1), the fractional change in microwave power absorbed by the solution can be converted to change in conductivity as a function of time. The conductivity is related to the concentration and the mobility of charges by

$$\Delta\sigma(t) = e\sum_i N_i(t)\mu_i \tag{9.5}$$

where e is the elementary charge and $N_i(t)$ is the number density of charge carriers of type i existing at time t. The mobility of the charges, μ, is a microscopic property of the charges that determines the drift speed v, of a charge in an electric field E, according to

$$v = \mu E \tag{9.6}$$

In order to convert the conductivity to the charge carrier mobility for charges moving along conjugated polymer chains in solution as shown in Figure 9.4, the concentration of charges has to be known as a function of time. These concentrations can be obtained by solving the set of differential equations that arise when all reactions that involve charged species are considered, with Eqs. (9.3) and (9.4) being the most important ones. The procedure by which this is done has been discussed in detail in previous work [17]. The quality of the fits made in this way is illustrated in Figure 9.4 for a range of concentrations of MEH-PPV in benzene. Ultimately, the important result from these fits is the charge carrier mobility, in this case for positive charges along the polymer chains. The (microwave) mobility determines the absolute height of the conductivity signal and was found to be 0.46 cm^2/Vs for MEH-PPV [17].

9.3 Mechanisms for Charge Transport along Conjugated Chains

The measured charge carrier mobility depends on the frequency of the probing electric field and is according to the work of Kubo *et al.* given by [29–31]

$$\mu_{ac}(\omega) = -\frac{e\omega^2}{2k_BT}\int_0^\infty \langle \Delta x^2(t) \rangle \cos(\omega t)\, dt \tag{9.7}$$

with e the elementary charge, ω the (radial) frequency of the probing electric field, k_B Boltzmann's constant, T the temperature, and $\langle \Delta x^2(t) \rangle$ the mean squared displacement of the charge. An implicit convergence factor $\exp(-\varepsilon t)$ ($\lim \varepsilon \to 0$) is understood in the integral [30].

According to this equation, the charge carrier mobility at a certain frequency of the probing electric field is fully specified by the mean squared displacement of the charge. This mean squared displacement can, in principle, have any functional form, the only requirement for Eq. (9.7) to hold is that the charges are in thermal equilibrium and the electric field is sufficiently weak to give rise to a drift velocity of charges that increases linearly with field strength.

If the charge moves by normal Gaussian diffusion, the mean squared displacement of charge carriers moving along an infinitely long polymer chain increases linearly with time

$$\langle \Delta x^2(t) \rangle = 2Dt \tag{9.8}$$

where D is the diffusion constant. In this special case, the mobility does not depend on the frequency and Eq. (9.7) reduces to the Einstein relation

$$\mu = \frac{e}{k_BT}D \tag{9.9}$$

Another limiting case for the mean squared displacement is ballistic motion where the value of $\langle \Delta x^2(t) \rangle$ increases quadratically with time. In this case, the real part of the mobility obtained from the Kubo relation in Eq. (9.7) is zero for all frequencies. This is easily understood by considering the definition of the mobility for a charge in a static electric field. The mobility is the ratio between the (constant) drift velocity of the charge and the applied electric field, $\mu = v/E$. This definition entails that there has to be dissipation of energy during the motion of the charge in order obtain a nonzero mobility (without dissipation the speed would not be constant). In a general case, the motion of the charge will be neither fully diffusive nor ballistic but somewhere in between. More specifically, in the case of a charge moving along a conjugated polymer chain, the motion of the charge can be almost ballistic on a short planar part of the chain. On longer distance scales, the charge will also encounter sections where the chain exhibits disorder in the form of structural fluctuations, leading to scattering and dissipation. Therefore, the motion over the full time range will be a combination of ballistic and diffusive motion. The frequency-dependent mobility can be derived from the Kubo relation, provided that the mean squared displacement as a function of time is known. The mean squared displacement of a charge on a polymer chain is not accessible through experiments, but can be obtained by numerical simulations.

Two limiting theoretical models for the description of charge transport can be distinguished, depending on the degree of disorder in the material. Delocalized charges in structurally ordered materials that undergo weak scattering on dynamic fluctuations can be described by a Drude-like model. Localized charges in disordered materials can be described by incoherent hopping models [32]. In the intermediate regime where partially delocalized charges move in a weakly disordered energy landscape, the charge transport can be described using a tight-binding model as will be discussed in Section 9.6.

9.4 The Meaning of the Mobility at Microwave Frequencies

The measurements described in Section 9.2 of this chapter have been performed for a wide variety of conjugated polymers. An overview of the microwave mobility values obtained for both π-conjugated and σ-conjugated polymers is given in Figure 9.5 [18]. As can be seen, the values cover a wide range for the different polymers. While there is a tendency for the values for the carbon-based backbones to be somewhat higher than for silicon, both classes contain some compounds with mobility values well in excess of 0.1 cm^2/Vs and others close to the limit of the sensitivity of the measurements ($\sim$0.01 cm^2/Vs).

In the case of polysilicon chains, a perfect all-trans arrangement of the sigma-bonds has been shown to be the optimal configuration for intrachain electronic coupling [33–36]. On the basis of this, the lower mobility values for the shorter alkyl-chain compounds, PAPS2 and PAPS4, are attributed to lower barriers to conformations other than all-trans, which results in greater conformational disorder and a decrease in the overall electronic coupling.

In the case of the carbon-based polymers, a perfectly coplanar arrangement of the aromatic/ethylinic moieties should be optimal for electronic coupling between neighboring p-orbitals and should favor rapid charge transport along the chain. This could explain the much lower mobility found for the polythiophene compound since the barrier to relative torsional rotation between neighboring units is much

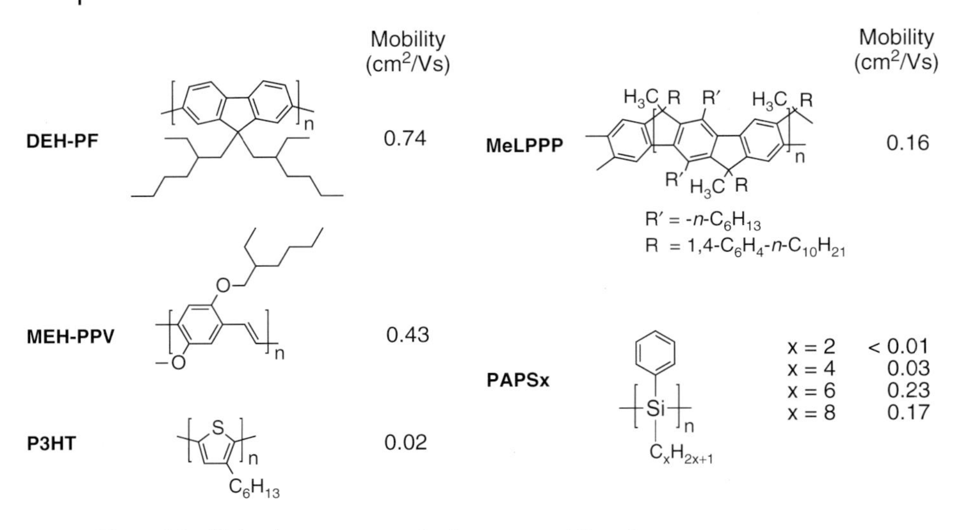

Figure 9.5 Molecular structure and microwave mobility of positive charges measured at 34 GHz for several different π- and σ-conjugated polymers.

lower than for PPVs [37]. Greater conformational disorder results in a smaller average electronic coupling, which can be interpreted as a larger effective mass of the charge carrier. On the basis of this, a lower mobility is predicted for the more disordered conjugated backbone of P3HT [37].

The large mobility found for DEH-PF would appear to be anomalous since rotation of the neighboring fluorene units about the connecting σ-bond might also be expected to occur. This compound and a similar 9,9′-dioctyl derivative have, however, been found to have an exceptional propensity for self-alignment indicating a high degree of order along the chains [38]. This characteristic would appear to be operative even in the dilute solutions and may explain the particularly high mobility observed.

An additional unexpected result is also that the mobility for the ladder polymer, MeLPPP, which should have the most rigid, coplanar backbone of all, was lower than that for MEH-PPV and DEH-PF. The explanation for this apparent anomaly may lie in the fact that MeLPPP has a branched structure with nonconjugated branching points every 20–30 monomer units. Therefore, the MeLPPP consists of a collection of relatively short conjugated segments in which the charge transport may be hampered by chain ends.

The wide variety of mobility values and unexpected results for some of the polymers have prompted us to examine the meaning of the charge carrier mobility at microwave frequencies in more detail. More specifically, we have considered the effects of limited chain length on the microwave mobility and the relation between the microwave mobility and DC mobility values that would be measured for a molecule between electrodes. As evident from the Kubo relation discussed in Section 9.3, the charge carrier mobility is, in general, dependent on the frequency of the oscillating electric field with which the charge transport is probed [31]. In order to gain insight into the effects of chain ends on the measured microwave mobility, we have considered the diffusive motion of a charge moving in one dimension along a conjugated polymer chain with infinitely high barriers at the ends (i.e., the chain ends). By analytically solving the diffusion equations that arise from these conditions, an expression can be obtained that relates the mean squared displacement of the charge to its diffusion coefficient and the chain length. This relation can be inserted in the Kubo relation, which yields the following equation for the complex charge carrier mobility, μ_{ac}, as a function of the frequency of the probing electric field, f [39, 40],

$$\mu_{ac}(f) = 8\mu_{dc} \sum_{k=0}^{\infty} \frac{[c_k]^{-2}}{\frac{k_B T \mu_{dc}}{ie(na)^2 2\pi f} [c_k]^2 + 1} \tag{9.10}$$

with $c_k = 2\pi(k + 1/2)$. In Eq. (9.10), k_B is Boltzmann's constant, T is the temperature, a is the length of a repeat unit, $i^2 = -1$, n is the number of units in the chain, e is the elementary charge, and μ_{dc} is the DC charge carrier mobility, that is, related to the diffusion coefficient, D, of the charge by the Einstein relation ($D = \mu_{dc} k_B T/e$). The real part of the mobility in Eq. (9.10) is due to motion of the charge in phase with the oscillating electric field. In this case, the field performs work on the charge by inducing a drift velocity and microwave power is absorbed by the charges. The imaginary part of the complex mobility is due to motion of charges with a velocity that oscillates out-of-phase with the microwave electric field. This leads to a phase shift of the microwaves without absorption of microwave power by the charges. In the remainder of this chapter, we will only consider the real part of the charge carrier mobility.

The microwave mobility measured for a specific material is determined by the intrachain DC mobility, the length of the polymer chain along which charge transport is probed, and the frequency of the microwaves used. The real part of the complex mobility as calculated with Eq. (9.10) is shown as a function of these three parameters in Figure 9.6. The values of the three parameters kept constant in the respective calculations are $f = 30$ GHz, $\mu_{dc} = 100$ cm^2/Vs, and $n \times a = 100 \times 7.5$ Å. In Figure 9.6(a), the AC microwave mobility is shown as a function of the intrachain DC mobility. For low-DC mobility (<1 cm^2/Vs), the microwave mobility is equal to the DC mobility; the motion of the charges is not significantly hindered by the chain ends during the oscillation period of the probing field. As the DC mobility exceeds a few centimeter square per volt second, the microwave mobility starts to deviate from the DC mobility. This occurs at the DC mobility where the charge carrier starts to encounter the ends of the polymer chain during the oscillation period, leading to a lower measured microwave mobility. This can be understood since a charge only contributes to the microwave mobility when its velocity is in phase with the microwave field. The latter effect becomes even clearer when the DC mobility exceeds tens of centimeter square per volt second. At these high values of the intrachain DC mobility, the chain ends strongly affect the high-frequency mobility, resulting in a decrease of the real part of the microwave mobility with increasing DC mobility.

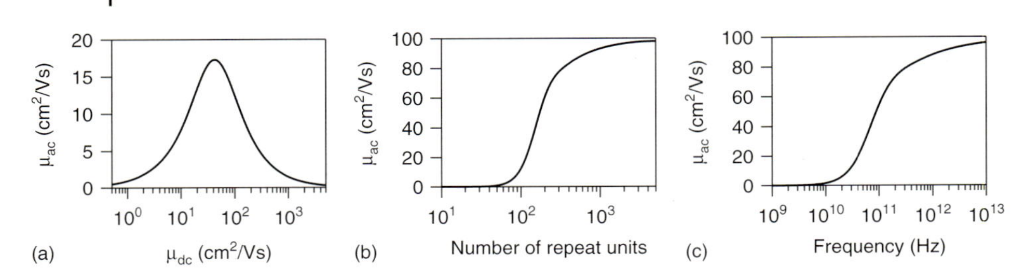

Figure 9.6 Real part of the high-frequency mobility for a charge carrier moving along a polymer chain as calculated with Eq. (9.7) as a function of (a) DC mobility ($f = 30$ GHz, $n = 100$), (b) number of repeat units per polymer chain ($f = 30$ GHz, $\mu_{dc} = 100$ cm^2/Vs), and (c) probing frequency ($\mu_{dc} = 100$ cm^2/Vs, $n = 100$).

According to the explanation above, lengthening of the polymer chains should have a strong effect on the measured microwave mobility. This is shown in Figure 9.6(b) where the calculated microwave mobility is plotted as a function of polymer chain length. For short chains (up to 100 repeat units), the motion of the charge is severely hindered by the chain ends, leading to a low mobility. When the polymer chain becomes longer, the microwave mobility increases (to reach the intrachain DC mobility of 100 cm^2/Vs) because the charge encounters the chain ends less often during the oscillation period of the microwaves. Ultimately, for infinitely long chains a charge can move freely without ever encountering a chain end. In this case, the microwave measured mobility is equal to the DC mobility of the charge along the chain.

A similar effect as reached by making the polymer chains infinitely long can be obtained by increasing the frequency of the oscillating electric field. At a higher frequency, the oscillation period is shorter and the charges encounter the chain ends less often. This is illustrated in Figure 9.6(c) where the microwave mobility is plotted as a function of the frequency for a fixed chain length and intrachain DC mobility (100 repeat units and a mobility of 100 cm^2/Vs). For a very low frequency, the microwave mobility is zero since the charge is always "waiting" at the chain ends for the microwave field to reverse direction. When the frequency is increased, the measured microwave mobility increases, until it reaches the same value as the intrachain DC mobility. It has experimentally been found that the mobility increases with the frequency in the range 10–30 GHz [41]. A frequency dependence of the charge mobility has also been demonstrated in the terahertz regime [42, 43].

The chain length and frequency dependence of the measured microwave mobility should be carefully considered as described below for several types of conjugated polymer. At the same time, measurement of the microwave mobility for different chain lengths or at varying frequency yields important information that makes it possible to derive values for the mobility at infinite chain length or high frequency. This infinite chain value then corresponds to the mobility as it would be measured for a charge moving along a chain between electrodes, in the absence of injection barriers.

9.5 Charge Transport along Ladder-Type PPP

Ladder-type poly-*para*-phenylene (LPPP) polymers consist of phenyl units that are locked into a mutually planar configuration by bridging carbon atoms, see Figure 9.7.

PR-TRMC measurements as described in Section 9.2 were performed on four samples of LPPP with varying average chain length [39]. The change in conductivity upon irradiation of solutions containing 0.315 mM (in monomer units) of the polymer is shown as a function of time for the four samples in Figure 9.8. The conductivity transients in this figure directly show the strong effect of the average chain length on the measured microwave mobility, as can be expected based on the discussion in Section 9.4. The microwave mobility was obtained from kinetic fits that are also shown in the same figure. From these fits, microwave mobilities of 0.025, 0.036, 0.10, and 0.24 cm^2/Vs were obtained for average chain lengths of 13, 16, 35, and 54 repeat units, respectively. This increase in microwave mobility with chain length directly indicates that the motion of charges is hindered by the chain ends for chain lengths up to at least 35 repeat units. The fact that we observe this chain length dependence for the microwave mobility means that the charge must diffuse over the entire length of the 35 units long polymer chain and encounter a chain end on a timescale of the order of one period of the oscillating electric field. From this, a minimum value for the intrachain DC mobility can be derived. For one-dimensional diffusion, the mean squared displacement as a function of time

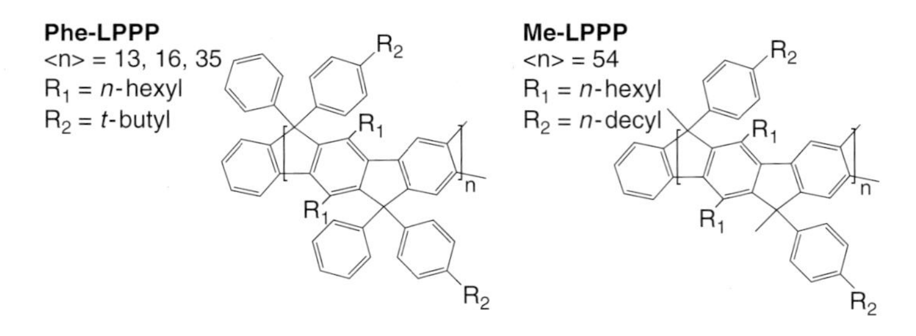

Figure 9.7 Molecular structures of ladder-type poly-*para*-phenylene (LPPP) derivatives.

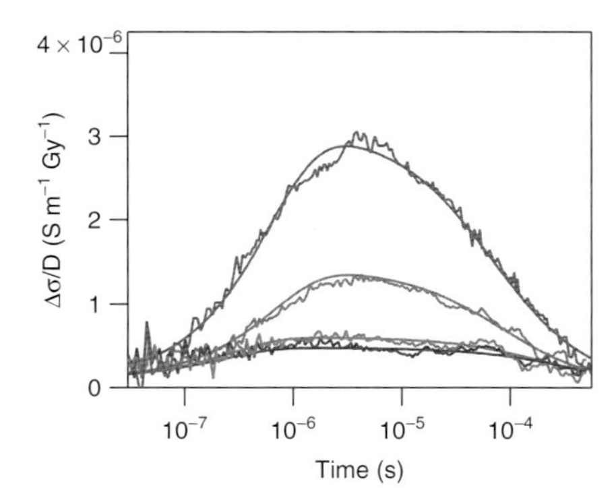

Figure 9.8 Radiation-induced conductivity due to positive charges on ladder-type poly-*para*-phenylene (LPPP) chains with number averaged length of 13, 16, 35, or 54 repeat units (from bottom to top).

is given by $\langle x^2(t)\rangle = 2Dt$, where the diffusion constant D is related to the intrachain mobility of the particle performing diffusive motion (μ_{intra}) and temperature (T) according to the Einstein relation $D = \mu_{\text{dc}} k_B T/e$. Using a displacement equal to a chain length of 35 repeat units (30 nm), and a migration time equal to one period of the microwave field (30 ps), a value of 5 cm^2/Vs can be deduced as a lower limit to the intrachain mobility for positive charges moving along LPPP chains.

To gain more accurate quantitative insight into the effects of limited chain length on the charge carrier mobility measured by microwave conductivity, we have compared the experimental data with values obtained using Eq. (9.10). The microwave mobility (at 34 GHz) was calculated as a function of the intrachain DC mobility, see Figure 9.9.

For a good comparison, it is essential to average the mobility obtained from Eq. (9.10) over the chain length distribution. For the LPPP chains considered here, this was a Flory distribution with $5 \leq n \leq 75$ [39]. The length of the LPPP chains is the product of the number of repeat units and the length of one repeat unit (8.3 Å). The results for the four average chain lengths are shown in Figure 9.9. The mobility follows the trend as shown for the hypothetical case in Figure 9.6(a); at low-intrachain DC mobilities, the microwave mobility is equal to the DC mobility, while at high-intrachain mobilities the motion of the charge is severely hindered by scattering at chain ends, resulting in a much lower value of the measured microwave mobility.

In order to describe the experimental data in terms of the one-dimensional diffusion model in Eq. (9.10), we need to determine the intrachain mobility that gives rise to the experimental values for the microwave mobilities for the four average chain lengths. Mobilities of the same order of magnitude as the experimental values (denoted by the horizontal lines in Figure 9.9) are obtained at low mobilities of a few tenths of square centimeter per volt second and at very high

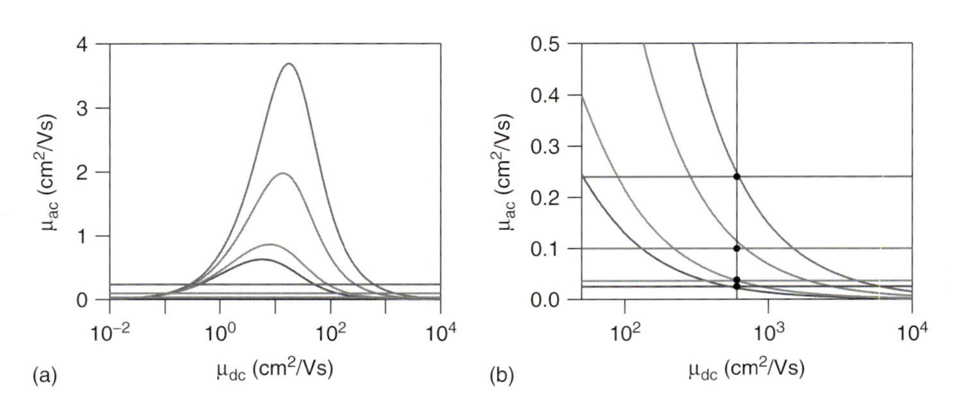

Figure 9.9 (a) Microwave mobility, μ_{ac}, as a function of intrachain DC mobility for average polymer chain lengths of 13, 16, 35, and 54 repeat units (from bottom to top). (b) The experimental chain length dependence is reproduced when assuming an intrachain mobility of 600 cm^2/Vs.

mobility values of a few hundred square centimeter per volt second. However, at a low-intrachain DC mobility, almost no chain-length dependence would be observed when the mobility is probed by the microwave field–contrary to the experimental results. Therefore, the experimentally observed increase in the high-frequency mobility with chain length cannot be reproduced with an intrachain mobility in the order of a few tenths of square centimeter per volt second. This agrees with the estimate made above that the intrachain mobility must exceed 5 cm^2/Vs in order to explain the chain-length dependence of the experimental results. The data in Figure 9.9(b) show that the chain-length dependence of the microwave mobility can be reproduced assuming an intrachain DC mobility close to 600 cm^2/Vs.

For an intrachain mobility of 600 cm^2/Vs, the charge carrier motion along the polymer chains used in this study is severely hindered by the chain ends. According to Eq. (9.10), this means that the measured microwave mobility should depend strongly on the frequency of the microwaves used. Therefore, we have also performed PR-TRMC measurements using a microwave frequency of 10.6 GHz, instead of 34 GHz. From these measurements a microwave mobility of 0.032 cm^2/Vs was obtained for the polymer with an average chain length of 54 repeat units. This value is significantly (7.5 times) lower than the mobility found for the same polymer at 34 GHz. Using Eq. (9.10) with $\langle n \rangle = 54$, $\mu_{dc} = 600$ cm^2/Vs, yields a microwave mobility of 0.031 cm^2/Vs at of 10.6 GHz, which is in very good agreement with the experimental value. From this, we can conclude that the frequency dependence of the microwave mobility confirms that the intrachain DC mobility is indeed very high.

From the measurements presented in this section, it becomes clear that the DC mobility of charges along chains of conjugated polymers can be very high, comparable to the mobility in some inorganic semiconductor materials. This shows that there are no fundamental limitations to applying organic semiconductors in the same way as their inorganic counterparts. More interestingly, the results show that conjugated polymers that have a rigid planar backbone are attractive candidates for application as wires in single-molecule devices. It should be noted that using these polymers in solid-state devices also requires high order on the supramolecular level. We have shown that a small amount of energetic disorder caused by interchain interactions reduces the mobility by a factor 20 [44].

9.6 Effect of Torsional Disorder on the Mobility

The LPPPs that were discussed in Section 9.5 are restricted to a planar backbone conformation by bridging carbon atoms, leading to a very high intrachain DC mobility. Most conjugated polymers exhibit more conformational freedom, especially in solution. Examples of the latter include derivatives of phenylene-vinylene (PV, see Figure 9.10), which have been widely studied due to their applicability in organic light-emitting diodes.

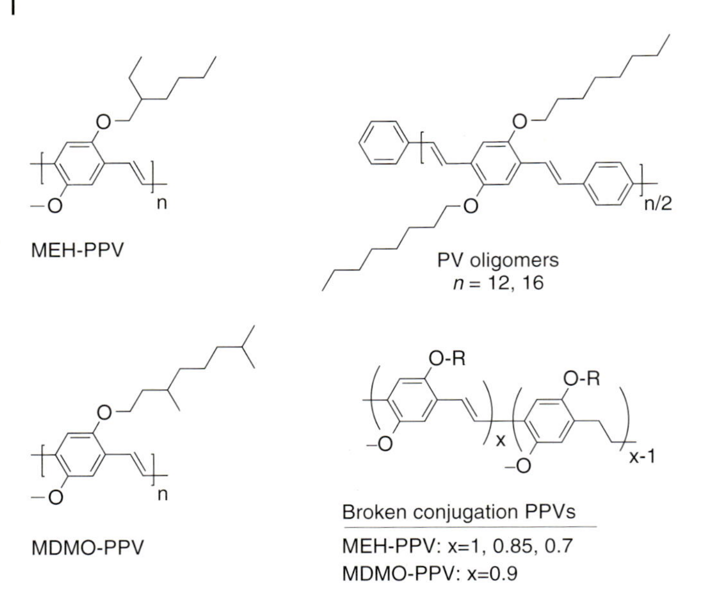

Figure 9.10 Molecular structures of phenylene–vinylene polymers and oligomers.

As shown above for the LPPPs, the chain-length dependence of the microwave mobility gives valuable information on the intrachain DC mobility. Therefore we have performed PR-TRMC measurements for a series of PPV derivatives that contain a well-defined fraction of saturated bonds that are statistically distributed along the polymer chains [28]. The introduction of saturated bonds breaks the conjugated pathway. This restricts the motion of charges over finite length between two saturated bonds. Additionally, we have considered two PV oligomers with a well-defined length [41].

The PR-TRMC transients obtained for the MEH-PPV derivatives with conjugation fractions of 100%, 85%, and 70% are shown in Figure 9.11. It is clear that introduction of saturated units in the PPV chain strongly reduces the conductivity (and hence the mobility) due to charges moving along the chains. The charge carrier mobility values obtained from the analysis of these transients are listed in

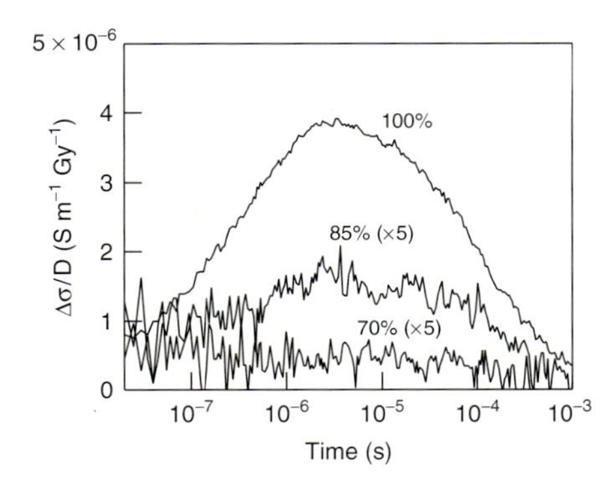

Figure 9.11 Radiation-induced conductivity for fully conjugated ($x = 1$) MEH-PPV and MEH-PPV derivatives containing 15% ($x = 0.85$) and 30% ($x = 0.7$) of saturated vinylene units.

Table 9.1 Experimental and calculated microwave charge carrier mobility for PV oligomers and polymers.

PV chain	μ_{exp} (cm²/Vs)	μ_{calc} (cm²/Vs)
PV12	0.018	0.038
PV16	0.029	0.072
MEH-PPV (70%)	<0.01	0.0029
MEH-PPV (85%)	0.04	0.018
MEH-PPV (100%)	0.46	0.46
MDMO-PPV (90%)	0.085	0.055

Table 9.1. The table also contains the mobility obtained for a 90% MDMO-PPV for comparison. The mobility of charges along the chains is seen to decrease by more than an order of magnitude upon going from 0.46 cm^2/Vs for the fully conjugated MEH-PPV to 0.04 cm^2/Vs for a derivative with 15% of conjugation breaks. After introduction of 30% of saturated units in the MEH-PPV backbone, the signal reduces even more and only an estimate of the upper limit of the mobility of 0.01 cm^2/Vs can be given. These results show that charge transport is severely limited by conjugation breaks; however, it is not immediately clear whether the charges move on a single conjugated segment. Although it has been found that charge transfer between neighboring conjugated segments occurs [28], this process will always be much slower than transport along conjugated parts. In fact, if the transport would be dominated by the transport over conjugation breaks, the mobility would be much lower than the lower limit that can be measured in our experiment. Therefore, the lowering of the mobility with increasing fraction of conjugation breaks can be attributed to the reduction of the average length of the conjugated segments. Similar to the situation for the ladder-type polymer, on shorter segments the charge is hindered in its motion by the chain ends, leading to a lower average mobility.

A very direct way to study the effects of finite conjugation length on the measured microwave mobility is to consider conjugated chains of a well-defined length. Therefore we have performed PR-TRMC measurements for the two PV oligomers shown in Figure 9.10. The results of these experiments are shown in Figure 9.12. This figure shows that even for a short chain consisting of 16 PV units, a microwave conductivity signal is observed, although the signal is more than an order of magnitude lower than that for the fully conjugated MEH-PPVs, see Table 9.1. For the oligomer consisting of 12 PV units, the conductivity is considerably lower.

9.6.1 Simulations of Charge Transport along Phenylene–Vinylene Chains

The analysis to obtain the intrachain DC mobility used for the ladder-type polymers cannot be used for PPV derivatives. The reason for this is that PPV is

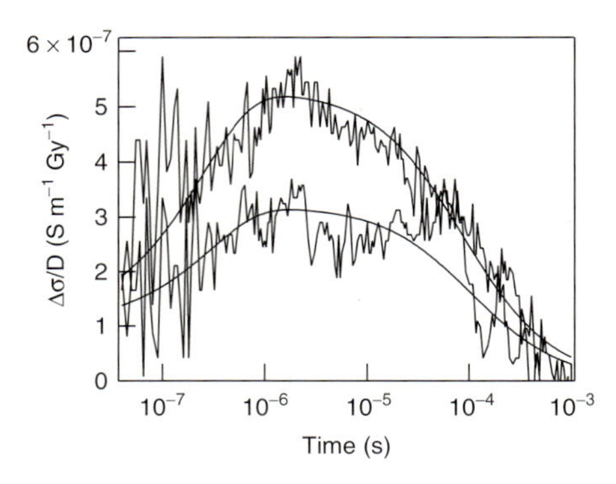

Figure 9.12 Radiation-induced conductivity for PV chains consisting of 12 (lower curve) and 16 (upper curve) repeat units. The smooth lines are kinetic fits to the transients.

not restricted to a planar conformation due to fluctuations of the torsional angle between adjacent phenyl and vinyl units along the polymer chain. Consequently, the polymer is broken up into sections that are relatively planar, separated by sites with a large dihedral angle. This will lead to a frequency dependence of the measured charge mobility even for infinitely long chains. Effects of torsional disorder have been taken into account in numerical simulations of charge transport based on the tight-binding approximation [41, 45]. In the model, the PPV chain is represented by an alternating sequence of vinylene and 2,5-methoxy-substituted phenylene units. The time-dependent wavefunction, $\Psi(t)$, of the charge moving along the polymer is taken to be a linear combination of the highest molecular orbitals, Φ_n, located on the individual repeat units according to

$$\Psi(t) = \sum_n c_n(t, n_0)\Phi_n \tag{9.11}$$

In situations with fast dephasing due to interactions of the charge with vibrations and solvent reorganization, the mean squared displacement of the charge may be calculated by initially localizing the charge on a single unit; that is, $c_n(t = 0, n_0) = \delta_{n,n_0}$ [45]. The time-dependent expansion coefficients of the basis functions on the molecular units in the PV chain ($c_n(t)$) are obtained by propagation of the wavefunction according to the time-dependent Schrödinger equation

$$i\hbar\frac{\partial\Psi(t)}{\partial t} = \hat{H}\Psi(t) \tag{9.12}$$

In the tight-binding approximation, the diagonal matrix elements of the Hamiltonian ($\hat{H}$) correspond to the site energies ($\varepsilon_{i,i}$), that is, the energy of the charge localized on a single-molecular unit in the PV chain. When only nearest-neighbor interactions are taken into account, the off-diagonal matrix elements of the Hamiltonian $J_{i,i\pm1}$ represent the electronic coupling or charge transfer integral between adjacent molecular units. The other off-diagonal matrix elements are zero in this approximation and the Hamiltonian matrix is given by

$$H = \begin{pmatrix} \varepsilon_{11} & J_{12} & 0 & \dots & 0 \\ J_{21} & \varepsilon_{22} & & & \\ 0 & & \ddots & & \\ \vdots & & & \ddots & \\ 0 & & & & \varepsilon_{NN} \end{pmatrix} \tag{9.13}$$

The mean squared displacement of the charge as a function of time can be expressed in terms of the time-dependent site coefficients and the positions of the molecular units

$$\langle \Delta^2(t) \rangle = \sum_{n,n_0} f(n_0) \left| c_n(t, n_0) \right|^2 (n - n_0)^2 a^2 \tag{9.14}$$

where $f(n_0)$ describes the initial distribution of the charge and $(n - n_0)a$ is the distance between the orbitals at sites n and n_0. $c_n(t, n_0)$ is the coefficient of the orbital at site n at time t for a state which was initially localized at n_0.

The values of the charge transfer integrals, J, in Eq. (9.13) depend strongly on the conformation of the polymer chain, particularly on the dihedral angle between neighboring phenylene and vinylene units. Therefore, the polymer conformation and temporal fluctuations of this conformation should be taken into account in the simulations of charge transport. In Figure 9.13(a), the torsion potential is shown as a function of the dihedral angle for di-methoxy-substituted phenylene unit with respect to the vinylene. The torsion potential is minimum for a close to planar configuration and maximum for a dihedral angle of 90°. The asymmetry in the potential energy surface around a dihedral angle of 90° results from the difference in steric hindrance due to the two methoxy substituents at inequivalent sites on the phenyl unit. The charge transfer integral between phenylene and vinylene as a function of dihedral angle is shown in Figure 9.13(b). As expected, J is maximum for a planar configuration and (almost) zero when the phenyl unit is perpendicular to the adjacent vinyl unit. The structural fluctuations are taken into account in the simulations by accounting for the rotational diffusion of the repeat units along

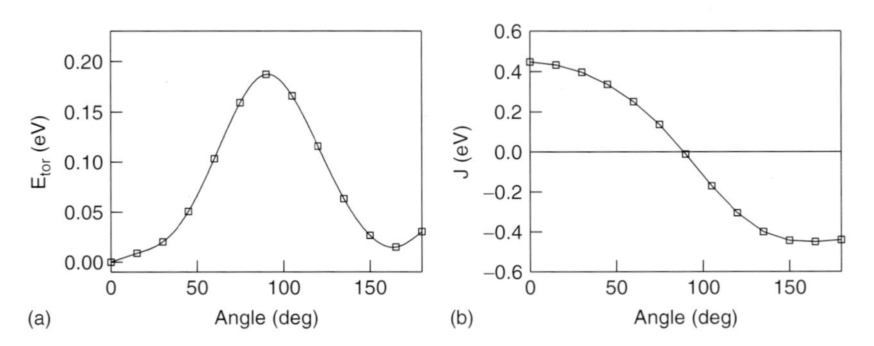

Figure 9.13 (a) Torsion potential (b) and charge transfer integral calculated as a function of the dihedral angle between neighboring vinylene and 2,5-methoxy-phenylene units.

the polymer chains. The phenylene units are propagated during the simulation by assuming rotational diffusion in the dihedral potential in Figure 9.13(a) and using a rotational diffusion time of 200 ps [41, 45].

9.6.2 Mean Squared Displacement of a Charge on Phenylene–Vinylene Chains

Using the methodology summarized above, we have simulated charge transport along PV chains of varying of lengths. Two illustrative graphs of the mean squared displacement as a function of time are shown in Figure 9.14. The calculated mean squared displacement for a positive charge on a PV chain consisting of 50 PV units is shown in Figure 9.14(a). Initially, the charge rapidly delocalizes along the chain, after which the mean squared displacement becomes constant. The mean squared displacement for an infinitely long PV chain is shown in Figure 9.14(b). At short times, the mean squared displacement is dominated by charges moving on the relatively planar parts of the chain, leading to the initial ($t < 5$ ps) fast increase in the mean squared displacement. After some time, the charge also encounters larger torsional angles along the polymer chain and the motion will be determined mostly by the largest angles where the charge transfer integral is small. Therefore, after a few picoseconds, the motion of charges becomes diffusive; that is, the mean squared displacement increases linearly with time and can be described by Eq. (9.6).

9.6.3 Mobility of Charges along Phenylene–Vinylene Chains

The mean squared displacement as a function of time for the PV chains as plotted in Figure 9.14 can be used to obtain the mobility of charges by applying Kubo's relation (Eq. (9.7)) as discussed in Section 9.4. We have numerically integrated the curves in Figure 9.14 according Eq. (9.7), to obtain the mobility as a function of frequency as shown in Figure 9.15. The frequency dependence of the mobility for the chain consisting of 50 PV units nicely illustrates the strong effect of the

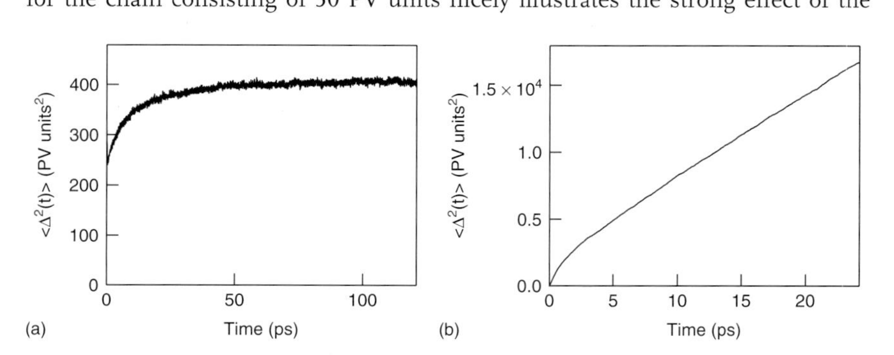

Figure 9.14 Mean squared displacement as a function of time calculated for a positive charge on (a) a chain consisting of 50 PV units and (b) an infinitely long PV chain.

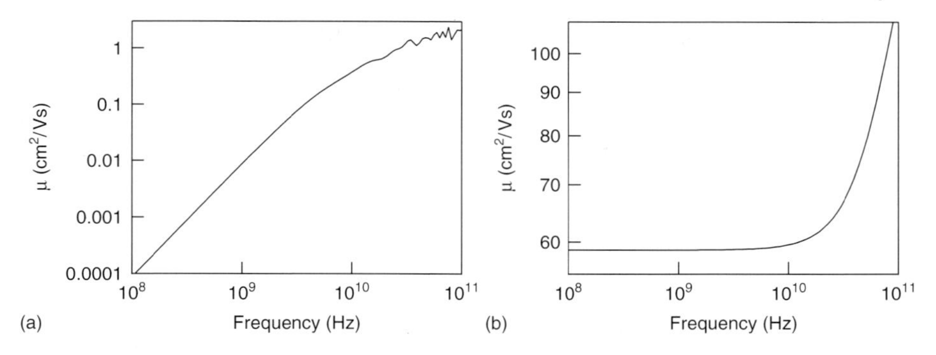

Figure 9.15 Charge carrier mobility as a function of frequency calculated for a charge on (a) a chain consisting of 50 PV units and (b) an infinitely long PV chain.

chain ends on the mobility, see Figure 9.15(a). In agreement with the discussion in Section 9.4, the mobility increases with the frequency of the probing electric field. This shows that at higher frequencies, the charge encounters chain ends less often during the oscillation period of the electric field.

For the PV chain that is infinitely long, the mobility of charge is independent of the frequency below 10 GHz, see Figure 9.15(b), and has a value of 59 cm^2/Vs. This regime corresponds to the diffusive motion of the charge at times exceeding 5 ps in Figure 9.14. At higher frequencies, the motion of charges is probed over a smaller distance, and the mobility is dominated by the motion of charges on relatively planar parts of the PV chain, that is, in between places where the dihedral angle between adjacent molecular units is large. This results in the frequency-dependent mobility at frequencies above 10 GHz and corresponds to the initial fast increase in the mean squared displacement in Figure 9.14(b).

9.6.4
Comparison of the Calculated Mobility with Experimental Data

To compare the calculated results with the experimental mobility values found for the oligomers PV12 and PV16, the mobility was calculated as a function of the number of PV units in the chain for a frequency of 34 GHz. At this frequency, the mobility of the charge carriers is strongly affected by the ends of the PV chain, as can be seen in Figure 9.15. For chains longer than 500 PV units, the mobility at 34 GHz is found to be independent of the chain length and is as high as 68 cm^2/Vs. This value is slightly higher than the DC mobility found for the infinitely long chain in Figure 9.15(b), due to the stronger limiting effect of large angles on the mobility at low frequencies. Taking into account the *ab initio* character of the charge transport calculations, the calculated values of the mobility for positive charges on PV12 and PV16 are in good agreement with the experimental results (see the inset of Figure 9.16 and Table 9.1).

For comparison of the calculated results with the experimental mobility values for the polymers with broken conjugation, the mobility for the different chain lengths

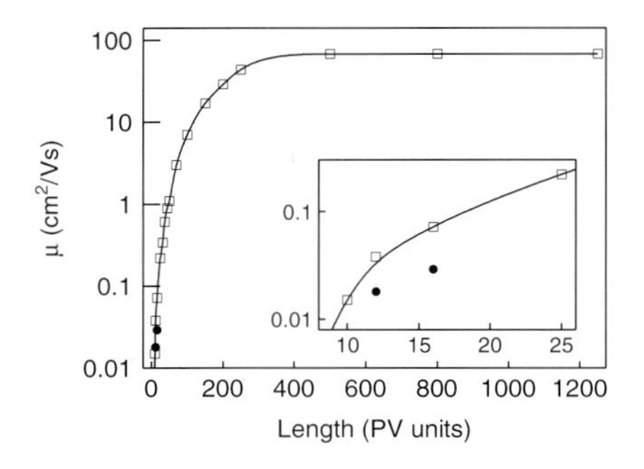

Figure 9.16 Charge carrier mobility as a function of chain length calculated for a positive charge. The solid dots indicate the experimental data (see inset).

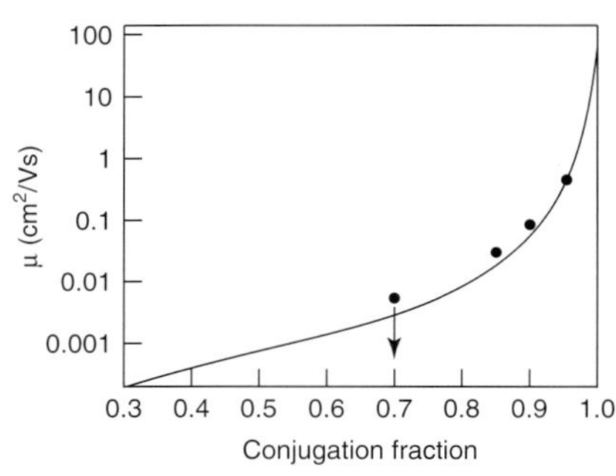

Figure 9.17 Charge carrier mobility as a function of conjugation fraction for broken conjugation PPVs. The solid dots indicate the experimental data.

in Figure 9.17 are weighted according to a Flory distribution, where the probability of segments consisting of m PV units is given by $P(m) = (1 - p)p^{m-1}$, p being the conjugation fraction. The mobility is found to decrease strongly with decreasing conjugation fraction, or equivalently, with increasing fraction of conjugation breaks. Introduction of only 5% of defects leads to a reduction in the mobility by 2 orders of magnitude (see Figure 9.17). At 34 GHz, a calculated mobility of 68 cm^2/Vs is found for a fully conjugated ($p = 1.00$) PPV chain, which is equal to the mobility found for the long (>500 units) PV oligomer chains, as expected. The calculated values for the mobility are in good agreement with the experimental results (see Figure 9.17 and Table 9.1). The mobility found experimentally for MEH-PPV ($p \sim 1$) can be reproduced by taking a conjugation fraction of $p = 0.955$. This conjugation fraction is reasonable since a small percentage of defects is inevitably induced during polymerization [46, 47].

The good agreement between the experimental and calculated microwave mobility values demonstrates that the theoretical model is reliable. Hence, a high DC mobility close to the calculated value of 59 cm^2/Vs for infinitely long PPV chains is to be expected when effects of conjugation breaks or chain ends are avoided.

9.7
Effect of Chain Coiling on the Mobility of Charges

In the two preceding sections, we have dealt with the motion of charges along polymer chains that exhibit a relatively linear conformation in solution, meaning that

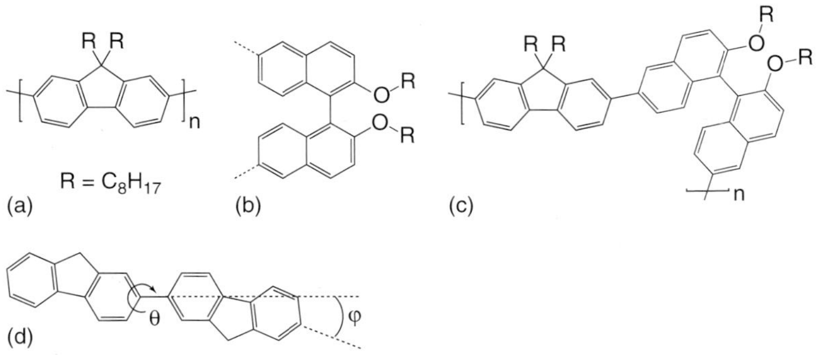

Figure 9.18 Molecular structure of polyfluorene and fluorene–binaphthyl copolymers.

the charge moves primarily in one dimension. The assumption of one-dimensional charge transport is valid for the ladder-type polymer LPPP and for PPV because these chains are relatively straight and rigid, and in addition because the (conjugation) length of the chains was relatively short. If the conjugation length becomes much larger or if the polymers chains are inherently coiled in solution, the three-dimensional chain conformation has to be taken into account explicitly when studying charge transport along such chains. An example of such an inherently coiled conjugated polymer is polyfluorene (PF) as shown in Figure 9.18(a). The coiled conformation of PF is a result of a nonzero bend angle, φ, and nonzero dihedral (torsional) angle, θ, between adjacent fluorene units, see Figure 9.18(d). The angle φ is 23.5°, while the dihedral angle deviates around 45° from a planar conformation. These two angles result in a persistence length of only 10 fluorene units for the fully perfect PF shown in Figure 9.18(a). Incorporation of binaphthyl units leads to an even more coiled chain conformation since the dihedral angle between the two naphthyl units is almost 90°. In fact, the incorporation of such binaphthyl units has been show to influence the formation of ordered phases in solid PF and thus markedly influences the chain conformation [48].

To gain insight into the effect of coiling, and to some extent conjugation breaks, on the charge transport properties of PF chain we have performed PR-TRMC measurements for polyfluorenes that contain varying fractions of binaphthyl units, as shown in Figure 9.19. The fractions of naphthyl units in the PFs studied are listed in Table 9.2, where also the chain length based on the number averaged molecular weight is given. Additionally, we have performed experiments for an alternating PF-binaphthyl copolymer.

The results of the PR-TRMC measurements are shown in Figure 9.19, together with kinetic fits to the conductivity transients. The mobility values obtained from the kinetic analysis are presented in Table 9.2. It should be noted that in this case the mobility that is quoted actually refers to the (isotropic) three-dimensional mobility, rather than the one-dimensional mobility for charge motion along a linear polymer chain [49]. The magnitude of the isotropic microwave mobility found here for holes on PF chains is comparable to values discussed above for other conjugated polymers in dilute solution. The mobility is found to decrease with increasing

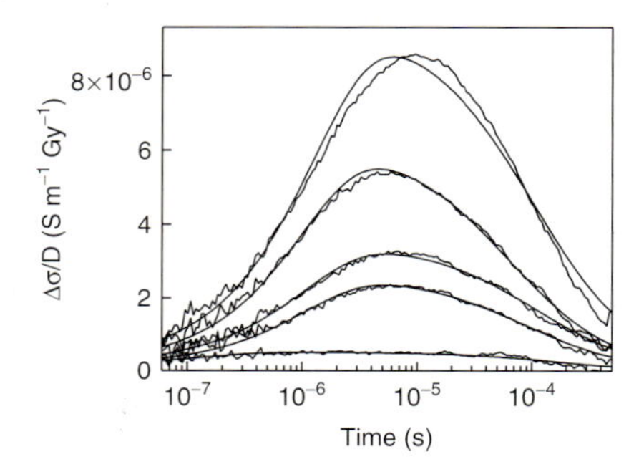

Figure 9.19 Radiation-induced conductivity for polyfluorene polymers with increasing fraction of binaphthyl incorporated in the backbone. The fractions of binaphthyl (from top to bottom): 0, 0.04, 0.09, 0.12, and an alternating block copolymer. The smooth curves represent kinetic fits to the experimental data.

fraction of BN incorporated in the PF backbone, from 0.23 for PF to 0.06 cm^2/Vs for PF with a BN fraction of 0.121, indicating that charge transport is less efficient when BN is present. Interestingly, the F-BN alternating copolymer shows a nonzero conductivity signal, indicating that charges can still move along these copolymers. This shows that BN does not break the conjugated pathway along the polymer backbone completely. It should be noted that the charge carrier mobility listed in Table 9.2 for the alternating F-BN copolymer may be considerably limited by the chain ends, since this polymer is much shorter than the random copolymers in Table 9.2. On the basis of these experimental results, it is not possible to conclude whether the presence of BN affects the mobility of charges on PF chains due to a modification of the chain conformation or because BN is merely a barrier to charge transport as a result of the close to perpendicular orientation of the naphthyl units. To gain further insight into the charge transport on coiled polymer chains, we have modeled the chain conformation of fluorene–binaphthyl copolymers and we have

Table 9.2 Average chain length, M_n, in monomer units, polydispersity index (pdi), experimental (μ_{exp}), and calculated (μ_{calc}) charge carrier mobility and the calculated persistence length (l_{calc}) of the polyfluorene derivatives that were studied.

Polymer	M_n (units)	pdi	μ_{exp} (cm^2/Vs)	μ_{calc} (cm^2/Vs)*	l_{calc} (units)
Stretched PF	–	–	–	18	∞
PF	493	2.4	0.23	3.2	11.7
PF + 0.04 BN	235	2.0	0.14	1.0	8.7
PF + 0.09 BN	409	2.2	0.08	0.50	6.5
PF + 0.12 BN	321	1.2	0.06	0.38	5.8
F–BN block copolymer	21	1.7	0.01	0.14[a]	2.1

[a] Mobility calculated for infinitely long polymer chains.

simulated the charge transport along these polymer chains in the same way as described above for PPVs. The most important difference between the calculations for PF and PPV is that in the former the three-dimensional nature of the polymer is explicitly taken into account.

As mentioned above, the chain conformation of PF is inherently coiled due to the bending angle ($\varphi = 23.5^\circ$) and the relatively large dihedral angle ($\theta = 45^\circ$) between neighboring units. An example of such a coiled chain conformation is shown in Figure 9.20(a), where the dihedral angles were sampled from a Boltzmann distribution using the calculated potential energy surface for mutual rotation of adjacent fluorene units [50]. The average persistence length calculated for PF calculated in this way is 12 fluorene units, which is close to the experimental value in solution of 10 units. For comparison, Figure 9.20(c) shows the chain conformation for a PF where all dihedral angles are 45°. Such a "stretched"

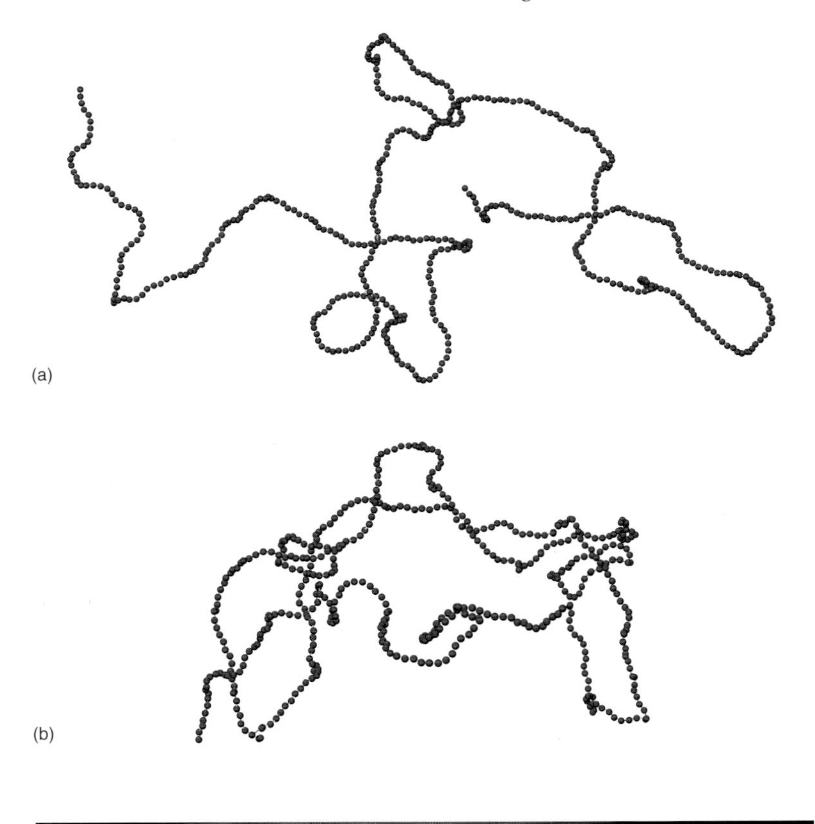

Figure 9.20 Chain conformation for (a) a 500-units-long fluorene chain, (b) a 500-unit-long polyfluorene chain containing a fraction of 0.0944 BN, and a stretched 100-unit-long polyfluorene.

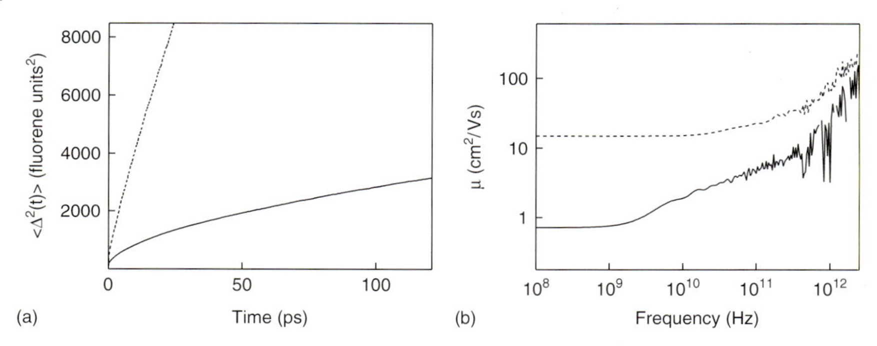

Figure 9.21 (a) Mean squared displacement as a function of time and (b) isotropic mobility as a function of frequency calculated for a hole along the contour of the polymer (dotted line) and for a hole moving in three-dimensional space (solid line).

PF adopts a helical conformation with a length of ~97% of the contour length. Introduction of BN units leads to even more coiled conformations, since BN fully prevents PF from adopting a planar conformation. The dihedral angle between the naphthyl units in BN is close to 90°. In Figure 9.20(b), a chain conformation is shown for a copolymer with a BN fraction of 0.09. The calculated persistence length for the experimentally studied PFs is listed in Table 9.2. It is evident that the persistence length drops rapidly with increasing BN fraction.

The mean squared displacement for an infinitely long fluorene chain is shown in Figure 9.21. The dotted line shows the mean squared displacement along the contour of the chain, whereas the full line shows the actual three-dimensional mean squared displacement of the charge in space, taking into account the coiled nature of the polymer. The motion of charges along the contour of the chain is determined by the electronic coupling (and hence by the dihedral angles) between adjacent fluorene units. On a short timescale, the mean squared displacement is dominated by charges moving on parts of the polymer chains with relatively high electronic coupling. On a longer timescale, various electronic couplings are encountered according to the distribution of dihedral angles. As a result, the motion of the charge occurs by normal Gaussian diffusion at longer times and the mean squared displacement along the contour of the chain increases linearly with time, according to Eq. (9.8). As a consequence of the coiled chain conformation of PF, the direction of the motion of the charge carrier changes in time and the actual mean squared displacement increases more slowly with time than the mean squared displacement along the contour of the polymer. Only after approximately 100 ps, when the charge has encountered parts of the polymer with various curvatures, the motion of the charge occurs by Gaussian diffusion with the mean squared displacement as described by Eq. (9.8).

As before, from the mean squared displacement as a function of time, the frequency-dependent charge carrier mobility can be obtained by numerically integrating the curves in Figure 9.21(a) according to Eq. (9.7). The three-dimensional mobility along the contour of the polymer chain as a function of frequency is calculated from the mean squared displacement along the contour of the polymer, whereas the spatial isotropic mobility is calculated from the mean squared displacement in three-dimensional space. The results of this integration are shown in Figure 9.21(b). At a very high probing frequency ($>10^{12}$ Hz), the motion of the charges is probed on a length scale that is smaller than the persistence length of PF. Because of this, the mobility along the contour of the polymer chain becomes equal to the mobility in three-dimensional space. The charge transport that is probed at this frequency is dominated by the motion on relatively planar parts of the chain. Upon decreasing the probing frequency, both the contour mobility and the spatial mobility decrease due to the larger angles that are encountered along the polymer chain during the oscillation period of the probing electric field. This situation is very similar to that for the PPVs discussed above. What is specific for the PF is that the spatial isotropic mobility exhibits a more pronounced decrease upon lowering the probing frequency than the contour mobility. This is a result of the coiling of the PF chains. The DC mobility of positive charges on coiled PF is found to be more than an order of magnitude lower than the DC mobility along the contour of the PF chains, which clearly illustrates the limiting effect of a coiled chain conformation on the mobility of charges.

Charge transport on PF–BN copolymers and on the PF–BN alternating copolymer was simulated, using the torsion potentials and electronic couplings between fluorene and naphthyl units [50]. The microwave mobilities calculated for these polymers are listed in Table 9.2. The mobility of holes on PF–BN copolymers is found to decrease by approximately a factor of 2 upon incorporation of a BN fraction of 0.05 in the polymer backbone. For the alternating copolymer, the calculated mobility is more than an order of magnitude lower than the mobility for PF. Incorporation of BN in the PF backbone reduces the charge carrier mobility because of two effects. The close to perpendicular orientation of two alkoxy-substituted naphthyls results both in a lowered electronic coupling and in a more strongly coiled chain conformation, as compared to PF.

9.7.1 Comparison of Experimental and Calculated Data

The calculated results are in qualitative agreement with the experimental results in the sense that the charge carrier mobilities are found to decrease with increasing BN fraction and that the mobility for the alternating copolymer is nonzero. However, the absolute values of the calculated mobilities are higher than the values determined experimentally with PR-TRMC on polymers in dilute solution. This difference can (in part) be attributed to the effect of the solvent, which is not taken into account in the charge transport simulations. The reorganization of the solvent molecules around a charge on a polymer chain may limit the efficiency of charge transport. Moreover, also the chain conformation is affected by the interaction between the polymer and the solvent. The conformation of PF in solution is more coiled (persistence length of 10 units) [51] than calculated on basis of the angles and dihedral angles (persistence length of 12 units). In addition, the finite length of the polymers used in the experimental study can lead to differences between the calculated and experimental mobilities in Table 9.2. The number averaged chain lengths listed in Table 9.2 were determined by GPC with a polystyrene reference. These chain lengths could be overestimated because of the difference in chain conformation of polystyrene and fluorene–binaphthyl copolymers. The calculated mobility is strongly affected by the (average) chain length as was shown above for PPV. For example, the calculated mobility for a Flory distribution of chain lengths with an average length of 150 fluorene units is a factor of 2 lower than for infinitely long fluorene chains. Equivalently, the calculated mobility is strongly affected by defects in the polymer backbone. Another possible explanation for the trend in the factor between the calculated and experimental mobility is the chain conformation. The calculated chain conformation with a large BN fraction might be closer to the actual chain conformation than for PF with no (or a small fraction of) BN incorporated in the backbone.

9.8
Supramolecular Control of Charge Transport along Molecular Wires

In the preceding sections of this chapter, we have shown that torsional disorder and three-dimensional conformation of the polymer chains have a strong influence on the mobility of charges. This shows that control over the microscopic conformation of molecular wires can lead to significant enhancements of the charge transport properties. An attractive way to control molecules on a nanometer scale is to exploit molecular self-assembly [52]. An example of a molecular wire in which the conformation can be controlled by such self-assembling properties is the conjugated porphyrin polymer shown in Figure 9.22. The polymer consists of zinc-porphyrins that are linked at the meso-position by butadiyne units. It was shown recently by photo-induced electron transfer measurements that such porphyrin wires can transfer charge efficiently over long distances [53]. EPR measurements on related ethyne-linked porphyrin oligomers indicate that they can mediate essentially barrier-less hole transport over distances of up to 75 Å. In solution, the conjugation length of these polymers is limited by the low barrier for rotation of one porphyrin with respect to its neighbors [54].

The molecular conformation of these polymers can be altered by adding bidentate ligands such as 4,4′-bipyridyl (Bipy), which leads to the formation of double-strand ladder-like assemblies, Figure 9.22 [55, 56]. In these ladder structures, the conjugation length is increased, amplifying the two-photon absorption [56] and nonlinear refraction [57].

We have studied these porphyrin polymers using the PR-TRMC method to establish the effect that the formation of the ladder structure has on the efficiency of charge transport along the chains. The result of such a PR-TRMC measurement for a dilute solution of the porphyrin polymers is shown in Figure 9.23(a). From the fit an isotropic mobility of 0.028 cm^2/Vs was found for single-strands of the

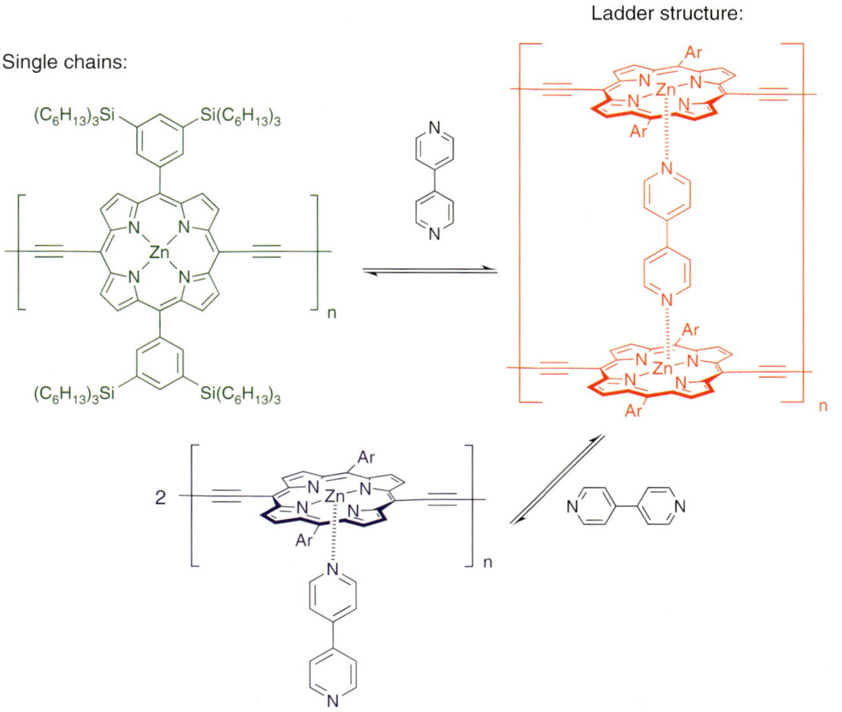

Figure 9.22 Formation of porphyrin polymer ladders.

porphyrin polymer. If it is assumed that the porphyrins adopt a straight rod-like conformation, then the one-dimensional mobility along the chains is three times this value; that is, 0.084 cm^2/Vs. The data shown in Figure 9.23 were obtained for a porphyrin polymer with a molecular weight corresponding to a number-average degree of polymerization of about 50 monomer units. Additional experiments on porphyrin polymers with average chain lengths of 10 and 37 monomer units yielded similar conductivity transients and mobility values. This is rather interesting since it was discussed above that the mobility (at 34 GHz) measured by the same method for LPPP and PPV derivatives strongly depends on the chain length. From those experiments, it was concluded that the measured mobility is limited by the chain ends and that the actual (DC or low frequency) mobility is considerably higher than the measured value at 34 GHz. Values of 600 and 60 cm^2/Vs were inferred for the LPPP and PPV polymers. In the present case, we find that the mobility does not depend on chain length, which leads to the conclusion that the DC mobility of the charge along the porphyrin chains is equal to the values measured at 34 GHz; that is, 0.084 cm^2/Vs.

Addition of Bipy to the solution of the porphyrin polymer leads to the formation of ladder structures as shown in Figure 9.22. In Figure 9.23(b), the result of a PR-TRMC measurement on an oxygen-saturated solution of the porphyrin

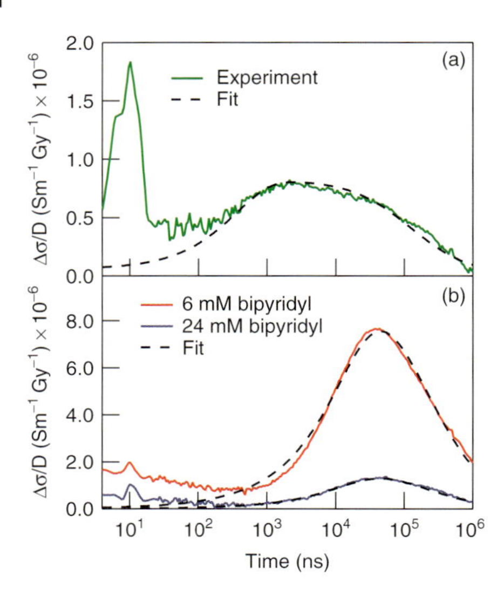

Figure 9.23 Change in microwave conductivity on irradiation of solutions of porphyrin polymer without (a) and with (b) 4,4′-bipyridyl.

polymer (0.25 mM in monomer units) containing a 6-mM concentration of Bipy is shown. Under these conditions, the polymer exists as the double-strand ladder (see Figure 9.22). Comparison with the transient in Figure 9.23(a) shows that the addition of Bipy leads to an increase in conductivity by roughly an order of magnitude. The isotropic microwave mobility obtained from fits to the transient for the solution containing the porphyrin ladders is 0.31 cm^2/Vs, corresponding to a one-dimensional mobility of 0.93 cm^2/Vs. This mobility is more than an order of magnitude larger than discussed above for isolated porphyrin wires, showing directly that planarization of the chains strongly improves charge transport. The mobility value of 0.93 cm^2/Vs is higher than typical DC mobilities for interchain transport in thin films of conjugated polymers. However, the mobility is lower than for intrachain transport along defect-free conjugated polymer chains.

Interestingly, the transfer of positive charges to the polymer is considerably slower in Figure 9.23(b) than in the absence of Bipy (Figure 9.23(a)). We attribute this to a stepwise charge transfer, in which the charges first transfer to Bipy and subsequently to the porphyrin chain. The slower increase in conductivity is explained by either a much lower diffusion coefficient of the $Bipy^+$ radical cation, due to the formation of cation–π complexes, or nondiffusion limited kinetics.

Increasing the concentration of Bipy above 6 mM leads to a decrease in the amplitude of the conductivity transients as shown in Figure 9.23(b), due to the dissociation of ladders to $\mathbf{1} \cdot \mathrm{Bipy}_n$ single strands (Figure 9.22) [57]. For the solution with a 24-mM concentration of Bipy, a mobility value close to that of the noncomplexed single-strand porphyrin polymer was obtained, and the amplitude of the conductivity transient becomes similar to that in Figure 9.23(a). This shows that the mobility of positive charges along porphyrin-based molecular wires can be increased by an order of magnitude by formation of the double-strand Bipy ladder complex.

9.9 Summary and Outlook

The mobility of charges moving along conjugated polymers in solution can be measured by using the PR-TRMC technique. We have shown that relatively high microwave mobility values are measured for a variety of π- and σ-conjugated polymers, ranging between 0.1 and 1 cm^2/Vs. It was shown that these already high values are severely limited by scattering of the charge at the ends of the chains, which implies that the actual intrachain mobility is even considerably higher. We have compared the experimental data obtained for conjugated chains of varying length or polymers containing well-defined conjugated lengths with theoretical models and conclude that the actual DC mobility can be up to 3 orders of magnitude higher than the measured microwave mobility. For fully planar conjugated chains of phenyl units, a mobility as high as 600 cm^2/Vs has been inferred from the analysis of PR-TRMC measurements. Introduction of torsional disorder was shown to decrease the mobility by an order of magnitude to 60 cm^2/Vs, while coiling of the polymer chains leads to a lowering of the mobility by another order of magnitude.

The measurements and calculations described in this chapter directly indicate the potential of conjugated polymers and oligomers for use as wires in single-molecule electronics, but also have implications for the understanding of the performance of these materials in thin film devices. In optoelectronic devices, the charge carriers have to move through the active material from one electrode to the other over a distance of the order of 100 nm. Polymers can consist of several hundreds of repeat unit and can be as long as a few hundred nanometers. If conjugated chains could be organized in such a way that the distance between the electrodes is bridged by a single extended polymer chain, as depicted in Figure 9.24(a), the mobility measured in a device would correspond to the one-dimensional mobility that was derived here for linear polymer; that is, ranging 600 cm^2/Vs for planar ladder-type chains to 60 cm^2/Vs for a chain with torsional disorder.

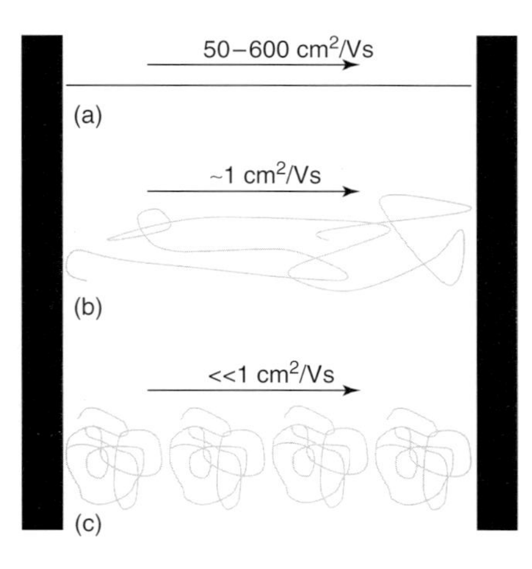

Figure 9.24 Schematic representation of conjugated polymer chains in devices in which the distance between the electrodes is bridged by (a) a single stretched chain, (b) a single coiled chain, and (c) multiple strongly coiled polymer chains.

Even if the polymer chains are not fully extended between the electrodes, much higher mobilities than generally obtained in device measurements can be expected. If a single-coiled PF chain is positioned between two electrodes (see Figure 9.24(b)), the transport of charge takes place with a mobility that corresponds to the actual DC mobility in three dimensions rather than the mobility along the chain. In other words, the charge moves along the coiled chain with a very high mobility, while the coiling reduces the actual displacement in space. As shown above, this leads to a significant lowering of the mobility to a value of the order of 1 cm^2/Vs.

At a certain degree of coiling, the spatial extent of the polymer chains becomes so small that a single polymer coil is not sufficient to bridge the gap between the two electrodes (see Figure 9.24(c)). This means that the charge transport has to take place through multiple polymer chains, which involves interchain charge transport steps. These interchain charge transfer events are the slowest steps in the charge transport and therefore limit the mobility that is obtained from measurements in devices.

This discussion illustrates the strong effect of the molecular organization in thin film polymer devices on the charge carrier mobility. Devices in which polymers adopt a regular, stretched structure are expected to exhibit a considerably higher mobility. This has been shown for poly(3-hexyl-thiophene) [58]. The results that we present here show that the intrinsic mobility in all conjugated polymers can be very high, if the chain conformation is controlled. This indicates that control over the molecular organization is a key issue in the processing of devices based on conjugated polymers.

References

1. Moore, G.E. (1965) *Electronics*, **38**, 114–117.
2. Moth-Poulsen, K. and Bjornholm, T. (2009) *Nat. Nanotechnol.*, **4**, 551–556.
3. Hush, N. (2003) *Ann. N.Y. Acad. Sci.*, **1006**, 1–20.
4. James, D.K. and Tour, J.M. (2005) *Top. Curr. Chem.*, **257**, 33–62.
5. Ratner, M.A., Davis, B., Kemp, M., Mujica, V., Roitberg, A., and Yaliraki, S. (1998) *Ann. N.Y. Acad. Sci.*, **852**, 22–37.
6. Chung, S.-J., Jin, J.-I., and Kim, K.-K. (1997) *Adv. Mater.*, **9**, 551–554.
7. van Hutten, P.F. and Hadziioannou, G. (2000) in *Semiconducting Polymers* (eds G. Hadziioannou and P.F. van Hutten), Wiley-VCH Verlag GmbH, Weinheim.
8. Kudernac, T., Katsonis, N., Browne, W.R., and Feringa, B.L. (2009) *J. Mater. Chem.*, **19**, 7168–7177.
9. Choi, S.H., Kim, B., and Frisbie, C.D. (2008) *Science*, **320**, 1482–1486.
10. McCreery, R.L. and Berggren, A.J. (2009) *Adv. Mater.*, **21**, 4303–4322.
11. Reed, M.A., Zhou, C., Muller, C.J., Burgin, T.P., and Tour, J.M. (1997) *Science*, **278**, 252–254.
12. Slowinski, K., Fong, K.Y., and Majda, M. (1999) *J. Am. Chem. Soc.*, **121**, 7257–7261.
13. Davis, W.B., Naydenova, I., Haselsberger, R., Ogrodnik, A., Giese, B., and Michel-Beyerle, M.E. (2000) *Angew. Chem. Int. Ed. Engl.*, **39**, 3649–3652.
14. Weiss, E.A., Ahrens, M.J., Sinks, L.E., Gusev, A.V., Ratner, M.A., and Wasielewski, M.R. (2004) *J. Am. Chem. Soc.*, **126**, 5577–5584.
15. Goldsmith, R.H., Sinks, L.E., Kelley, R.F., Betzen, L.J., Liu, W., Weiss, E.A., Ratner, M.A., and Wasielewski, M.R. (2005) *Proc. Natl. Acad. Sci. USA*, **102**, 3540–3545.

16. Hoofman, R.J.O.M., de Haas, M.P., Siebbeles, L.D.A., and Warman, J.M. (1998) *Nature*, **392**, 54–56.
17. Grozema, F.C., Hoofman, R.J.O.M., Candeias, L.P., de Haas, M.P., Warman, J.M., and Siebbeles, L.D.A. (2003) *J. Phys. Chem. A*, **107**, 5976–5986.
18. Grozema, F.C., Siebbeles, L.D.A., Warman, J.M., Seki, S., Tagawa, S., and Scherf, U. (2002) *Adv. Mater.*, **14**, 228–231.
19. Warman, J.M. (1982) in *The Study of Fast Processes and Transient Species by Electron Pulse Radiolysis* (eds J.H. Baxendale and F. Busi), Reidel, Dordrecht, pp. 433–533.
20. Warman, J.M. (1982) in *The Study of Fast Processes and Transient Species by Electron Pulse Radiolysis* (eds J.H. Baxendale and F. Busi), D. Reidel Publishing Company, Dordrecht, pp. 129–161.
21. Warman, J.M., de Haas, M.P., Dicker, G., Grozema, F.C., Piris, J., and Debije, M.G. (2004) *Chem. Mater.*, **16**, 4600–4609.
22. Luthjens, L.H., Hom, M.L., and Vermeulen, M.J.W. (1978) *Rev. Sci. Instrum.*, **49**, 230–235.
23. Schouten, P.G., Warman, J.M., and de Haas, M.P. (1993) *J. Phys. Chem.*, **97**, 9863–9870.
24. Warman, J.M., de Haas, M.P., and Hummel, A. (1973) *Chem. Phys. Lett.*, **22**, 480–483.
25. Infelta, P.P., de Haas, M.P., and Warman, J.M. (1977) *Radiat. Phys. Chem.*, **10**, 353–365.
26. Itoh, K. and Holroyd, R. (1990) *J. Phys. Chem.*, **94**, 8850–8854.
27. Huang, S.S.-S. and Freeman, G.R. (1980) *J. Chem. Phys.*, **72**, 1989–1993.
28. Candeias, L.P., Grozema, F.C., Padmanaban, G., Ramakrishnan, S., Siebbeles, L.D.A., and Warman, J.M. (2002) *J. Phys. Chem. B*, **107**, 1554–1558.
29. Kubo, R. (1957) *J. Phys. Soc. Jpn.*, **12**, 570–586.
30. Scher, H. and Lax, M. (1973) *Phys. Rev. B*, **7**, 4491–4502.
31. Dyre, J.C. and Schrøder, T.B. (2000) *Rev. Mod. Phys.*, **72**, 873–892.
32. Grozema, F.C. and Siebbeles, L.D.A. (2008) *Int. Rev. Phys. Chem.*, **27**, 87–138.
33. Miller, R.D. and Michl, J. (1989) *Chem. Rev.*, **89**, 1359.
34. Teramae, H. and Takeda, K. (1989) *J. Am. Chem. Soc.*, **111**, 1281–1285.
35. Mintmire, J.W. (1989) *Phys. Rev. B*, **39**, 13350–13357.
36. Van der Laan, G.P., de Haas, M.P., Hummel, A., Frey, H., and Möller, M. (1996) *J. Phys. Chem.*, **100**, 5470–5480.
37. Grozema, F.C., van Duijnen, P.T., Berlin, Y.A., Ratner, M.A., and Siebbeles, L.D.A. (2002) *J. Phys. Chem. B*, **106**, 7791–7795.
38. Lieser, G., Oda, M., Miteva, T., Meisel, A., Nothofer, H.-G., Scherf, U., and Neher, D. (2000) *Macromolecules*, **33**, 4490–4495.
39. Prins, P., Grozema, F.C., Schins, J.M., Patil, S., Scherf, U., and Siebbeles, L.D.A. (2006) *Phys. Rev. Lett.*, **96**, 146601.
40. Prins, P., Grozema, F.C., Schins, J.M., and Siebbeles, L.D.A. (2006) *Phys. Stat. Sol. B*, **243**, 382–386.
41. Prins, P., Grozema, F.C., and Siebbeles, L.D.A. (2006) *J. Phys. Chem. B*, **110**, 14659–14666.
42. Hendry, E., Schins, J.M., Candeias, L.P., Siebbeles, L.D.A., and Bonn, M. (2004) *Phys. Rev. Lett.*, **92**, 196601.
43. Hendry, E., Koeberg, M., Schins, J.M., Nienhuys, H.K., Sundstrom, V., Siebbeles, L.D.A., and Bonn, M.A. (2005) *Phys. Rev. B*, **71**, 125201.
44. Prins, P., Grozema, F.C., Schins, J.M., Savenije, T.J., Patil, S., Scherf, U., and Siebbeles, L.D.A. (2006) *Phys. Rev. B*, **73**, 045204.
45. Prins, P., Grozema, F.C., and Siebbeles, L.D.A. (2006) *Mol. Simul.*, **32**, 695–705.
46. Roex, H., Adriaense, P., Vanderzande, D., and Gelan, J. (2003) *Macromolecules*, **36**, 5613–5622.
47. Becker, H., Spreitzer, H., Ibrom, K., and Kreuder, W. (1999) *Macromolecules*, **32**, 4925–4932.
48. Rabe, T., Hoping, M., Schneider, D., Becker, E., Johannes, H.H., Kowalsky, W., Weimann, T., Wang, J., Hinze, P., Nehls, B.S., Scherf, U., Farrell, T., and

Riedl, T. (2005) *Adv. Funct. Mater.*, **15**, 1188–1192.

49. Prins, P., Candeias, L.P., Van Breemen, A.J.J.M., Sweelssen, J., Herwig, P.T., Schoo, H.F.M., and Siebbeles, L.D.A. (2005) *Adv. Mater.*, **17**, 718–723.

50. Prins, P., Grozema, F.C., Galbrecht, F., Scherf, U., and Siebbeles, L.D.A. (2007) *J. Phys. Chem. C*, **111**, 11104–11112.

51. Grell, M., Bradley, D.D.C., Long, X., Chamberlain, T., Inbasekaran, M., Woo, E.P., and Soliman, M. (1998) *Acta Polym.*, **49**, 439–444.

52. Hoeben, F.J.M., Jonkheijm, P., Meijer, E.W., and Schenning, A.P.H.J. (2005) *Chem. Rev.*, **105**, 1491–1546.

53. Winters, M.U., Dahlstedt, E., Blades, H.E., Wilson, C.J., Frampton, M.J., Anderson, H.L., and Albinsson, B. (2007) *J. Am. Chem. Soc.*, **129**, 4291–4297.

54. Winters, M.U., Karnbratt, J., Eng, M., Wilson, C.J., Anderson, H.L., and Albinsson, B. (2007) *J. Phys. Chem. C*, **111**, 7192–7199.

55. Taylor, P.N. and Anderson, H.L. (1999) *J. Am. Chem. Soc.*, **121**, 11538–11545.

56. Drobizhev, M., Stepanenko, Y., Rebane, A., Wilson, C.J., Screen, T.E.O., and Anderson, H.L. (2006) *J. Am. Chem. Soc.*, **128**, 12432–12433.

57. Screen, T.E.O., Thorne, J.R.G., Denning, R.G., Bucknall, D.G., and Anderson, H.L. (2002) *J. Am. Chem. Soc.*, **124**, 9712–9713.

58. Sirringhaus, H., Brown, P.J., Friend, R.H., Nielsen, M.M., Bechgaard, K., Langeveld-Voss, B.M.W., Spiering, A.J.H., Janssen, R.A.J., Meijer, E.W., Herwig, P., and de Leeuw, D.M. (1999) *Nature*, **401**, 685–688.

Part IV
Exciton Transport through Conjugated Molecular Wires

Charge and Exciton Transport through Molecular Wires. Edited by L.D.A. Siebbeles and F.C. Grozema

ISBN: 978-3-527-32501-6

10
Structure Property Relationships for Exciton Transfer in Conjugated Polymers

Trisha L. Andrew and Timothy M. Swager

10.1
Introduction

Conjugated polymers (CPs) are useful materials that combine the optoelectronic properties of semiconductors with the mechanical properties and processing advantages of plastics. In general, CPs in their neutral state are wide band-gap semiconductors with direct band gaps [1]. Many CPs have an extremely large absorption cross-section ($\sigma \approx 10^{-15}$ cm^2) because the $\pi \rightarrow \pi^*$ transition is allowed and the quasi one-dimensional electronic wavefunctions have a high density of states at the band edge [2]. Additionally, a CP can exhibit strong luminescence depending on the system. The luminescence efficiency is primarily related to the delocalization and polarization of the electronic structure of the CP [1].

A vast number of studies on oligomers confirm that the electronic states in a CP have limited delocalization, and the electronic structure of a given CP is often determined by 7–13 repeating units. This is particularly prevalent in systems containing aromatic rings since the aromatic character localizes the electronic wavefunctions. As a result of this localization, a CP's band gap is largely determined by its local electronic structure [1].

The emission of CPs is dominated by energy migration to local minima in their band structures. For example, the emission from electroluminescent devices occurs from regions with greatest conjugation [3] and the emission from complex ladder polymers can be dominated by defect sites present in low concentration [4]. However, a fundamental understanding of the relative mechanisms of energy migration in these systems remains elusive. This inherent difficulty is a result of the fact that CPs have disordered dynamic conformations that produce variable electronic delocalization, both within a given polymer and between neighboring polymer chains [5]. To improve this situation, it is necessary to design polymers with specific structures and properties intended to test proposed mechanisms of energy migration.

In this chapter, we review the photophysical properties of an exciton transport in a series of poly(*p*-phenyl ethynylene)s (PPEs, see Figure 10.1). First, the use of energy migration in PPEs to create signal gain in chemical sensors is discussed.

Charge and Exciton Transport through Molecular Wires. Edited by L.D.A. Siebbeles and F.C. Grozema

ISBN: 978-3-527-32501-6

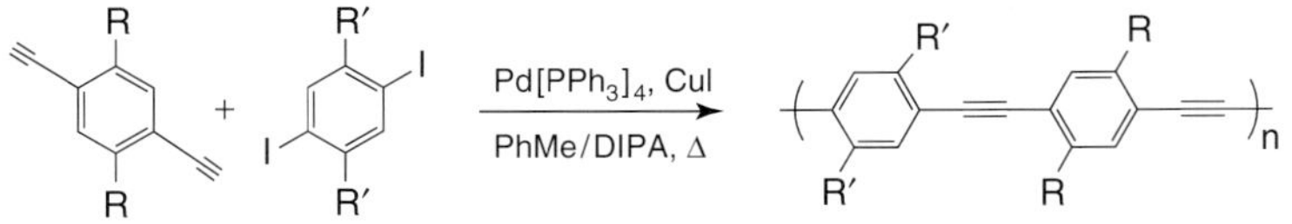

Figure 10.1 A general synthetic route to poly(*p*-phenylene ethynylene)s (PPEs).

Next, we detail the importance of dimensionality and molecular design in directing excitations and the effect of excited-state lifetime modulation on energy migration in PPEs. The ability to extend electronic delocalization and induce well-defined CP conformations in liquid crystal (LC) solutions is also discussed. We restrict our analysis to primarily PPEs, as details relating to energy migration in other common luminescent polymers, such as poly(phenylene)s and poly(*p*-phenylene vinylene)s (PPVs), will be addressed in later chapters.

10.2 Signal Gain in Amplifying Fluorescent Polymers

Rigid rod CPs, such as PPEs (Figure 10.1), may be thought of as "molecular wires" with well-defined lengths proportional to molecular weight. Therefore, PPEs can be used to interconnect, or wire in series, receptors to produce fluorescent chemosensory systems with sensitivity enhancements over single receptor analogs [6]. In a PPE with a receptor attached to every repeat unit, the degree of polymerization defines the number of receptor sites, n. If energy migration is rapid with respect to the fluorescence lifetime, then the excited state can sample every receptor in the polymer, thereby allowing the occupation of a single binding site to dramatically change the entire emission of the PPE. In the event that a receptor site is occupied by a quencher, the result is an enhanced deactivation of the excited state [6]. For isolated polymer chains in solution, the sensitivity may be enhanced by as much as n times over single-molecule receptors; however, larger effects may occur in the solid state wherein interpolymer energy transfer may also occur.

This concept is demonstrated by studying the fluorescence-quenching responses of model compounds **1–2** and their corresponding PPEs, **P1–2** (see Figure 10.3). The operative interaction that leads to quenching of fluorescence in these systems is the formation of a pseudorotaxane between a *bis*(*p*-phenylene)-34-crown-10 (BPP) moiety and paraquat, PQ^{2+}, a well-known electron transfer quenching agent (see Figure 10.2) [7].

Electron transfer fluorescence quenching by PQ^{2+} can occur by either dynamic (collisional) or static (associated complex) processes. The dependence of the fluorescence intensity on the quencher concentration follows the Stern–Volmer relationship, whose general form is shown in Eq. (10.1) [8]:

$$F_0/F = (1 + K_D[Q])\exp(V[Q]) \tag{10.1}$$

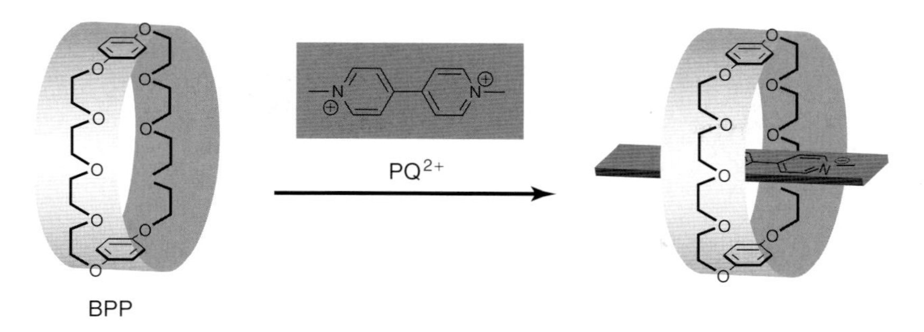

Figure 10.2 The formation of a pseudo-rotaxane between a *bis*(*p*-phenylene)-34-crown-10 (BPP) and paraquat (PQ^{2+}).

In this equation, F_0 and F are the fluorescence intensities in the absence and presence of the quencher, respectively, K_D denotes the dynamic quenching constant, and V represents the static quenching constant. When $[Q]$ or V is very small, the contribution from static quenching can be approximated by a linear function of quencher concentration and Eq. (10.1) simplifies to

$$F_0/F = (1 + K_D[Q])(1 + K_S[Q]) \tag{10.2}$$

where K_S now denotes the static quenching constant. Furthermore, if either a static or dynamic process dominates the quenching response, Eq. (10.2) can be further simplified to include only one linear term:

$$F_0/F = 1 + K_{SV}[Q] \tag{10.3}$$

where K_{SV} is either the dynamic (K_D) or static (K_S) quenching constant. When the fluorophore and the quencher form a simple one-to-one dark complex, K_{SV} is equivalent to the association constant, K_a. However, for systems with more complex species, the quenching profile may deviate from the linear function and the more general form of the Stern–Volmer relationship (i.e., Eq. 10.1) must be used [8].

In static quenching, diffusion rate of the quencher is not a factor and the fluorescence lifetime of the fluorophore, τ, is independent of $[Q]$. However, for purely dynamic quenching, the excited state is quenched by a collision with the quencher and thus the lifetime is truncated with added quencher [8]. As a result, monitoring the changes in the lifetime of the fluorophore with added quencher represents the conventional practice for determining the dynamic quenching constant independent of the static quenching process. The correlation of lifetime with quencher concentration can be expressed as

$$\tau_0/\tau = 1 + K_D[Q] \tag{10.4}$$

The structures of the model compounds and polymers used to investigate amplified fluorescence quenching are shown in Figure 10.3, along with their quenching constants with PQ^{2+} [9]. Comparing **1** and **P1**, a 16-fold enhancement in the dynamic quenching constant is observed upon transitioning from a small molecule to a CP (35 vs. 574 M^{-1}, respectively), even in a system lacking the BPP receptor.

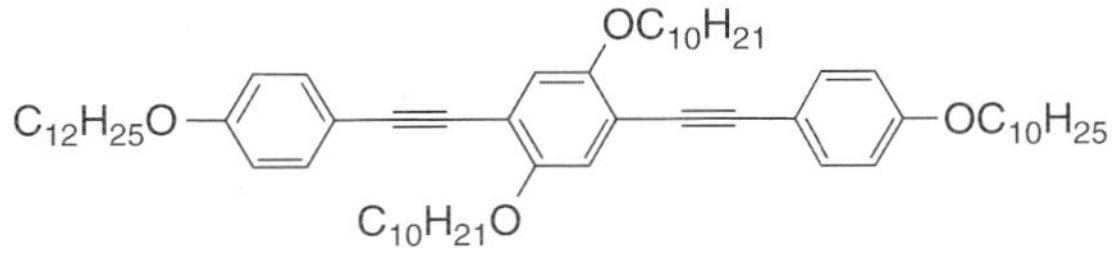

$OC_{10}H_{21}$
$C_{12}H_{25}O$
$OC_{10}H_{25}$
$C_{10}H_{21}O$

1

$K_{SV} = 35\ M^{-1}$ (K_D)

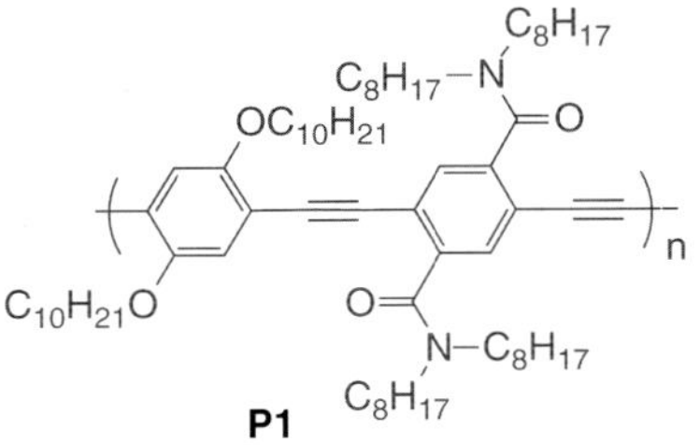

C_8H_{17}
$C_8H_{17}-N$
$OC_{10}H_{21}$
O
O
n
$C_{10}H_{21}O$
$N-C_8H_{17}$
C_8H_{17}

P1

$K_{SV} = 574\ M^{-1}$ (K_D)

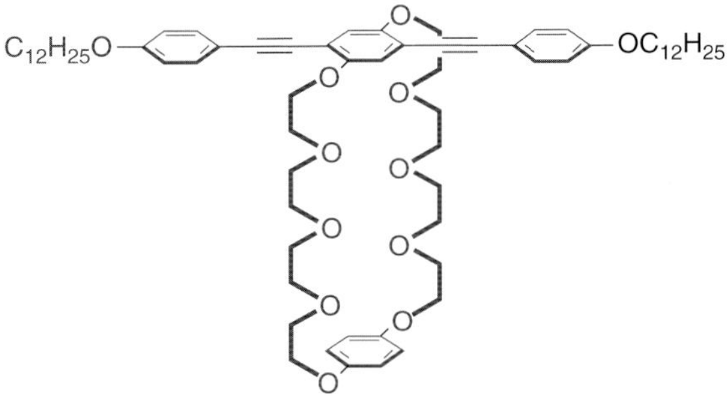

$C_{12}H_{25}O$
$OC_{12}H_{25}$
O O O O O O O O O O

2

$K_{SV} = 1600\ M^{-1}$ (K_S)

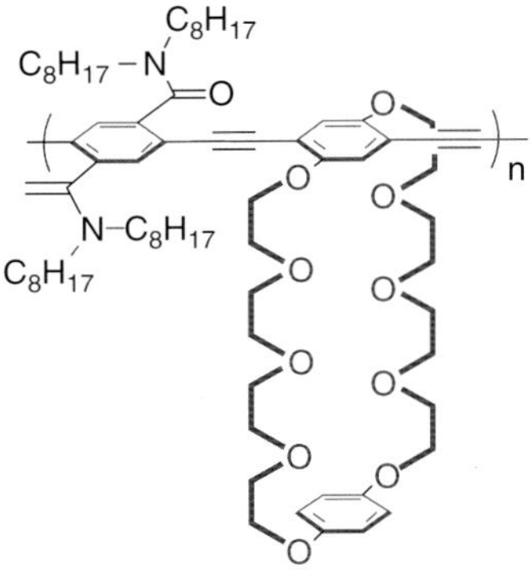

C_8H_{17}
$C_8H_{17}-N$
O
n
$N-C_8H_{17}$
C_8H_{17}
O O O O O O O O O O

P2

$K_{SV} = 105\,000\ M^{-1}$ (K_S)

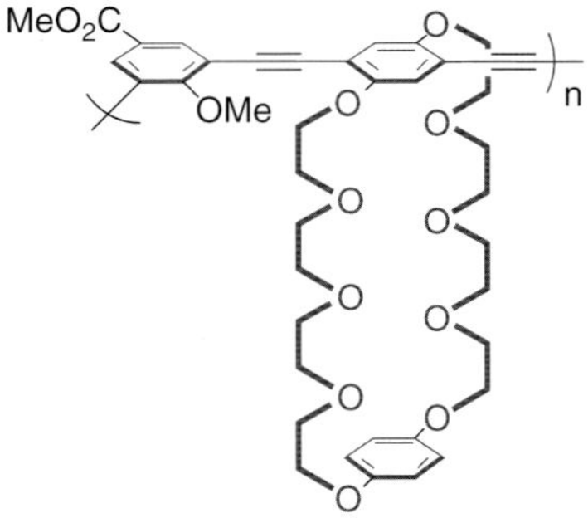

MeO_2C
OMe
n
O O O O O O O O O O

P3

$K_{SV} = 10\,600\ M^{-1}$ (K_S)

$C_{10}H_{21}$
S S S
n
O O O O O O O O O O

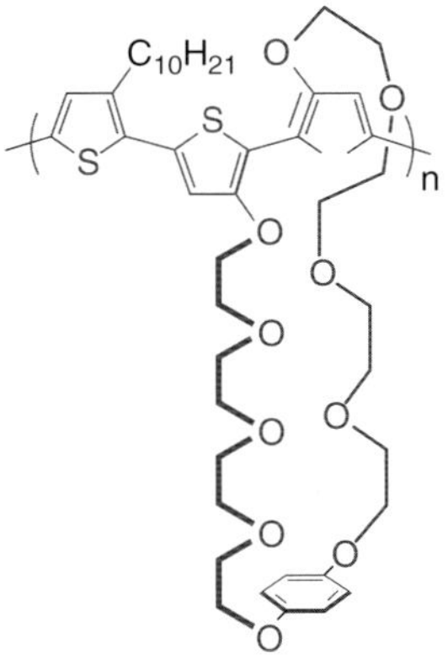

P4

$K_{SV} = 5340\ M^{-1}$ (K_S)

Figure 10.3 Structures of the model compounds and polymers used to investigate amplified fluorescence quenching and their associated quenching constants, K_{SV}.

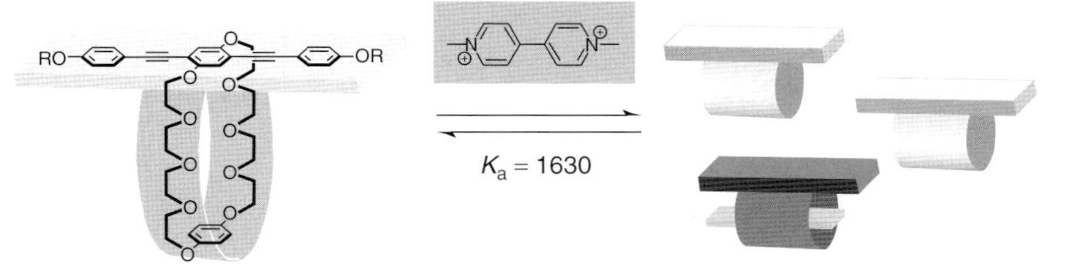

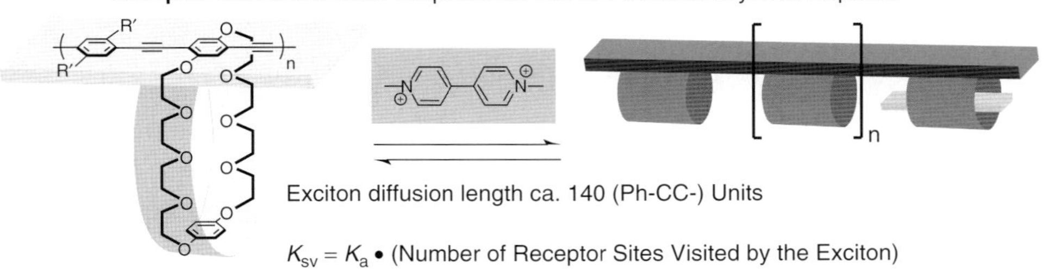

Figure 10.4 Schematic explaining the concept of signal amplification due to a collective system response.

This enhancement results from the extended electronic structure of the polymer, which produces a mobile delocalized excited state and a larger effective size. It is important to note that this enhancement occurs in spite of the fact that the lifetime of **P1** (0.5 ns) is shorter than that of **1** (1.2 ns). When the BPP receptor is introduced, addition of PQ^{2+} results in the formation of a charge transfer (CT) complex with either an associated red-shifted absorption onset or the growth of a new CT band and static quenching dominates. A 66-fold increase in K_S value is observed for CP **P2** relative to small molecule **2**, in agreement with our model for signal amplification due to a collective system response (see Figure 10.4).

The degree of enhancement resulting from energy migration is determined by the radiative lifetime and the mobility of the excitations in the polymer. Longer lifetimes and higher mobilities will produce longer average diffusion lengths. For isolated polymers in solution, if this diffusion length exceeds the length of the polymer, then an increase in molecular weight will produce greater enhancements. Accordingly, lower molecular weight analogs of **P2** display smaller values of K_S than higher molecular weight analogs: $K_S = 105\,000\ M^{-1}$ for M_n 122 500 but $K_S = 75\,000\ M^{-1}$ for M_n 31 100. However, it must be noted that the value of K_S remains largely unaffected after the molecular weight of the polymer exceeds ca. 65 000. This result reveals that the exciton was not able to visit the entire length of the higher molecular weight polymers because of its limited mobility and finite lifetime (there is always competitive relaxation to the ground state). Therefore, one can conclude that the exciton diffusion length in a PPE is approximately 140 (Ph–CC–) units [9].

Additionally, the *para*-linked polymer **P2** is observed to be more effective at energy migration than both its *meta*-linked analog, **P3**, and a poly(thiophene) analog, **P4**. This difference exists in spite of the fact that the lifetime of **P3** (1.88 ns) is about a factor of 3 longer than that of **P2** (0.64 ns). The observation that excitations in **P3** have longer lifetimes indicates that energy migration is slower in this system relative to **P2**.

It can be argued that the greater tendency for energy migration in *para*-linked **P2** over *meta*-linked **P3** might be expected based on delocalization; however, greater delocalization is not a guarantee of superior performance. This fact is illustrated by polymer **P4**, which displays a K_S value of only 5340 M^{-1}. This result indicates that **P4** is less effective at energy migration than **P2**, even though poly(thiophene)s display much greater bandwidths (delocalization) relative to PPEs.

10.3
Directing Energy Transfer within CPs: Dimensionality and Molecular Design

10.3.1
Solutions vs Thin Films

As stated before, the emission of CPs is often dominated by energy migration to local minima in their band structures. For example, selective emission from states associated with anthracene end groups has been demonstrated in solutions of PPEs (see Figure 10.5) [10]. As is characteristic for most PPEs, **P5** displays a broad absorption band centered at 446 nm and a relatively sharp emission spectrum with vibronically resolved (0,0) and (0,1) bands centered at 478 and 510 nm, respectively. In the case of **P6**, however, where terminal anthracene units are present, the solution emission spectrum is dominated by a single band at 524 nm, which corresponds

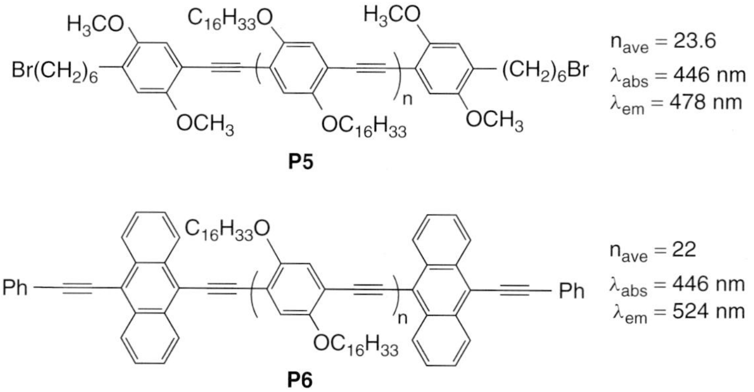

Figure 10.5 Structures of the PPEs used to demonstrate efficient energy transfer to low-energy anthracene end groups.

to emission from the anthracene end group. The fluorescence quantum yields of **P5** and **P6** are roughly comparable (0.35 and 0.28, respectively), the emission spectra of both polymers are insensitive to the excitation wavelength and their corrected excitation spectra match the absorption spectra, thus confirming that the 524 nm band of **P6** results from excitation of the bulk material and subsequent energy migration to the lower energy end groups. By comparison of the emission intensities of the 524 nm band and a residual 474 nm band (that corresponds to fluorescence from the PPE backbone) in the emission spectrum of **P6**, it was concluded that the energy transfer from the polymer backbone to the anthracene end-groups proceeded with >95% efficiency in solution.

Energy transfer in a related PPE, **P7**, (see Figure 10.6) however, was found to be more sensitive to the physical state of the system. In solutions of **P7**, emission from both the polymer backbone and anthracene end-group can be observed in an approximately 2 : 1 ratio. On the other hand, thin films of **P7** *exclusively* display emission from the anthracene end groups at 492 nm. Therefore, transitioning from a 1D solution system to 3D thin films enables intra- and interpolymer energy transfer, which results in more extensive exciton migration to energy minima.

Although PPEs superficially appear to have rigid-rod structures, materials with higher degrees of polymerization exhibit coiled solution structures with persistence lengths of approximately 15 nm [11]. For isolated polymer chains in dilute solutions, the migration of excitations along the polymer backbone follows the random walk statistics of 1D diffusion. Therefore, given that the exciton diffusion length in PPEs is ca. 140 (Ph–CC–) units (see Section 10.2), 1D exciton transport requires

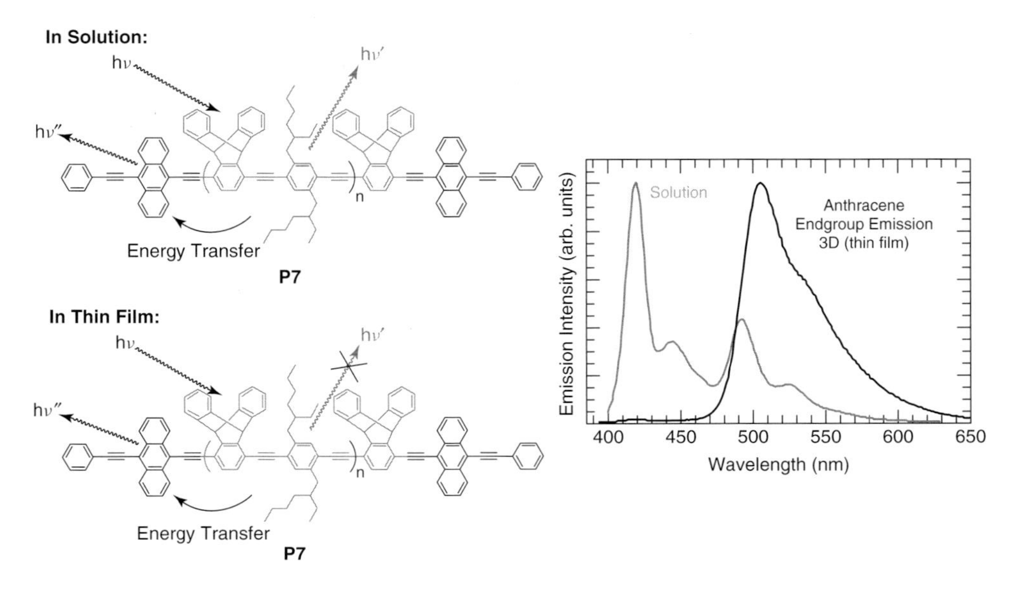

Figure 10.6 Structure and emission spectrum of an anthracene end-capped PPE that displays notably different behavior in isotropic solutions and thin films.

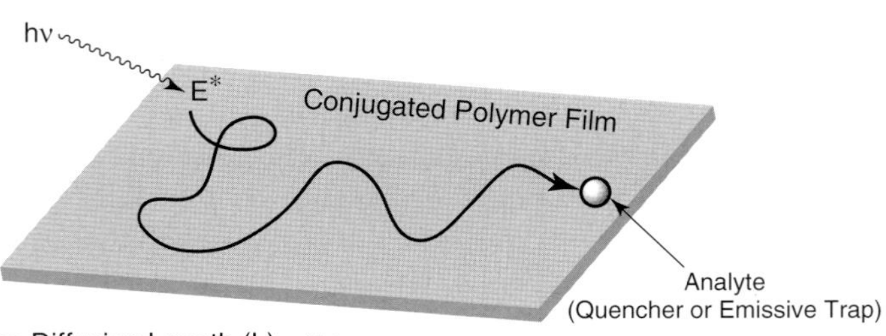

Exciton Diffusion Length (L) = v • τ
v ~ energy transfer rate (effective mass, delocalization, ε, solid state structure, energy topology)
τ ~ lifetime (electronic structure, chain-chain interactions, quenchers)

Figure 10.7 Factors that influence three-dimensional exciton migration in thin films of conjugated polymers.

$(140)^2$ hops to travel 140 linear hops. Although this means that the exciton can theoretically sample approximately 20 000 phenylethynyl repeat units, 1D random walks do not provide the optimal pathway for energy migration (and thus signal amplification) because an excitation necessarily retraces portions of the polymer backbone multiple times. Hence, it is necessary to enable 2D and 3D random walks of the excitations. This increased dimensionality decreases the probability of an excitation retracing a given segment of the polymer and thereby produces a larger amplification in sensory schemes.

It is for this reason that thin films of CPs can serve as unparalled, highly sensitive chemosensors (exemplarily for 2,4,6-trinitrotoluene (TNT) [12]). In general, increases in the diffusion length of the exciton within the CP thin film will enhance the sensitivity of the chemosensor. As shown in Figure 10.7, the exciton diffusion length (L) is provided by the product of the energy transfer rate (v) and lifetime (τ) of the exciton. The energy transfer rate, v, is dependent in part, on the extent of delocalization in the polymer, the effective mass of the exciton, and the energy surface topology of the CP thin film. The lifetime of the exciton is largely defined by the photophysics of the polymer repeat unit and can be further influenced by the presence of interchain interactions and quenchers.

Within thin films, individual polymer chains electronically couple, thus encouraging interpolymer energy transfer. The efficiency of intermolecular energy migration depends on facile dipolar Förster-type processes, which are optimal when the transition dipoles of the donor and acceptor groups are aligned. As a result, films of aligned polymers with extended chain conformations provide an ideal situation for energy migration. Such optimal polymer conformations are best achieved by forming monolayer or multilayer films of PPEs prepared by the Langmuir–Blodgett (LB) deposition technique [13].

For example, a striated multipolymer system composed of three different PPEs with tailored absorption and emission maxima designed to have large spectral overlap between a donor emission and an acceptor absorption can be precisely fabricated using the LB technique (see Figure 10.8) [14]. In this system, spectral overlap encourages energy transfer from **P8** to **P2** and from **P2** to **P9**. Polymers

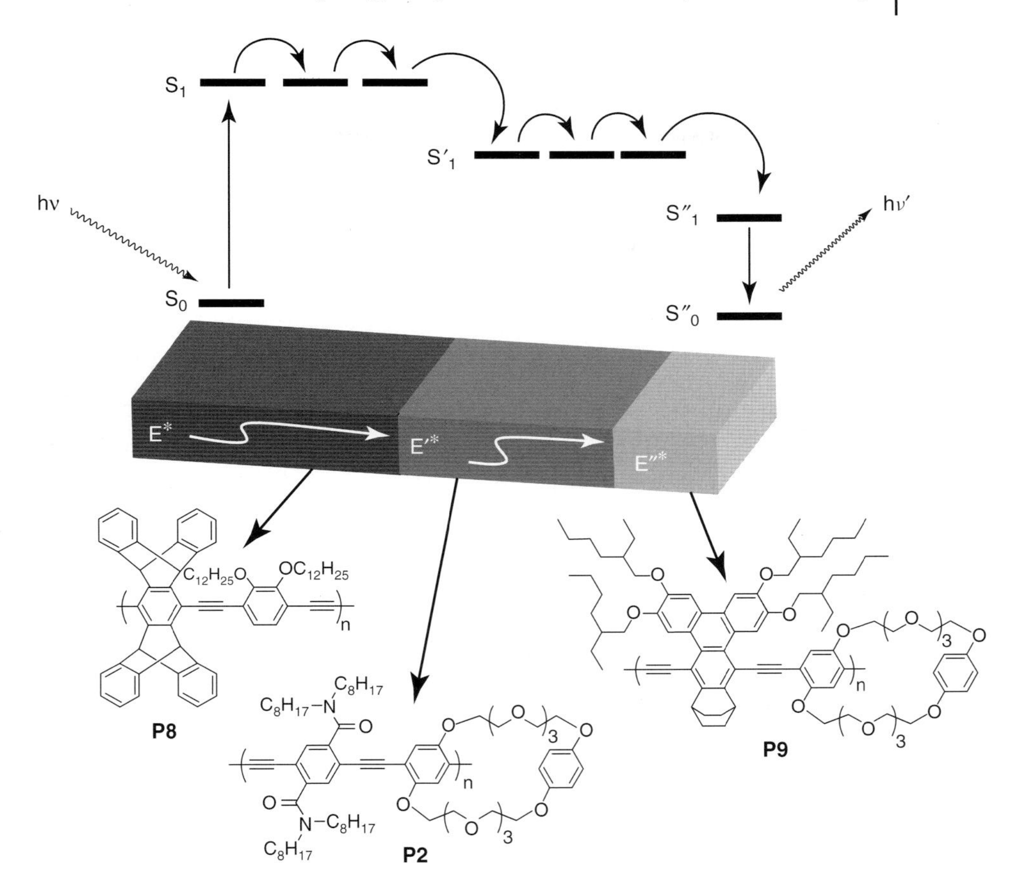

Figure 10.8 A striated multipolymer system that demonstrates vectorial energy transfer normal to the substrate. (Adapted from Ref. [14].)

P2 and **P9** are also nonaggregating and amphiphilic, thus allowing manipulations at the air–water interface. The trilayer assembly was created by first spin-coating the shortest-wavelength polymer, **P8**, on a glass substrate. Next, 16 LB layers of **P2** were coated over this spin-cast film and finally a single monolayer of **P9** was coated, thus providing a composite film where the band gap decreases directionally from the substrate to the polymer–air interface. Excitation of the three-component film at 390 nm (λ_{max} of **P8**) resulted in an emission spectrum consisting of three peaks: two small peaks at 423 and 465 nm, which are attributed to emission from **P8** and **P2**, respectively, and a third, dominant peak at 512 nm that is a result of energy transfer from **P8** and **P2** to **P9** and subsequent emission from **P9**. The observation that most of the excitation energy is transferred from **P8** through 16 layers of **P2** to **P9** demonstrates that energy can be efficiently moved in the *z*-direction, thereby concentrating the energy at the film–air interface. Direct excitation of **P9** at 490 nm resulted in a peak at 512 nm of much lower fluorescence intensity than the peak resulting from excitation at 390 nm; the difference in fluorescence intensity is directly proportional to the difference in optical density at 390 nm versus 490 nm, again confirming the efficiency of energy transfer.

10.3.2
Aggregates

Continuing with the aforementioned practice of using anthracene moieties as low energy emissive traps, PPEs **P10** and **P11** (see Figure 10.9) incorporating anthryl units were synthesized and investigated (polymer **P11** is simply a polyelectrolytic analog of **P10)** [15, 16]. In contrast to polymers **P6** and **P7**, which contained anthracene moieties as end groups, polymers **P10** and **P11** are random copolymers containing small concentrations (1–9%) of an anthryl comonomer. However, the optical effects of incorporating anthryl moieties into the polymer backbone are largely similar to those observed with **P6** and **P7**: new, intense, long-wavelength emission bands that are sensitive to the dimensionality of the system are observed.

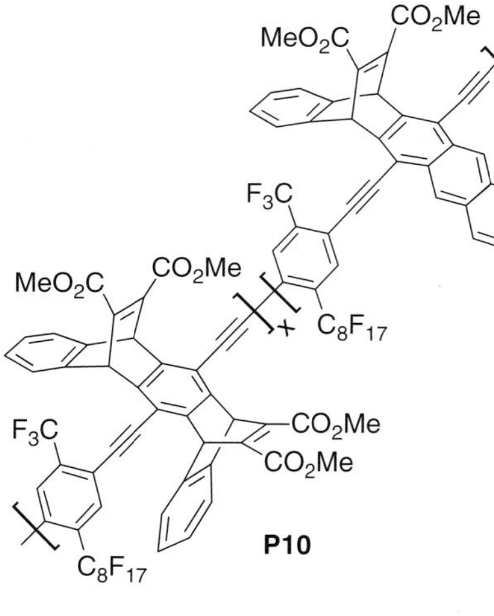

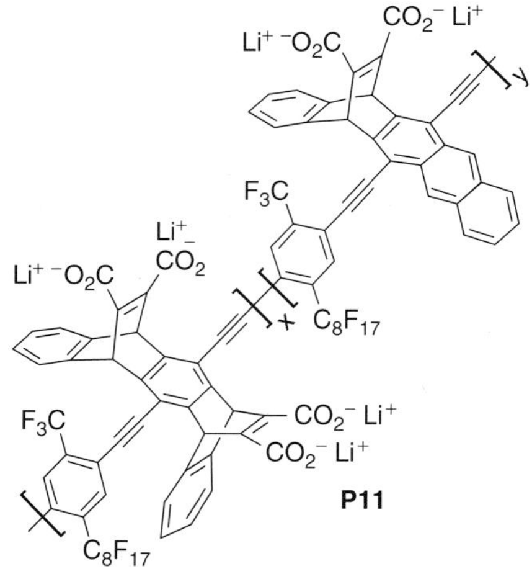

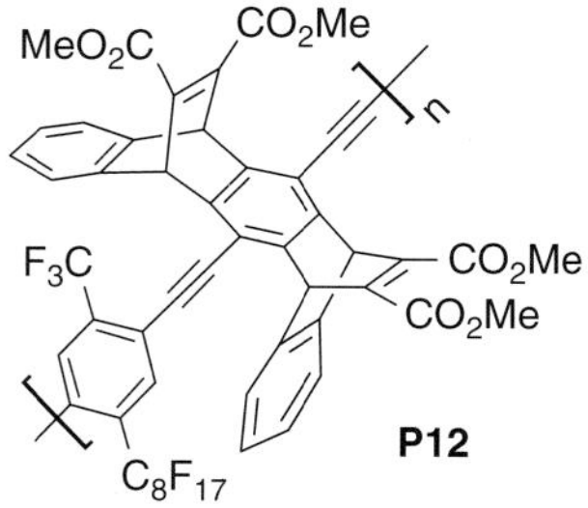

Figure 10.9 Structures of anthryl-incorporated PPEs and a corresponding PPE lacking low-energy trap sites.

For example, the effect of anthracene incorporation can best be appreciated upon comparing the emission spectra of **P10** with that of **P12**, which is an anthryl-free analog of **P10**. PPE **P12** displays the characteristic two-band emission spectrum of PPEs, with an emission maximum at 433 nm in solutions and in thin films. Solution emission spectra of **P10** exhibit these same two bands, but also display a third, green emission band centered between 500 and 520 nm (depending on the percent of anthryl comonomer) that greatly increases in intensity in thin films of **P10**. Furthermore, **P10** has a long (1.5–1.9 ns) excited-state lifetime, as compared to **P12** (0.44 ns).

As discussed before, the green band emission from **P10** is much more pronounced when the polymer is in its film state than when it is dissolved in dilute solutions because of the enhanced exciton migration present in CP films (3D) relative to that present in dilute polymer solutions (1D). In dilute solutions only, *intra*chain exciton migration is possible because individual polymer chains are isolated from one another. However, in the film state, chains of **P10** are aggregated within close proximity to each other such that *inter*chain exciton migration becomes possible. If the low-energy exciton trap sites are emissive, such as the anthryl defect sites in **P10** and **P11**, then they can dramatically alter the emission spectra of CPs in their film state.

To further investigate the effects of exciton migration on luminescence properties, absorption and fluorescence spectroscopy were conducted on PPE solutions in various degrees of aggregation. By adding a poor solvent (i.e., a solvent in which **P10** is in a collapsed or aggregated state) to a PPE solution dissolved in a good solvent (i.e., a solvent in which the polymer is in an expanded and well-dissolved state), one can study the polymers in various degrees of aggregation. In dilute tetrahydrofuran (THF) solution, **P10** was well dissolved and individual polymer chains were isolated; therefore, only *intra*chain exciton migration is possible and the small concentration of emissive exciton traps was not noticeable in the fluorescence spectra. Thus, THF solutions of **P10** appear fluorescent blue, as characterized by the sharp emission band around 432–434 nm. However, in a 50 : 50 THF : H_2O cosolvent mixture, **P10** was present in the aggregated state, held together by hydrophobic and $\pi-\pi$ interactions. Upon aggregation, *inter*chain exciton migration became significant, so the emissive exciton traps noticeably altered the fluorescence spectra, exhibiting a dominant green emission band around 513 nm. If the ratio of the anthryl comonomer in **P10** is increased, the ratio of green to blue emission (I_{green}/I_{blue}) in the aggregated state also increases.

Furthermore, the fluorescence color change of **P10** dispersed in a solid poly(vinyl alcohol) (PVA) matrix was investigated. PVA is a water-soluble polymer that has been widely used to make water-permeable hydrogels. In order to disperse **P10** in PVA, a THF solution of **P10** was quickly added into an aqueous solution of PVA and the resulting, precipitated polymer blend was crosslinked with glutaric dialdehyde before isolation. Upon washing the **P10**/PVA blend with THF, it became fluorescent blue and remained so even after drying. Subsequently, submerging this blend into pure water for 2 min causes it to become fluorescent green. The observed blue-to-green fluorescence color change was attributed to the water-induced aggregation

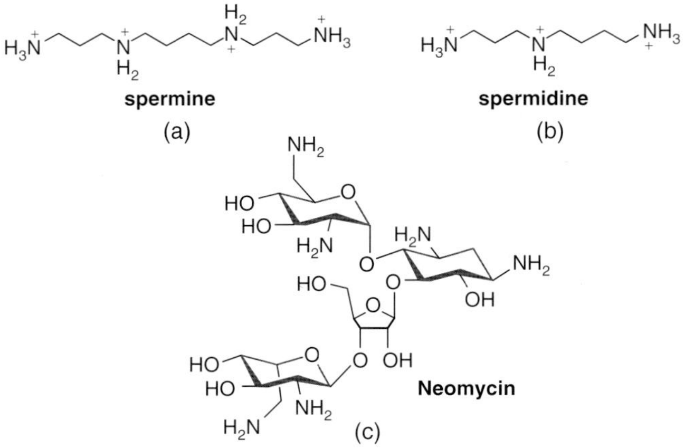

Figure 10.10 Structures of (a) spermine, (b) spermidine, and (c) neomycin.

of the PPE chains within the PVA matrix. Surprisingly, rewashing this fluorescent green blend in THF did not restore blue fluorescence – this was probably due to the difficulty of separating individual PPE chains once they become strongly aggregated in 100% water.

The phenomenon of enhanced exciton trapping in PPE aggregates was exploited to make aggregation-based sensors for nonquenching multicationic analytes [16]. Nonquenching analytes are described as analytes that cannot participate in direct quenching of the inherent fluorescence intensity of a CP via electron transfer or energy transfer due to incompatible redox and spectral properties, relative to the photoexcited CP. Examples of such analytes are biologically relevant small-molecules, such as multicationic spermine and spermidine, and neomycin (see Figure 10.10). It was found that spermine, spermidine, and neomycin induced the formation of tightly associated aggregates in ethanol solutions of the polyelectrolyte **P11**, which was accompanied by a visually noticeable blue-to-green fluorescence color change (Figure 10.11). Dicationic and monocationic amines were not observed to affect this change, thus demonstrating that a conjugated polyelectrolyte sensor relying on nonspecific, electrostatic interactions may still attain a certain level of selectivity.

Although CP aggregates represent another option to enable 3D exciton migration (other than thin films), it must be noted that, with few exceptions, the strong electronic interactions between chains of CPs that accompany aggregation dramatically lower their quantum efficiency. This is due to the phenomenon of self-quenching, which can be generally described as any interaction between an excited molecule, M^*, and a ground-state molecule of the same type, M, that leads to fluorescence quenching of M^* [17]. Therefore, it is generally true that the design principles for maintaining high quantum yields in CPs have been diametrically opposed to those for the optimization of charge and exciton transport, which encourage greater interpolymer contact [18, 19].

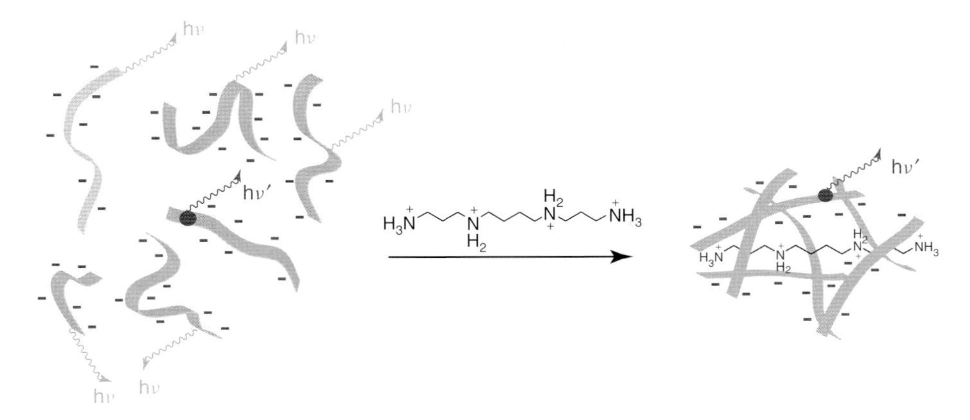

Figure 10.11 Schematic description of aggregation-based sensors for nonquenching, multicationic analytes. (Adapted from Ref. [16].)

However, it *is* possible to produce strongly interacting polymer chains with 3D electronic interactions while maintaining high luminescence efficiency (Figure 10.12). Specifically, it has been proposed that an oblique orientation between neighboring transition dipole moments of CPs will prevent self-quenching [20]. Based on an exciton-coupling model, a parallel orientation of polymer chains is expected to result in cancellation of transition dipoles to give a forbidden S_0-S_1 transition, but coupled chromophores with oblique organizations should exhibit an allowed S_0-S_1 transition [21]. Therefore, by incorporating specific chemical structures within the repeat unit of a CP that enforce an oblique arrangement of chain segments, highly luminescent CP aggregates can be accessed. Following this concept, we will describe two PPE systems that exhibit a highly emissive

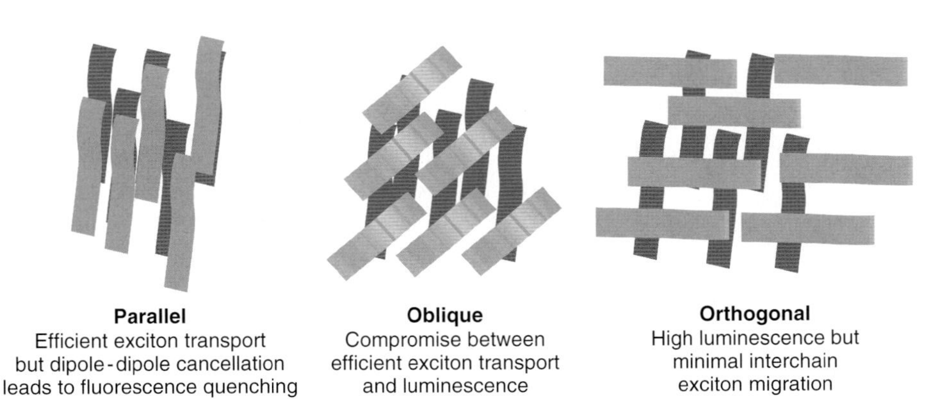

Figure 10.12 Three possible orientations of strongly interacting polymer chains and their corresponding optical and charge-transport properties.

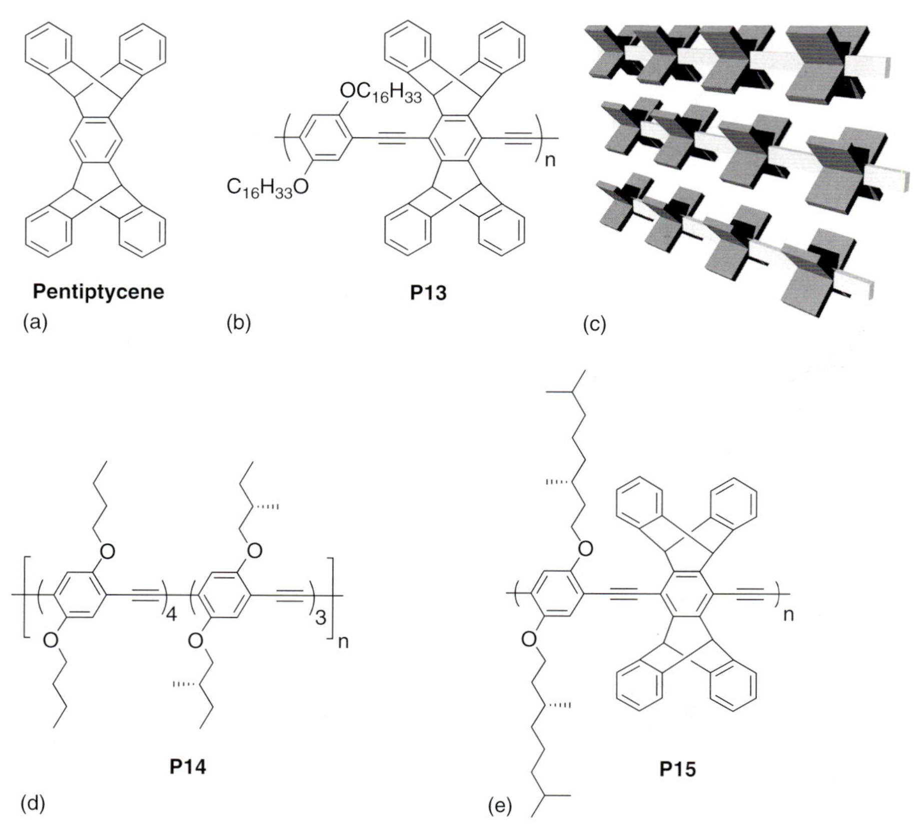

Figure 10.13 (a–c) Structures of pentiptycene, a pentiptycene-incorporated PPE and a simplified representation of the three-dimensional structure of pentiptycene-incorporate PPEs. (d, e) Structures of two PPEs used in aggregation studies.

aggregated phase and discuss the role of specific chemical structures in enabling oblique packing of chromophores.

Pentiptycene (see Figure 10.13) displays a rigid, 3D structure, which, upon inclusion within a CP backbone, effectively prevents π-stacking or excimer formation between individual chains [18]. In comparison with analogous PPEs lacking a pentiptycene comonomer, thin films of **P13** display enhanced fluorescence quantum yield and stability. Moreover, thin films of **P13** exhibit exceptionally high sensitivity as artificial fluorescent chemosensors for the vapors of nitroaromatic compounds, such as TNT and 2,4-dinitrotoluene (DNT) [12]. Essentially, the pentiptycene moiety imparts a porosity to solid-state structures (Figure 10.13(c)) that prevents direct electronic interaction between polymer chains (thus inhibiting self-quenching) while still allowing for strong dipole–dipole interactions (thus enabling 3D exciton migration).

Given the rigid structure of pentiptycene-incorporated PPEs, however, it was not initially anticipated that oblique aggregates of such polymers could be formed. Therefore initial investigations [22] into fabricating obliquely aligned PPE aggregates were carried out with **P14**. Chiral side chains were introduced into the repeat unit with the expectation that chirality, coupled with the normal twisting of polymer backbones, will yield self-assembled, ordered aggregates. Although enantiomerically pure **P14** was found to initially form chiral aggregates in 40% methanol/chloroform (see Figures 10.14(a)–(c)), the fluorescence intensity of these aggregates was strongly decreased relative to isolated polymer chains in dilute chloroform solutions. Moreover, the preliminary chiral structures thus formed ultimately rearranged to favor a stronger aggregate with coincident alignment of polymer chains at methanol concentrations higher than 50%. The resultant organization gave a low or nonexistent dihedral angle between polymer chains and, as expected, the fluorescence quantum yield dropped to <5% of its original value.

Therefore, in order to stabilize a strongly aggregated chiral and emissive organization of polymers while preventing aggregated chains from achieving a collinear structure, pentiptycene-containing structures were investigated [22]. Although pentiptycene-containing PPEs were not expected to form aggregates, it was initially found that addition of methanol (30%) to solutions of **P13** yielded aggregates with significantly quenched emission ($\Phi = 0.21$). These aggregates were much slower to assemble than those of **P14** and it was hypothesized that the polymer chains assemble into an interlocking structure, in which the polymer chains are constrained in the clefts between the pentiptycene groups (see Figures 10.14(d)–(f)). Considering that such an interlocking structure would prevent a coincidence of strongly interacting polymer chains, the aggregation behavior of an enantiomerically pure, pentiptycene-containing PPE, **P15**, was investigated. On the basis of solvent-dependent circular dichroism and absorbance spectroscopy, **P15** was indeed found to form restricted chiral aggregates in a poor solvent (methanol) yet still retaining the majority of its fluorescence intensity ($\Phi = 0.61$).

This unique aggregated state of P15 also showed sensitivity enhancements toward nitroaromatics. In solutions, fully aggregated **P15** was 15-fold more sensitive to fluorescence quenching by TNT and DNT than fully solvated **P13**. In addition, spun-cast films of aggregated **P15** displayed a fourfold increase in sensitivity toward TNT vapor (75% fluorescence quenching within 10 s) over optimized thin films of **P13**. The increased sensitivity of the fluorescent, chiral aggregates is proposed to derive from both an improved exciton diffusion length in the 3D-coupled chiral grids and an extension of the polymer conjugation length in the highly organized aggregated structure.

In addition to pentiptycene moieties encouraging an oblique packing of PPE chains, other 3D structures, such as cyclophanes, were also found to yield emissive PPE aggregates [23]. Specifically, spun-cast samples of polymer **P16** (Figure 10.15) displayed a visible, strong yellow emission that could be assigned to fluorescence from aggregated main chains. Notably, **P16** has a very low solution fluorescence quantum yield ($\Phi = 0.06$) due to electron transfer quenching of the polymer excited state by the amine residues. However, the aggregated phase of **P16** has a quantum

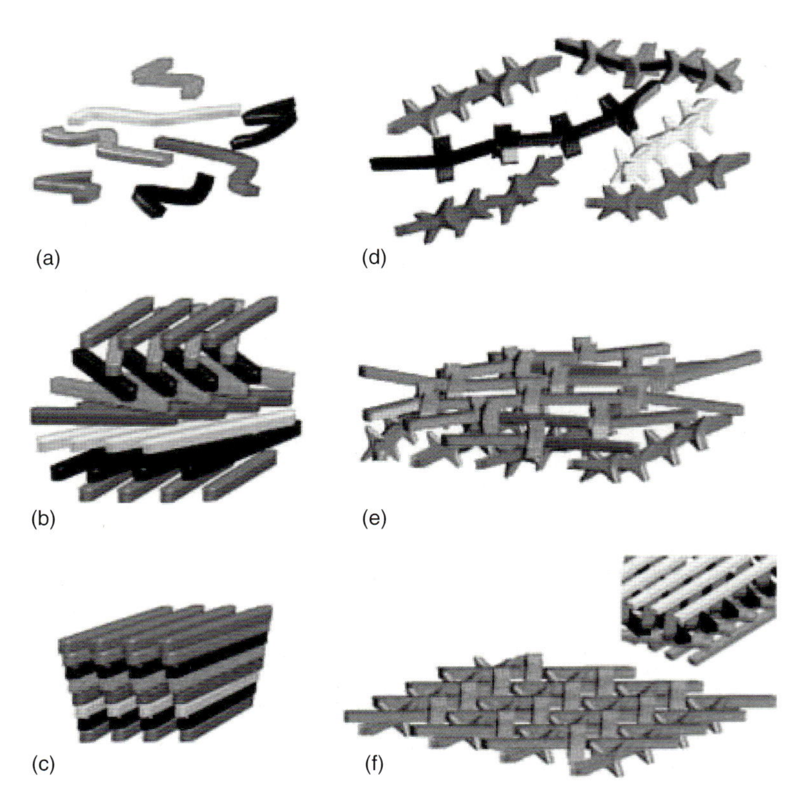

Figure 10.14 (a–c) Representation of the solution aggregation behavior of **P14** with increasing methanol concentrations. (d–f) Representation of the solution aggregation behavior of **P15** with increasing methanol concentration. Top: **P14** (a) and **P15** (d) dissolved in neat chloroform. Both polymers are highly solvated and there are no interactions between polymer chains. Middle: Aggregation of **P14** occurred and the rigid-rod PPEs form a lamella structure within each plane (b). The chiral side chains guide the polymers into a chiral macrostructure as depicted. The formation of the optically active macrostructure is guided by the influence of the chiral side chains. Polymer **P15** aggregates analogously to **P14**, but because of the presence of the pentiptycene groups, a slightly irregular interlocked structure results (e). The macrostructure of **P15** is shown in a two-layer graphic for simplification and clarity. Bottom: The initial chiral macrostructure of **P14** has been "untwisted" (c), which is favorable as it maximizes $\pi-\pi$ stacking in the edge-on conformation. The dihedral angle assumes a very small value, affording a very weak dichroic signal and low fluorescence quantum yield. Polymer **P15** self-assembles into a tighter structure (f) by incorporating the polymer into the clefts of the pentiptycene groups. Because the "untwisting" motion observed in **P14** is hindered, **P15** is able to maintain its optically active structure and its high fluorescence quantum yield. The inset illustrates the anticipated chiral gridlike structure. (Reprinted with permission from Ref. [22].)

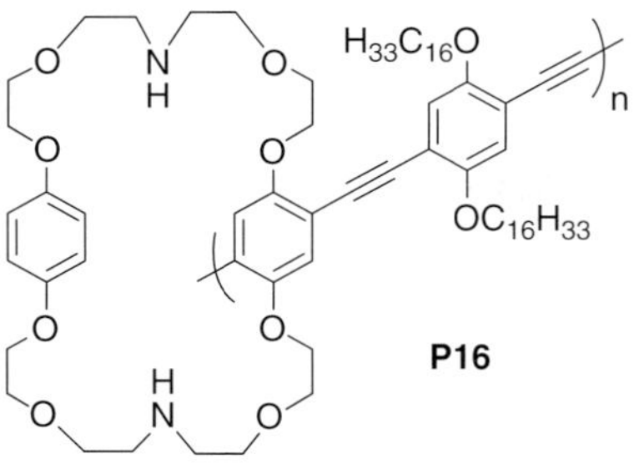

Figure 10.15 Structure of a PPE that forms a highly emissive aggregate.

yield 350% ($\Phi = 0.21$) of its solution value. This observation is unique because most other examples of CP aggregates display fluorescence quantum yields that either match (at best), or are only a few percent of the solution values. In this case, it is likely that the system displays a disordered structure and that isolated oblique aggregates are a minority species. The strong emission is a result of the fact that these aggregates are low-energy species and that energy migration results in a disproportionate emission intensity.

10.3.3
Kinetics of Energy Migration in Thin Films

In order to ascertain the photophysical and energy transport properties of PPEs, highly aligned LB multilayers of **P2** that were surface modified with luminescent traps (acridine orange, AO) were fabricated and investigated [24]. The LB deposition technique produced highly anisotropic films of **P2** with a well-defined thickness. The film thickness increases linearly with the number of layers transferred, thereby producing a well-defined geometry and distance for which to study energy migration processes.

The fact that energy migration is present in **P2** could be readily seen in wavelength-dependent lifetime measurements on spin-cast films. The excited-state lifetime was observed to increase when monitored at progressively longer wavelengths ($\lambda = 460$, 475, and 495 nm). These lifetime characteristics are consistent with a model that describes the polymer as a continuous distribution (usually Gaussian) of site energies. In this model, each state corresponds to a CP segment that is interrupted by chain defects (conformational or chemical), with the longer segments having lower energy, and energy migration is described as incoherent hopping of excitations to lower energy states. Emission from high-energy states (i.e., shorter wavelength of emission) should exhibit a faster decay rate due to energy transfer to lower energy chromophores within the system, consistent with what was observed for spin-cast films of **P2**. Evidence for intrachain energy migration in **P2** is likewise provided by fluorescence depolarization measurements, which will be discussed in a later section.

Emissive trapping sites were deposited selectively on the film surfaces by dipping LB films into methanol solutions of AO, which was chosen because its emission and absorption spectra are well separated from those of **P2** and its absorption spectrum has good overlap with the emission of **P2**. Additionally, the solubility of AO was almost orthogonal to that of **P2**, which allowed LB films of **P2** to be dip-coated in solution of AO with varying concentrations. AO was found to selectively localize at the film's surface, as evidenced by the fact that the ratio of AO fluorescence intensity between films of different thicknesses examined immediately after dipping or after extended periods of time remained constant. Polarization measurements showed that the AO transition dipole was principally aligned parallel to the polymer chain.

In the simplest case of energy transfer to AO from a monolayer of **P2**, the average lifetime of an excitation (τ) in an infinite 1D chain with randomly distributed efficient quenching traps should be $\tau = 1/(2WC^2)$, where C is the trap concentration and W is the hopping rate between neighboring sites. Therefore, the steady-state transfer rate should be proportional to C^2. In contrast, it was found that the degree of energy transfer to AO traps from a monolayer of **P2** was linearly dependent on the concentration of AO. This observation can be explained by either 1D energy migration with inefficient trapping or 2D transport. Considering that monolayer films of **P2** organize into highly aligned structures that could potentially allow excitations to undergo efficient interpolymer energy transport, the latter explanation of 2D transport is more likely.

If the number of LB layers was varied, an increase in the AO emission with increasing polymer layers was observed, up to 16 layers. At low concentrations of the AO trap, the AO fluorescence had a linear dependence on AO concentration, similar to the monolayer system described above. This last point leads to the conclusion that, at low AO concentration, the steady-state energy transfer rate in a monolayer film is less than $1/\tau$, where τ is the polymer's excitation lifetime. Moreover, the fact that the relative fluorescence of AO increases with increasing numbers of polymer layers is a direct indication of a transition to a 3D energy migration topology. The observation of saturation behavior in films with higher numbers of layers is a manifestation of the diffusion length for energy migration. The increase in the efficiency of energy migration to surface traps with increasing film thickness may, at first glance, seem counterintuitive since the concentration of AO relative to that of **P2** is actually smaller in thick films. However, this increased trapping efficiency is a direct result of the fact that 3D exciton migration necessarily creates a more efficient trapping process.

X-ray measurements on the monolayer and multilayer films revealed that the thickness per layer is 11 Å. Since the bimolecular Förster radius for most organic compounds is 20–60 Å, dipole–dipole excitation transfer between polymers must be involved as part of the mechanism for energy transfer to the AO trap.

In order to model the various energy transfer processes in LB films of **P2** surface modified with AO, rate constants for each possible energy transfer and decay process for an N-layer system were assigned. Figure 10.16 outlines these rate constants and their associated processes for a three-layer system. Since PPEs have a relatively large band gap and a narrow bandwidth, excitations were assumed to exist as strongly bound excitons. Assuming a steady-state population of all excited species, a set of balanced equations can be formulated (see Figure 10.16) and, ultimately, the relative intensity of AO fluorescence versus number of LB layers can

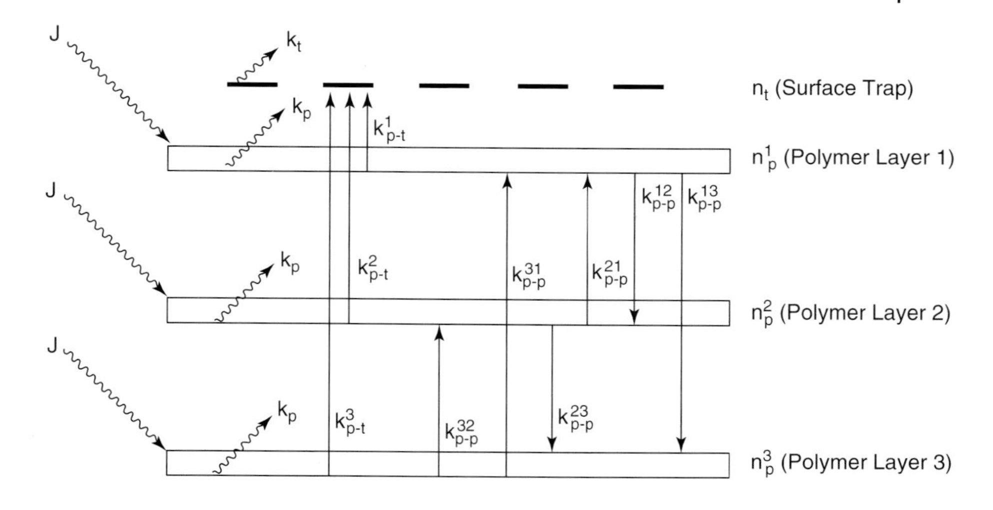

J : Intensity of steady state excitation

$n_{p,t}$: Excitation population in polymer layers and traps

$k_{p,t}$: Decay rates of polymer and trap. $k_P = 1/\tau_P$, $k_t = 1/\tau_t$

k^1_{p-t} : Rate constant for energy migration from polymer layer 1 to the trap. $k_{p-t} \sim C$, where C is the concentration of traps.

$k^{2,3}_{p-t}$: Rate constant for direct energy transfer to the trap from polymer layers 2 and 3. ℓ is the thickness of the layer $k^2_{p-t} \sim \frac{C}{\ell^6}$, $k^3_{p-t} \sim \frac{C}{(2\ell)^6}$

k^{ij}_{p-t} : Rate constant for energy transfer between polymer layers i and j $k^{ij}_{p-p} \sim \frac{1}{(\ell(i-j))^6}$

A steady-state assumption for n_t, n^1_p, n^2_p, and n^3_p gives:

$$-n_t k_t + n^1_p k^1_{p-t} + n^2_p k^2_{p-t} + n^3_p k^3_{p-t} = 0$$

$$-n^1_p k_p - n^1_p k^1_{p-t} - n^1_p k^{12}_{p-p} - n^1_p k^{13}_{p-p} + n_p k^{21}_{p-p} + n^3_p k^{31}_{p-p} + J = 0$$

$$-n^2_p k_p - n^2_p k^2_{p-t} - n^2_p k^{21}_{p-p} - n^2_p k^{23}_{p-p} + n^1_p k^{12}_{p-p} + n^3_p k^{32}_{p-p} + J = 0$$

$$-n^3_p k_p - n^3_p k^3_{p-t} - n^3_p k^{32}_{p-p} - n^3_p k^{31}_{p-p} + n^1_p k^{13}_{p-p} + n^2_p k^{12}_{p-p} + J = 0$$

Figure 10.16 Schematic representation and rate constants for a three-layer Langmuir–Blodgett assembly of **P2** with an emissive trap (acridine orange) placed at the film surface. The equations resulting from a steady-state excitation population of the layers and the trap are shown. (Adapted from Ref. [24].)

be modeled. Such modeling has determined, first, that the rate of energy transfer between layers exceeds $6 \times 10^{11}\ s^{-1}$. This high rate results in a uniform excitation population throughout all the layers of the LB films. Additionally, the model confirms the saturation of AO fluorescence intensity with increasing LB layers.

Therefore, it is clear that an optimal thickness will exist in sensor schemes requiring exciton trapping at the polymer surface. However, it must be pointed out that additional enhancements in energy migration may be possible by creating multilayer structures that provide vectorial energy transport in a specific direction. For example, in the striated, three-component film depicted in Figure 10.8, energy was preferentially transferred to the surface by utilizing layers of sequentially decreasing band gap and, in this way, the 16-layer energy transfer limitation for PPEs was overcome.

10.4
Lifetime Modulation

As discussed above, a thorough understanding of the mechanisms underlying energy migration in CPs is necessary to design its enhancement. The high efficiency of energy transfer in most conjugated systems relative to systems with pendant chromophores suggests that strongly coupled electronic intrachain (Dexter-type) processes may increase transport in these systems over those provided solely by the dipole–dipole (Förster-type) interactions that govern weakly interacting chromophores. Discrepancies between the two mechanisms allow the determination of which process dominates in a given system [25].

As derived by Förster, the dipole–dipole approximation yields a transition probability (k_{ET}):

$$k_{ET} = \frac{\kappa^2 J 8.8 \times 10^{-28}\,\text{mol}}{n^4 \tau_0 R_{\text{DA}}^6} \tag{10.5}$$

where κ is an orientation factor, n is the refractive index of the medium, τ_0 is the radiative lifetime of the donor, R_{DA} is the distance (cm) between donor (D) and acceptor (A); and J is the spectral overlap (in coherent units $\text{cm}^6\,\text{mol}^{-1}$) between the absorption spectrum of the acceptor and the fluorescence spectrum of the donor. Therefore, a weakly allowed transition, as manifest in a long radiative lifetime, should discourage purely coulombic energy transfer.

Electron exchange effects contributing to the energy transfer described by Dexter account for shorter range processes that result from direct wavefunction overlap of interacting molecules. In this case, the transition probability is described by

$$k_{ET} = KJ\exp(-2R_{\text{DA}}/L) \tag{10.6}$$

where K is related to specific orbital interactions, J is the spectral overlap, R_{DA} is the donor–acceptor distance, and L is the van der Waals radii distance between donor and acceptor. This process is often termed electron exchange because molecules must be almost within the van der Waals radii of each other to interact. In the specific case of CPs, chromophores are directly conjugated and, therefore, one might expect the Dexter mechanism to dictate the overall efficiency of energy migration, at least within the polymer backbone.

To determine the dominant *intra*chain energy migration mechanism in PPEs, the unique oscillator strength independence of the Dexter mechanism was invoked to guide the design of polymers with long radiative lifetimes. Long radiative lifetimes translate into reduced oscillator strengths of D* to D and A to A* transitions, which, according to the Förster mechanism (Eq. (10.5)), would result in a severely truncated rate of energy transfer. However, because the Dexter electron exchange mechanism does not depend on the oscillator strength, longer lifetimes affected by less allowed transitions can serve to increase energy transfer by providing more time for the excitation to migrate before radiative decay [25].

10.4.1 Triphenylene-Incorporated PPEs

Long-lifetime PPEs can be accessed by incorporating structures with extended aromatic cores, such as triphenylene, dibenzo[*g,p*]chrysene and benzothiophene, into the backbone of the polymer. Triphenylene has a well-known symmetrically forbidden ground-state transition and, therefore, exhibits a long excited-state lifetime. Although incorporation into a CP will decrease the triphenylene's symmetry, it was hypothesized that the strong aromatic structure would dominate the photophysics of the resulting polymer. In order to determine the general effect of triphenylene incorporation, a family of triphenylene-based poly(*p*-phenyl ethynylene)s (TPPEs) was synthesized along with chemically similar phenylene analogs (see Figure 10.17) [26]. Polymers were size-selected by gel permeation chromatography to ensure comparison of similar chain lengths and the excited-state lifetimes were measured in the frequency domain in methylene chloride solutions. It was found that triphenylene incorporation universally extended the excited-state lifetime of targeted PPEs without severely compromising quantum yield. The combination of Φ and τ data demonstrate that the enhanced lifetimes are principally due to differences in radiative rates and not differences in nonradiative rates.

In PPEs where the excitation is more localized, the lifetime-enhancing effect of the triphenylene moiety was more pronounced. An example is **P23**, consisting of a triphenylene monomer and a biphenyl monomer. Because biphenyl planarizes in the excited state, a large Stokes shift is observed in the resulting polymer. This process serves to localize the excitation and the radiative decay rate becomes more competitive with energy transfer. Consequentially, the lifetime of **P23** is about three times longer than its phenylene analog, **P24**. This effect is also observed in *meta*-linked PPE, **P27**: the *meta* linkage disrupts conjugation, thus localizing the excitation and resulting in one of the longest lifetimes observed for TPPEs. As expected, PPEs with a larger triphenylene component demonstrated more pronounced lifetime enhancement relative to their phenylene analogs (**P19** vs. **P20**).

In addition, electrostatic variation in TPPEs was found to lead to excited-state interactions. The long lifetime (4 ns) observed for **P25** most probably arises from an exciplex that is formed between the triphenylene moiety and its electron-deficient tetrafluorinated comonomer. This proposal is also supported by the broad, red-shifted emission spectra recorded for **P25**. Notably, these features were not observed for the phenylene analog, **P26**, thus suggesting that the flat, electron-rich nature of the triphenylene is necessary to induce exciplex formation.

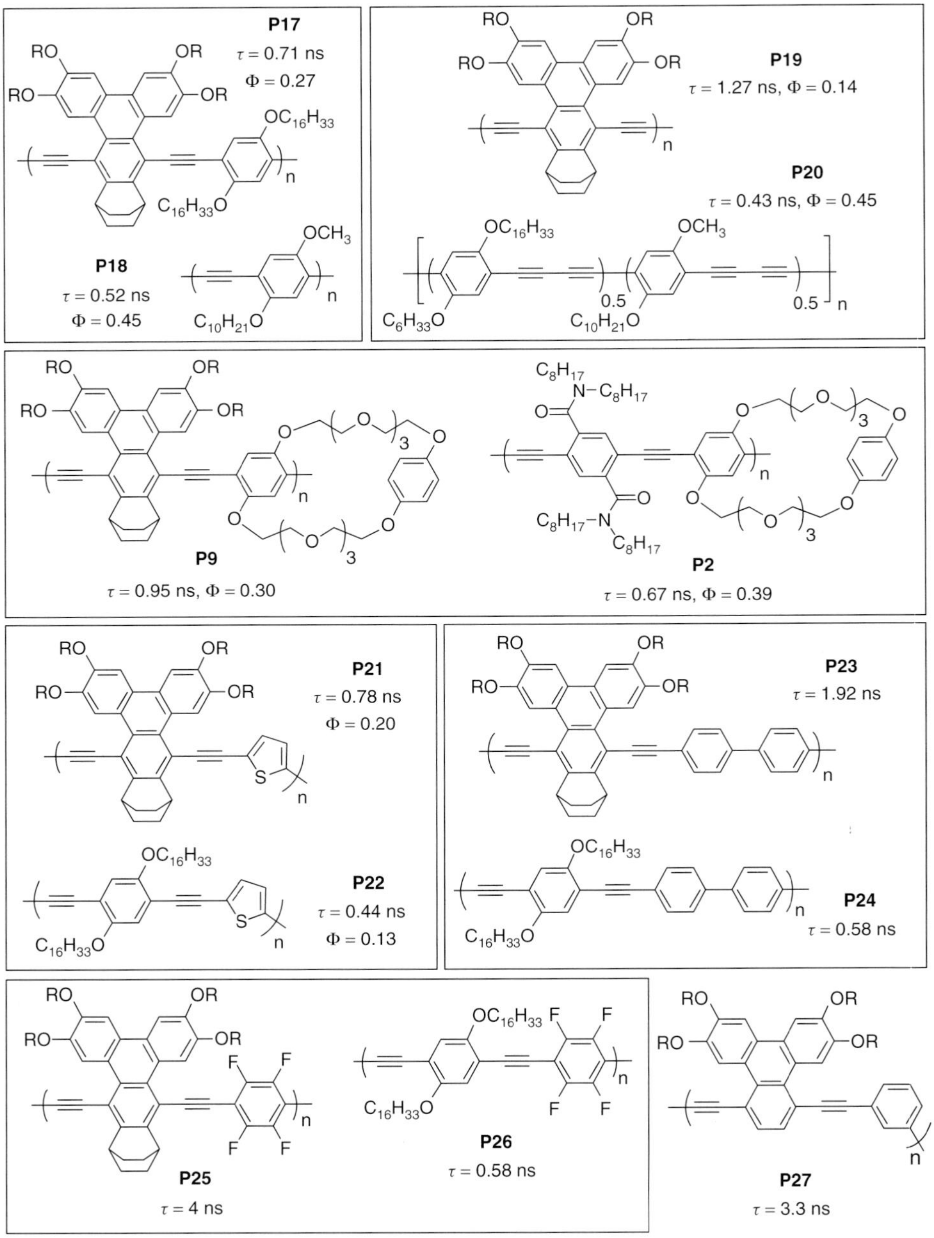

P17 $\tau = 0.71$ ns $\Phi = 0.27$

P18 $\tau = 0.52$ ns $\Phi = 0.45$

P19 $\tau = 1.27$ ns, $\Phi = 0.14$

P20 $\tau = 0.43$ ns, $\Phi = 0.45$

P9 $\tau = 0.95$ ns, $\Phi = 0.30$

P2 $\tau = 0.67$ ns, $\Phi = 0.39$

P21 $\tau = 0.78$ ns $\Phi = 0.20$

P22 $\tau = 0.44$ ns $\Phi = 0.13$

P23 $\tau = 1.92$ ns

P24 $\tau = 0.58$ ns

P25 $\tau = 4$ ns

P26 $\tau = 0.58$ ns

P27 $\tau = 3.3$ ns

Figure 10.17 Structures of triphenylene-incorporated PPEs and their corresponding phenylene PPEs, and the photophysical properties of all polymers considered.

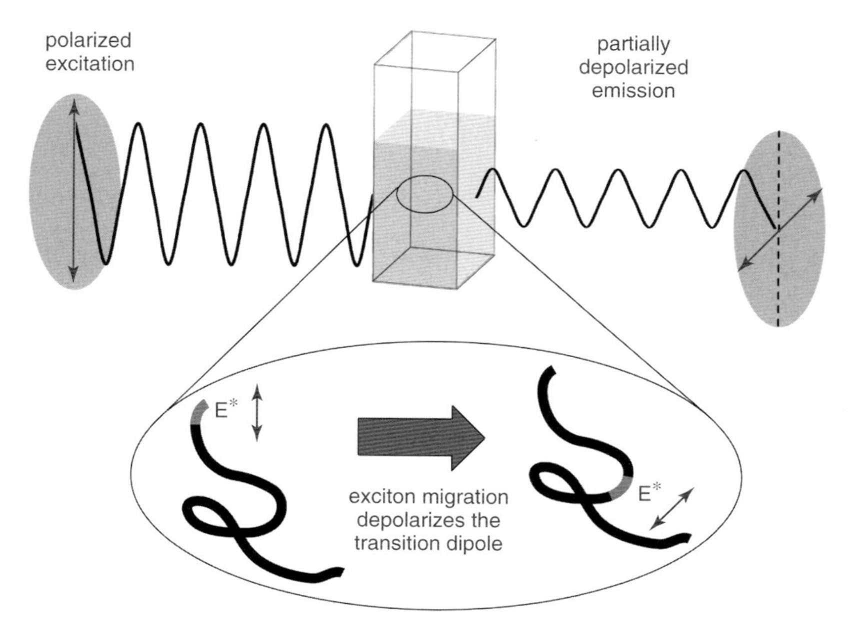

Figure 10.18 A simplified pictorial of depolarization due to energy migration in conjugated polymers (CPs). The excitation beam is vertically polarized and therefore only vertical transition dipoles are initially excited on the CP chain. Vertically polarized excitons on the polymer chain migrate. As they move over a disordered polymer chain, they lose their initial polarization. Thus, the emission of the polymer is depolarized relative to the excitation beam. This amount of measured depolarization directly indicates the extent of energy migration in the CP. (Adapted from Ref. [25].)

The relationship between excited-state lifetime and energy migration can be investigated through fluorescence depolarization measurements [25]. A simplified pictorial depiction of depolarization due to energy migration in CPs is shown in Figure 10.18. The excitation beam is vertically polarized and, therefore, only vertical transition dipoles are initially excited on the CP chain. Vertically polarized excitons on the polymer chain can migrate and, as they move over a disordered polymer chain, can lose their initial polarization. The emission of the CP is thus depolarized relative to the excitation beam. Therefore, the amount of measured depolarization directly indicates the extent of energy migration in the CP.

Since all polymers studied were high molecular weight materials, they can be considered rotationally static over the emission lifetime of the polymer. Therefore, energy migration is the major contributor to the fluorescence depolarization in CPs, and the exciton loses more of its initial polarization as it diffuses along a disordered polymer chain. The polarization value, P, was determined from the standard equation

$$P = \frac{I_{\parallel} - GI_{\perp}}{I_{\parallel} + GI_{\perp}} \tag{10.7}$$

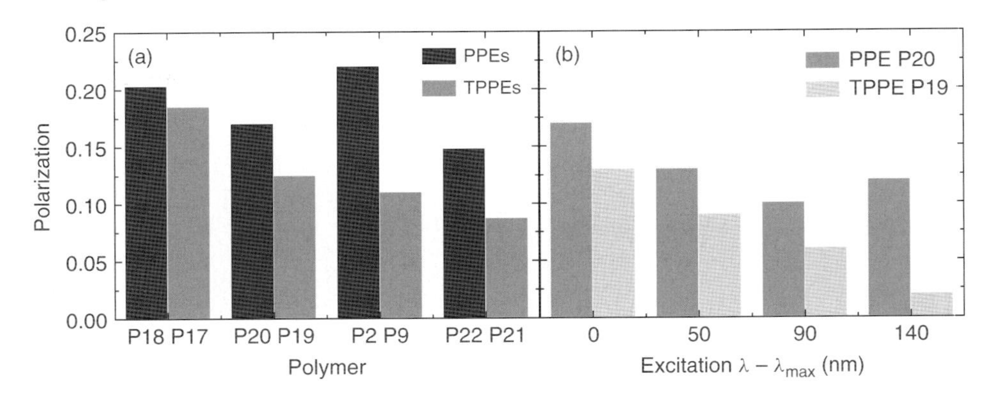

Figure 10.19 Polarization values for a family of TPPEs and their corresponding PPEs. Gray bars are polarization values for TPPEs and black bars are polarizations for PPEs. Both are excited at λ_{max} and in every instance, TPPEs exhibit greater depolarization, indicating more extensive energy migration. (Adapted from Ref. [26].)

where $I_{||}$ and $I_{\perp}$ are the intensities of emissions detected parallel and perpendicular to the polarization vector of the incident light, respectively, and G is an instrumental correction factor. Theoretically, the highest value of P for a randomly oriented, isolated, fixed chromophore with coincident transition dipoles for absorption and emission is 0.5.

Concurring with lifetime data, depolarization was found to be universally more pronounced in TPPEs than their phenylene analogs (Figure 10.19(a)). In the case of polymers **P21** and **P22**, the kinked thiophene linkage resulted in a much greater depolarization than in other polymers studied. Greater polarization loss per migration step is expected in a kinked structure. Polymer **P27**, with its localizing biphenyl monomer, retained one of the highest polarization values. Polymer **P25**, however, displayed the highest polarization value among the TPPEs. The energy minimum formed by its exciplex probably quickly traps the wandering excitation, thus reducing energy migration.

Additionally, the fluorescence depolarization of a subset of polymers as a function of excitation energy was studied in order to separate depolarization owing to energy migration from that due to absorption/emission dipole alignment (Figure 10.19(b)). If energy migration is indeed present, then polarization values should decrease as excitations move to shorter wavelengths. Measurements were performed on materials selected for similar chain length, all above the small molecular weight regime. As excitation energy is increased, it was found that both the TPPEs and their phenylene analogs display lower P values, consistent with population of higher energy excitons that readily lower their energy by migration to lower energy states. However, polarization continued to significantly decrease with shorter wavelengths of excitation in TPPEs but only leveled off in PPEs, indicating that radiative rates of emission are not competitive with energy migration in TPPEs as they are in PPEs.

If the Förster mechanism dominated, then the enhanced radiative rates in PPEs would encourage more extensive energy migration (and therefore greater fluorescence depolarization) as compared to the TPPEs; however, the opposite phenomenon is observed, thus lending credence to the claim that the Dexter mechanism is the dominant intramolecular energy transport process in these systems.

10.4.2
Chrysene-Incorporated PPEs

It was also possible that the increased fluorescence depolarization observed for the TPPEs is due to a reduced persistence length as compared to analogous PPEs (see Figure 10.20). Steric interactions between the ethynylenes and proximate CH bonds of the TPPEs could conceivably result in a more coiled structure that could be responsible for the more rapid fluorescence depolarization in the TPPEs. To address this concern, polycyclic dibenzo[*g,p*]chrysene-based poly(*p*-phenyl ethynylene)s (CPPEs) were investigated (Figure 10.21) [27]. CPPEs lack any complicating steric factors and have a more rigid structure that should increase the persistence length and yield less coiled polymers compared to PPEs.

The photophysical properties of the CPPEs were found to be similar to those observed in TPPEs: that is, longer excited-state lifetimes and more extensive energy migration. Polymers **P29** and **P31** displayed lifetimes greater than 2 ns, while most PPEs have sub-nanosecond lifetimes. Polarization studies confirmed the presence of enhanced exciton migration in CPPEs [25]. For all chain lengths and at all excitation wavelengths, the polarization values of CPPEs were about half of those in the corresponding PPEs. Additionally, polarization data as a function

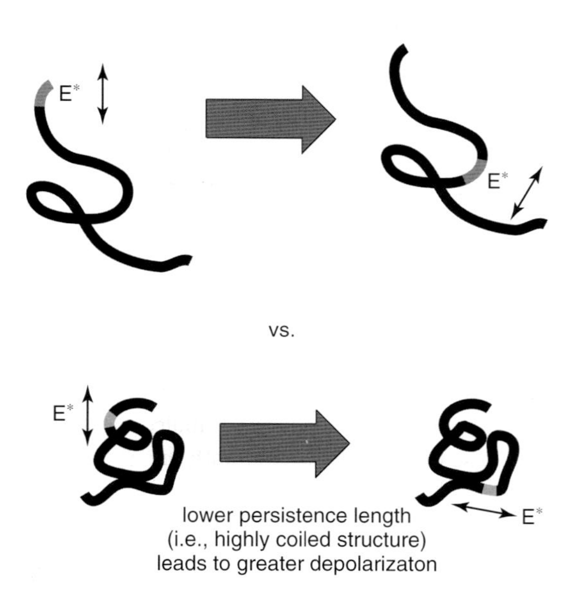

Figure 10.20 Possible persistence-length dependence of fluorescence depolarization.

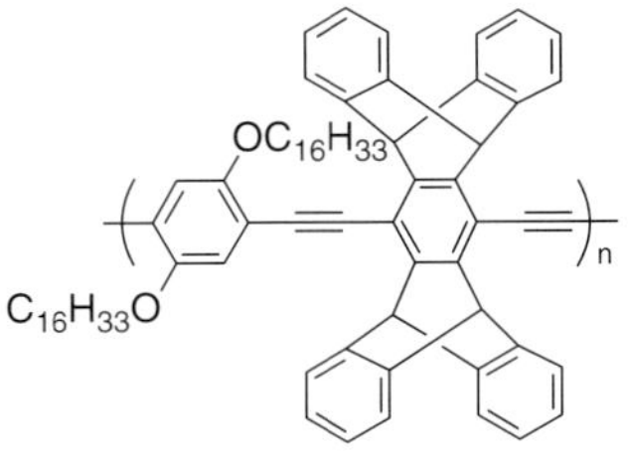

P28
$\tau = 0.51$ ns
$\Phi = 0.50$

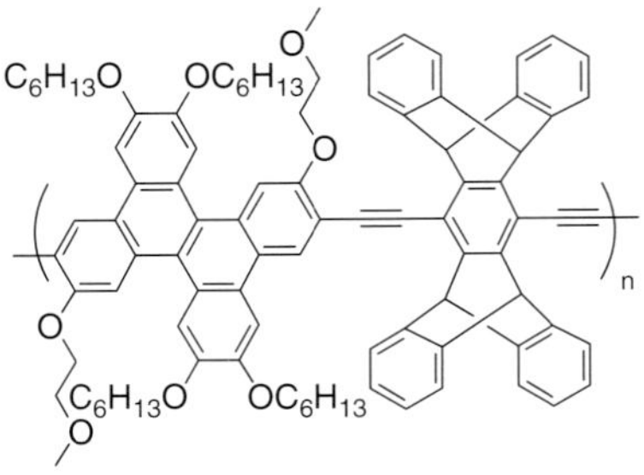

P29
$\tau = 5.5$ ns
$\Phi = 0.29$

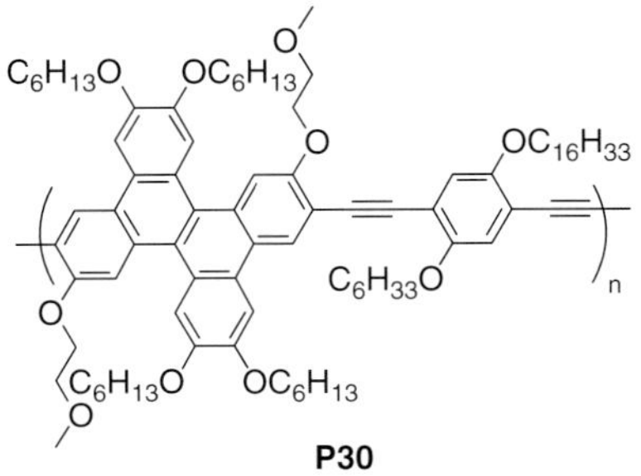

P30
$\tau = 1.4$ ns
$\Phi = 0.35$

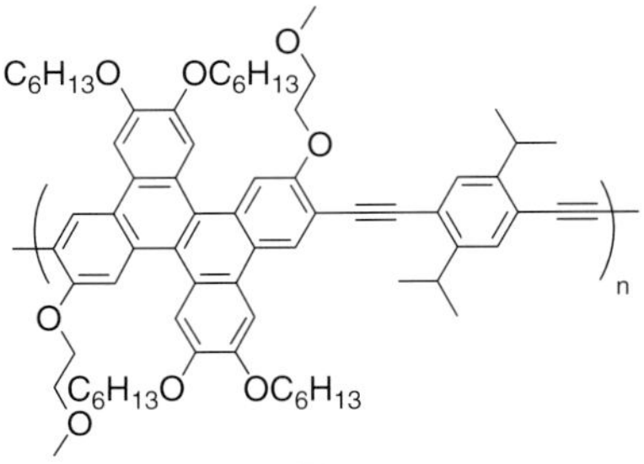

P31
$\tau = 2.6$ ns
$\Phi = 0.25$

Figure 10.21 Structures of chrysene-incorporated PPEs.

of excitation wavelength discounted dipole displacement as a main contributor to depolarization. Chain length dependent studies on **P29** revealed that radiative decay did not supersede energy migration even for the longest chain lengths ($n \sim 220$). Therefore, the CPPEs, similar to the TPPEs, allowed for greater intrachain exciton migration than PPEs because of the fact that energy migration is not truncated by radiative deactivation of the excited state.

10.4.3
Thiophene-Based Model Compounds and PPEs

Lastly, polymers with pendant thiophenes further illuminate lifetime extension (Figure 10.22) [28]. Sulfur incorporation benefits materials properties in part owing to the larger radial extension of its bonding. This promotes cofacial electronic interactions between stacked molecules that could enhance energy transfer. Cyclized and noncyclized versions of each model compound and polymer were investigated to assess the effects of imposed symmetry and rigidity on the photophysics of the material. With both *meta* and *para* linkages represented, the family of polymers investigated allowed for exploration of the effects of different degrees of aromatization as well as changes in conjugation pathways.

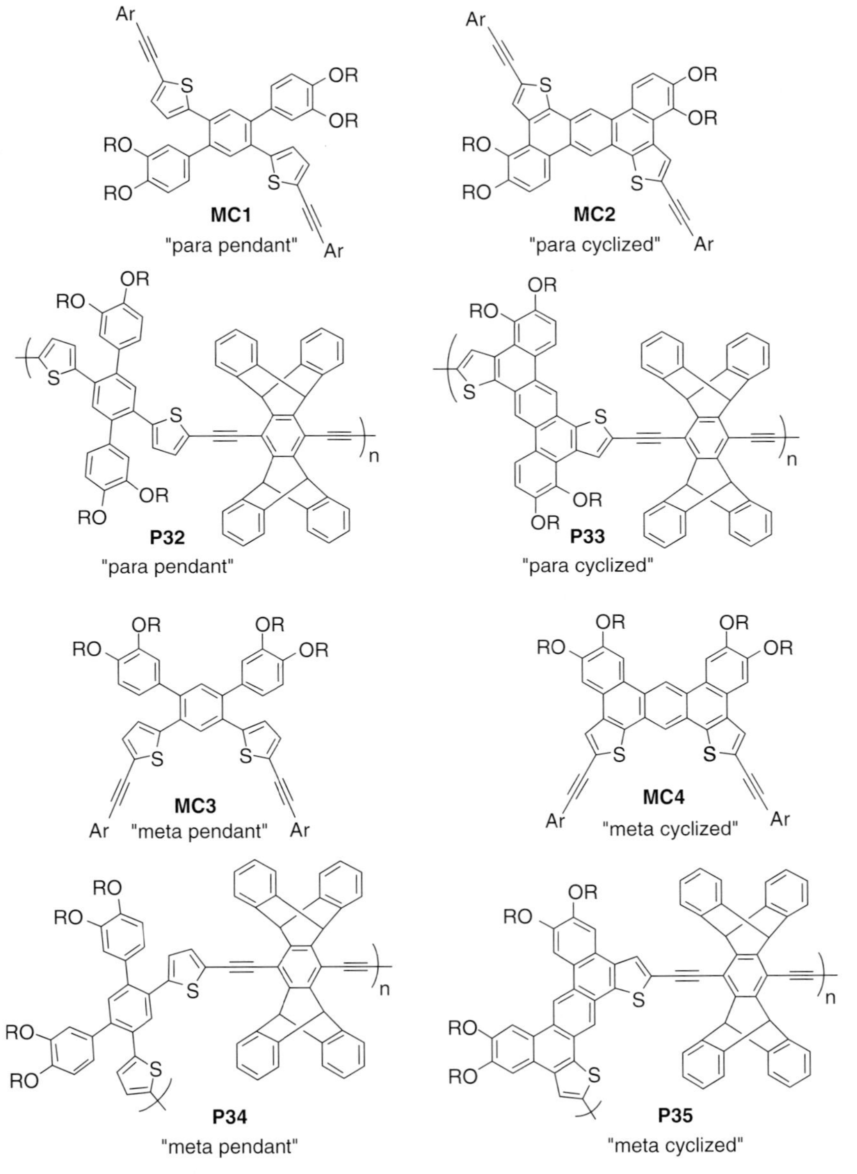

Figure 10.22 Structures of thiophene-based model compounds and PPEs.

To predict the behavior of such thiophene-containing polymers, model compounds **MC1–4** provided systems with precisely defined conjugation lengths, allowing separation of planarization effects from effective conjugation-length variations. The cyclized compounds **MC2** and **MC4** showed a sharpening of vibronic structure concomitant with a decrease in Stokes shift as degrees of freedom were reduced. The oscillator strength of the (0,0) transition was significantly reduced in the *meta* system **MC4** (log $\varepsilon = 3.63$) when compared with the *para* isomer **MC2** (log $\varepsilon = 4.69$). Accordingly, *meta* **MC4** had a longer excited-state lifetime (5.00 ns) than the *para* **MC2** (1.12 ns).

While a sharpening of emission spectra was observed, no significant wavelength shift in either system occurred upon cyclizing (aromatizing) either **MC1–MC2** or **MC3–MC4**. This suggested that there is planarization in the excited state of **MC1** and **MC3** to allow for greater delocalization. Additionally, the lifetime of the both systems was found to increase upon cyclization: the *meta* system displayed a ninefold increase in excited-state lifetime upon cyclization (0.58 ns for **MC3** to 5.0 ns for **MC4**) while the *para* system showed a more modest increase (0.8 ns for **MC1** to 1.12 ns for **MC2**).

Consistent with other CPs, the photophysical properties of the model compounds **MC1–4** were reflected in the corresponding polymers. The absorption spectra of rigid, aromatized **P33** and **P35** displayed sharper vibronic structure and a decrease in Stokes shift when compared with the noncyclized **P32** and **P34**. The *meta*-cyclized polymer **P35** displayed much lower oscillator strength at the band edge than its *para* analog, **P33** – which was predicted by the corresponding model compounds. Aromatization effects only slightly shifted the emission maxima of both the *meta* and *para* systems. As in **MC1** and **MC3**, this may attest to excited-state planarization in the flexible systems.

Lifetime trends in the model compounds were also consistent with the related polymers. In accordance with the diminished oscillator strength, the *meta* polymers **P34** and **P35** displayed a lifetime discrepancy (0.30 vs 1.06 ns, respectively) similar to that of their model compounds, **MC3** and **MC4** (0.58 vs 5.00 ns, respectively). The *para* polymers also mimicked the model systems. However, in this case, the model compounds **MC1** and **MC2** exhibited comparable lifetimes (0.80 vs 1.12 ns, respectively) before and after cyclization. As a result, the *para* polymers shared almost identical lifetimes (**P32**: 0.57 ns, **P33**: 0.61 ns). Both these examples correspond to the previously observed trend, which suggests that monomer photophysics critically influences the photophysics of the resulting polymers.

Polarization experiments revealed that energy migration is not enhanced without lifetime enhancement [25]. In other words, the *meta*-cyclized polymer **P35** displayed the greatest fluorescence depolarization, with polarization values reaching near zero. This was due to both the enhanced lifetime of the polymer and the curved architecture of the polymer chain, which causes significant depolarization even when energy has migrated through a single hop. In contrast, since a large lifetime enhancement was not obtained upon aromatizing the *para*-pendant polymer **P32** to its cyclized analog, **P33**, a large depolarization was not observed for either of the two *para* polymers. Together, the polarization measurements on these thiophene systems underscore an important distinction: simply rigidifying the polymer backbone is not enough to extend lifetime and enhance energy migrations. One must carefully consider chromophore photophysics when attempting to impart these properties into CPs because the excited-state behavior of polymers is essentially encoded by the choice of monomers.

10.5 Conformational Dependence on Energy Migration: Conjugated Polymer – Liquid Crystal Solutions

The achievement of complete control over the conformation of CP single chains and their assembly into functional structures is paramount to the thorough understanding and optimization of energy transfer and conductivity in CPs. Inconveniently, high molecular weight PPEs have finite persistence lengths and exist as flexible coils (as opposed to rigid rods) in solution [11]. Consequently, the disorder displayed in solution is often transferred to the solid-state structures of PPEs, and there is a general lack of long-range molecular order due to conformational disorder in the polymer main chain. The many structural defects in the solid state ultimately result in diminished electronic delocalization and limit the ability to study the intrinsic properties of these materials.

As one potential solution to this conundrum, liquid crystals (LCs) represent an ideal means to produce ordered arrays of molecular wires. Columnar LCs with extended aromatic cores have long been considered 1D conductors; however, there is limited electronic coupling between aromatic cores due to limited overlap of the π-orbitals, particularly in the liquid crystalline state [29]. An alternate strategy to exploit the long-range order of LCs and assemble electronic materials is to dissolve CPs into an LC host. In this way, the strong intramolecular electronic coupling of the CP and the organizational ability of the LC host can work together to form a highly organized electronic material.

Revisiting the rigid, 3D pentiptycene scaffold mentioned earlier, it is worth noting that the 3D nature of pentiptycene and its analog triptycene (Figure 10.23) also has important organizational influences. Specifically, the addition of triptycene moieties into the backbone of a polymer can either be used to redirect [30] or enhance [31] molecular alignment in LCs and stretched polymers. This property results from the natural tendency of host–guest mixtures to lower their energy by minimizing the amount of free volume. Therefore, fluorescent dyes, PPEs, and PPVs containing triptycene groups can theoretically be aligned along the nematic director in LCs and thus achieve significant ordering with high-order parameters (S) and dichroic ratios (D) [31].

Accordingly, nematic liquid crystalline solutions of the highly emissive, triptycene-incorporated PPV, **P36**, and the triptycene-incorporated PPE, **P37**

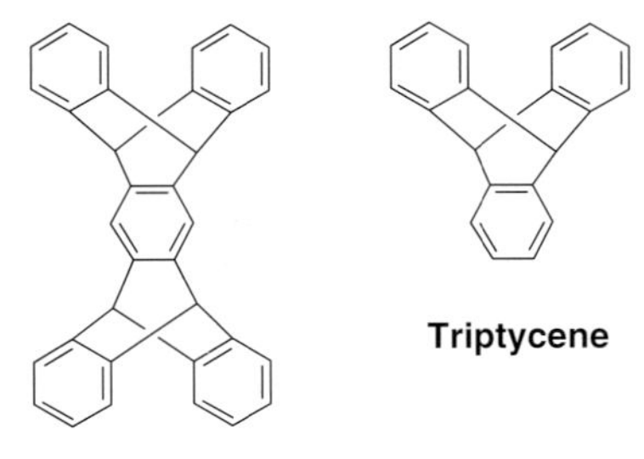

Figure 10.23 Common iptycene scaffolds.

(Figure 10.24) were investigated [32]. Solutions of **P36** and **P37** in **6CHBT** (Figure 10.25) were loaded into LC cells with rubbed internal polyimide surfaces that gave a homogenous alignment of the nematic LC. Polarized absorption spectroscopy of these test cells with the polarizers aligned parallel (0°) or perpendicular (90°) to the nematic director were used to calculate the order parameters, *S*. The liquid crystalline solvent (a wide variety of nematics were acceptable, but **6CHBT** and **5PCH** were primarily used) had the important feature that it created an extended CP chain conformation that was highly aligned (*S* ranged between 0.7 and above 0.8). Additionally, the polymers were found to have greatly enhanced conjugation lengths in nematic LC solutions. This could readily be observed by comparing the absorption spectra of **P37** in a CH_2Cl_2 solution and an LC solution: **P37** in a LC solvent displayed an absorption spectrum that was red shifted and had a comparatively abrupt band edge relative to its absorption spectrum in a CH_2Cl_2 solution (Figure 10.24). Both these features suggest that the CPs' long axes aligned with the director of the nematic LC and that the polymer chains were in a highly extended conformation as opposed to the typical random coil present in isotropic solutions.

Further proof that the polymers formed true solutions in nematic LCs was provided by demonstrating that the polymers could be reoriented with the nematic host by application of electric fields (Figure 10.26). Under an applied field (9 V), the nematic director and the polymer backboned aligned normal to surface of the LC test cells. This resulted in a dramatic reduction (75–80%) in the polymer absorption and complete loss of polarization. These results are due to the realignment of the CP's transition dipole (that is coincident with the polymer's long axis) to match the direction of the electric field (normal to the surface of the test cell), which minimized the projection of the transition dipole along the electric vector of the incident light beam. The reorientation of the CPs was also readily apparent by visualizing the polymer's fluorescence in the presence (fluorescence OFF) and absence (fluorescence ON) of an applied voltage (Figure 10.26). In all cases, the polarized fluorescence was rapidly recovered upon removal of the voltage.

To better illustrate the conformational dependence of energy migration in CPs, LC solutions of **P38**, which is a pentiptycene-incorporated PPE end-capped with low-energy anthracene trapping sites, were investigated (Figure 10.27) [33]. As explained earlier, the introduction of anthracene end groups solicits efficient energy transfer from the polymer backbone to these trapping sites and thus, site-selective, green emission from the polymer termini can be observed if significant exciton migration is operative. Therefore, dissolving **P38** in **6CHBT** allowed for the study of the rate of intrachain exciton migration under conditions of increased conjugation length and high alignment. These studies revealed that the order imposed by the nematic LC solvent increased the energy transfer efficiency to the low-energy anthracene termini. This process was accompanied by a significant increase in the fluorescence quantum yield. The liquid crystalline phase was found to be a necessary requirement for this phenomenon, as when the temperature of the system was increased above the nematic–isotropic transition temperature of

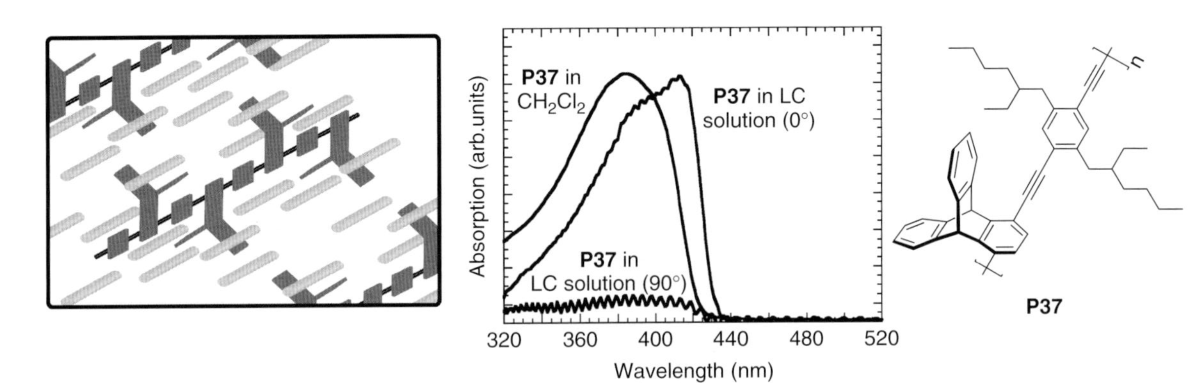

Figure 10.24 Comparison of the absorption spectrum of PPE **P37** in methylene chloride and a nematic liquid crystal solvent (**6CHBT**). The liquid crystal imposes a planarized conformation that is responsible for the steep band edge and red shift of the absorption spectrum. Note that the polarized absorption reveals that the polymer backbone is aligned parallel to the liquid crystal director. (Adapted from Ref. [32].)

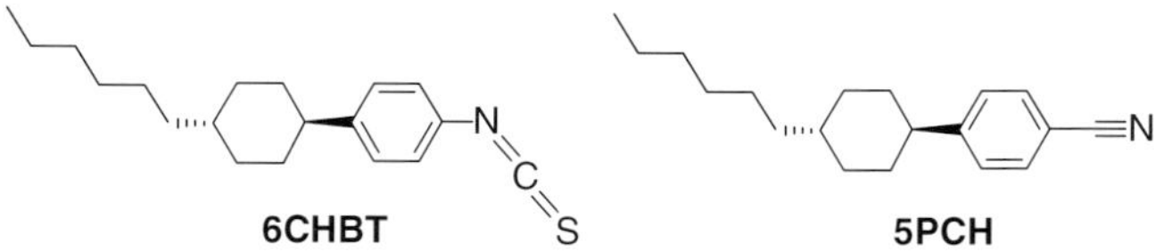

6CHBT 5PCH

Figure 10.25 Structures of some nematic liquid crystals.

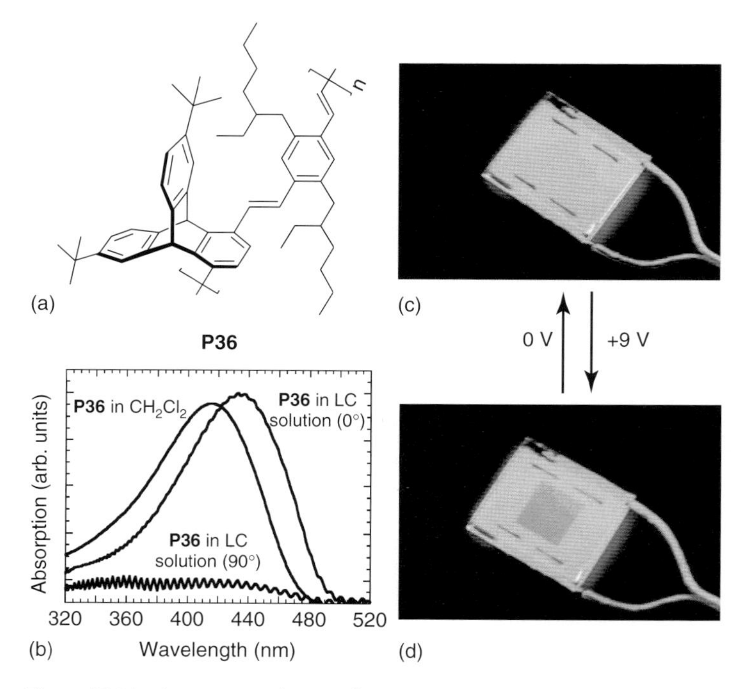

Figure 10.26 A nematic solution of PPV **P36** in **6CHBT**. The red-shift in the absorption spectrum of the liquid crystal solution is clearly apparent, and the alignment is apparent from the ratio of the spectra taken at 0° and 90° to the nematic director. The periodic signal at 90° is due to interference effects from the rubbed polyimide coatings of the test cell. The pictures document the fluorescence behavior of a **P36/6CHBT** nematic solution in a test cell. (c) A test cell is shown with no applied voltage, and the nematic director and polymer are aligned with the short axis of the cell. (d) 9 V is applied between the ITO pads on the top and bottom of the test cell that realigns the liquid crystal and polymer normal to the glass slides. This results in a dramatic reduction in the polymer's absorption and emission and complete loss of polarization.

the LC host, a dramatic reduction of the energy transfer efficiency and fluorescence quantum yield was observed.

Structure–property relationships that govern the extent of conformational enhancement achievable in PPE–LCs mixtures were investigated using PPEs **P39–41** (Figure 10.28) [34]. These PPEs contain more elaborate iptycene scaffolds introduced to create polymers displaying greater order and enhanced solubility in LCs at high molecular weights (high molecular weight versions of **P36** and **P37** were

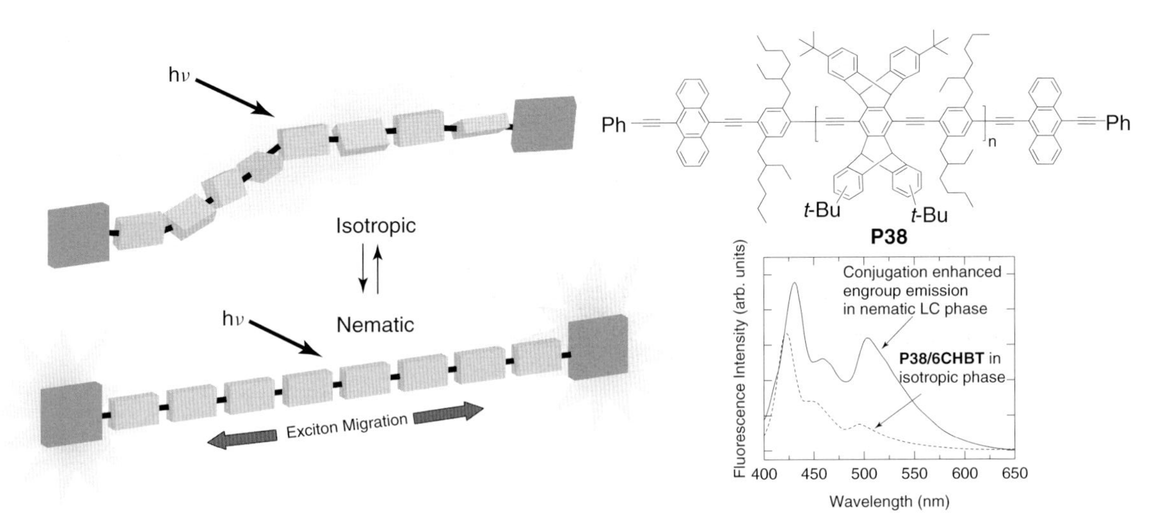

Figure 10.27 Conformational enhancement of energy transfer to low-energy trap sites located at polymer termini with nematic liquid crystals. (Adapted from Ref. [32].)

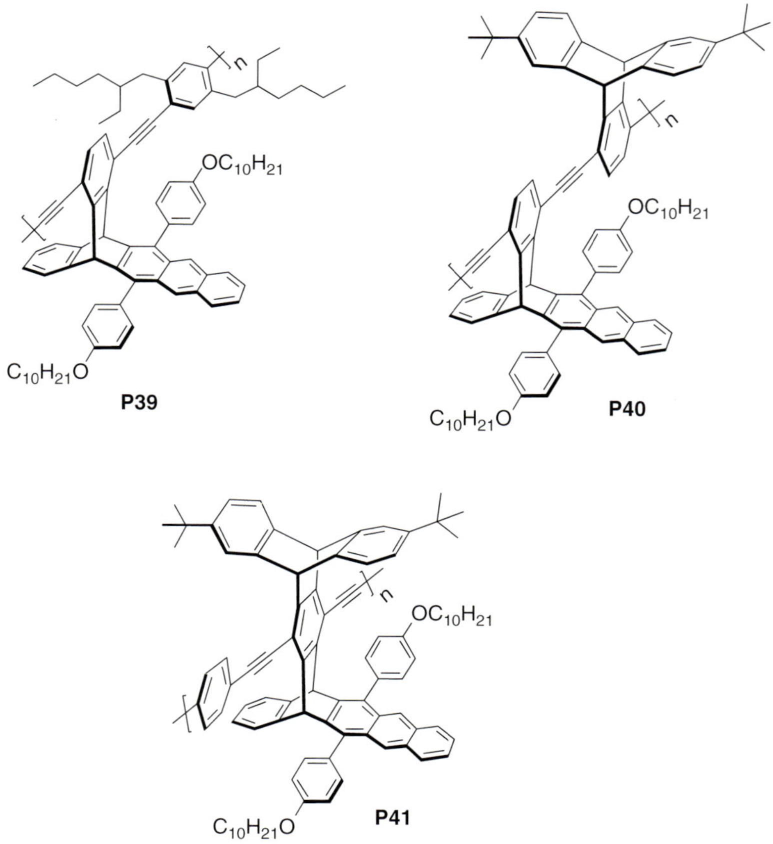

Figure 10.28 Structures of PPEs containing elaborate iptycene scaffolds.

found to be poorly soluble in nematic LCs). As expected, **P39** and **P40** displayed higher order parameters ($S_A = 0.86$ and 0.81, respectively) than both **P36** and **P37** ($S_A = 0.69$ and 0.73, respectively). Similar to observations made with **P36** and **P37**, mixtures of **P39** and **P40** in MLC-6884 (which has a negative dielectric anisotropy and a nematic phase at room temperature) displayed the same absorption red-shift and band-sharpening relative to isotropic solutions, thus indicating conjugation-length enhancement in LC solutions. However, **P41** did not exhibit any signs of conjugation-length enhancement, and hence it appears that steric crowding in this material restricts the large degree of interaction with the LC solvent necessary to promote planarization. Consistently, only low molecular weight versions of **P41** were soluble in the LC. In contrast, **P39**, with the least amount of steric congestion, showed the greatest amount of conjugation-length enhancement and the highest order parameter. Therefore, it is reasonable to conclude that the role of steric congestion about the polymer main chain plays an important role in determining the order parameter and conjugation-length enhancement in PPE–LC mixtures.

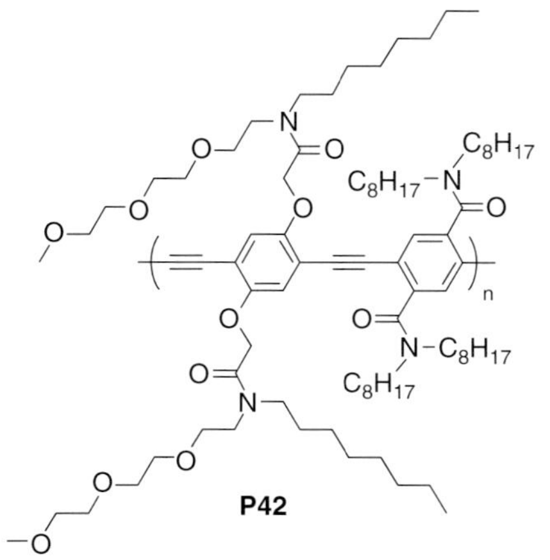

Figure 10.29 Structure of an amphiphilic PPE.

Additionally, molecular weight dependent studies with **P39** revealed that order parameters as high as $S = 0.90$ can be achieved using samples with M_n greater than ca. 20 000.

Lastly, conformational and conjugation-length enhancements in PPEs are not only restricted to nematic LC solvents. Lyotropic LCs composed of water–potassium dodecanoate–decanol were also found to affect the same changes in PPEs (e.g., **P42**, Figure 10.29), provided the repeat unit of the polymer contained amphiphilic side chains [35].

10.6 Conclusions

The ability of CPs to function as electronic materials is dependent on the efficient transport of excited states (excitons) along the polymer chain. Facile exciton migration in PPEs allows energy absorbed over large areas to be funneled into traps created by the binding of analytes, resulting in signal amplification in sensory devices. The energy migration in CPs can occur both *intra*molecularly and *inter*molecularly. In the case of dilute solutions, the *intra*molecular process dominates in the form of a 1D exciton random walk along isolated chains. Much higher efficiency can be reached in polymer aggregates and in solid films, where the energy migration occurs as a 3D process by both *intra*molecular and *inter*molecular pathways. The interplay between these two pathways has been a topic of contention, with intrachain migration being sometimes considered slow and inefficient as compared to its interchain counterpart. However, based on signal amplification of fluorescence quenching and efficient energy transfer to low-energy emissive traps in solution, we have shown that the intramolecular exciton migration in isolated polymer chains may indeed be very efficient.

A detailed understanding of intramolecular energy transfer in CPs can be elusive and is complicated by the conformational complexities that are typically associated with CPs in solutions and in thin films. LB monolayers constitute exceptions, as the rate of energy transfer can be investigated in PPEs assembled into discrete multilayers with precise control of polymer conformation and alignment. Kinetic analyses on LB multilayers surface modified with emissive traps indicated that energy transfer was much faster in the plane defined by each layer of the polymer chains as compared to the direction normal to the chains, thereby suggesting that *intra*molecular energy transfer is faster than the *inter*molecular process.

Fluorescence depolarization studies conducted on a family of triphenylene-incorporated PPEs with long lifetimes point to the through-bond Dexter energy transfer mechanism as being the dominant energy transfer pathway for *intra*molecular exciton diffusion. Extending the lifetime of a CP was universally found to increase the degree of *intra*molecular energy migration. Inclusion of structures with extended aromatic cores into the repeat unit of a PPE generally leads to an increase in the excited-state lifetime of PPEs. In addition, introduction of features that tend to localize excitations – such as biphenyl moieties, kinked polymer backbones, or exciplexes – into the polymer backbone also causes dramatic increases in excited-state lifetime. Generally, the photophysics of the chromophore monomer dictate the excited-state behavior of the corresponding CPs.

Emissive polymer films with modest to high quantum yields of fluorescence usually have limited electronic interaction between polymer chains, and in this case interchain energy migration is generally accepted to occur through the dipole-induced dipole mechanisms. The 3D nature of energy migration in films usually leads to longer exciton diffusion lengths, but often is accompanied by formation of low-emissive intermolecular species, resulting in diminished emission quantum yields. However, incorporation of rigid, 3D scaffolds, such as iptycenes and cyclophanes, can encourage an oblique packing of the chromophore units of a CP, thus allowing the formation of electronically coupled aggregates that retain high quantum yields of emission.

The rigid iptycene scaffolds also act as excellent structural directors that encourage complete solvation of PPEs in an LC solvent. LC–PPE mixtures display both an enhanced conformational alignment of polymer chains and extended effective conjugation lengths relative to isotropic solutions.

References

1. McQuade, D.T., Pullen, A.E., and Swager, T.M. (2000) Conjugated polymer-based chemical sensors. *Chem. Rev.*, **100**, 2537–2574.
2. McGehee, M.D. and Heeger, A.J. (2000) Semiconducting (Conjugated) polymers as materials for solid-state lasers. *Adv. Mater.*, **12**, 1655–1668.
3. Samuel, I.D.W., Crystall, B., Rumbles, G., Burn, P.L., Holmes, A.B., and Friend, R.H. (1993) The efficiency and time-dependence of luminescence from poly (*p*-phenylene vinylene) and derivatives. *Chem. Phys. Lett.*, **213**, 472–478.
4. Goldfinger, M.B. and Swager, T.M. (1994) Fused polycyclic aromatics via electrophile-induced cyclization reactions: application to the synthesis of graphite ribbons. *J. Am. Chem. Soc.*, **116**, 7895–7896.

5. Schwartz, B.J. (2003) Conjugated polymers as molecular materials: how chain conformation and film morphology influence energy transfer and interchain interactions. *Annu. Rev. Phys. Chem.*, **54**, 141–172.
6. Zhou, Q. and Swager, T.M. (1995) Methodology for enhancing the sensitivity of fluorescent chemosensors: energy migration in conjugated polymers. *J. Am. Chem. Soc.*, **117**, 7017–7018.
7. Allwood, B.L., Spencer, N., Shahriari-Zavareh, H., Stoddart, J.F., and Williams, D.J. (1987) Complexation of Paraquat by a bisparaphenylene-34-crown-10 derivative. *J. Chem. Soc., Chem. Commun.*, 1064–1066.
8. Lakowicz, J.R. (2006) *Principles of Fluorescence Spectroscopy*, 3rd edn, Springer, New York, pp. 9–12.
9. Zhou, Q. and Swager, T.M. (1995) Fluorescent chemosensors based on energy migration in conjugated polymers: the molecular wire approach to increased sensitivity. *J. Am. Chem. Soc.*, **117**, 12593–12602.
10. Swager, T.M., Gil, C.J., and Wrighton, M.S. (1995) Fluorescence studies of poly (*p*-phenyleneethynylene)s: the effect of anthracene substitution. *J. Phys. Chem.*, **99**, 4886–4893.
11. Cotts, P.M., Swager, T.M., and Zhou, Q. (1996) Equilibrium flexibility of a rigid linear conjugated polymer. *Macromolecules*, **29**, 7323–7328.
12. Yang, J.-S. and Swager, T.M. (1998) Fluorescent porous polymer films as TNT chemosensors: electronic and structural effects. *J. Am. Chem. Soc.*, **120**, 11864–11873.
13. Wegner, G. (1992) Ultrathin films of polymers: architecture, characterization and properties. *Thin Solid Film.*, **216**, 105–116.
14. Kim, J., McQuade, D.T., Rose, A., Zhu, Z., and Swager, T.M. (2001) Directing energy transfer within conjugated polymer thin films. *J. Am. Chem. Soc.*, **123**, 11488–11489.
15. Satrijo, A., Kooi, S.E., and Swager, T.M. (2007) Enhanced luminescence from emissive defects in aggregated conjugated polymers. *Macromolecules*, **40**, 8833–8841.
16. Satrijo, A. and Swager, T.M. (2007) Anthryl-doped conjugated polyelectrolytes as aggregation-based sensors for nonquenching multicationic analytes. *J. Am. Chem. Soc.*, **129**, 16020–16028.
17. Turro, N.J. (1991) *Modern Molecular Photochemistry*, University Science Books, California, pp. 357–359.
18. Yang, J.-S. and Swager, T.M. (1998) Porous shape persistent fluorescent polymer films: an approach to TNT sensory materials. *J. Am. Chem. Soc.*, **120**, 5321–5322.
19. Sato, T., Jiang, D.-L., and Aida, T. (1999) A blue-luminescent dendritic rod: poly(phenyleneethynylene) within a light-harvesting dendritic envelope. *J. Am. Chem. Soc.*, **121**, 10658–10659.
20. Bredas, J.-L., Cornil, J., Beljonne, D., dos Santos, D.A., and Shuai, Z. (1999) Excited-state electronic structure of conjugated oligomers and polymers: a quantum chemical approach to optical phenomena. *Acc. Chem. Res.*, **32**, 267–276.
21. Kasha, M. (1976) in *Spectroscopy of the Excited State* (ed B. Di Bartollo), Plenum, New York, pp. 337–363.
22. Zahn, S. and Swager, T.M. (2002) Three-dimensional electronic delocalization in chiral conjugated polymers. *Angew. Chem. Int. Ed.*, **41**, 4226–4230.
23. Deans, R., Kim, J., Machacek, M.R., and Swager, T.M. (2000) A poly(*p*-phenyleneethynylene) with a highly emissive aggregated phase. *J. Am. Chem. Soc.*, **122**, 8565–8566.
24. Levitsky, I.A., Kim, J., and Swager, T.M. (1999) Energy migration in a poly(phenylene ethynylene): determination of interpolymer transport in anisotropic Langmuir–Blodgett films. *J. Am. Chem. Soc.*, **121**, 1466–1472.
25. Rose, A., Tovar, J.D., Yamaguchi, S., Nesterov, E.E., Zhu, Z., and Swager, T.M. (2007) Energy migration in conjugated polymers: the role of molecular structure. *Philos. Trans. R. Soc. A*, **365**, 1589–1606.
26. Rose, A., Lugmair, C.G., and Swager, T.M. (2001) Excited-state lifetime modulation in triphenylene-based conjugated polymers. *J. Am. Chem. Soc.*, **123**, 11298–11299.

27. Yamaguchi, S. and Swager, T.M. (2001) Oxidative cyclization of bis(biaryl)acetylenes: synthesis and photophysics of dibenzo[g,p]chrysene-based fluorescent polymers. *J. Am. Chem. Soc.*, **123**, 12087–12088.
28. Tovar, J.D., Rose, A., and Swager, T.M. (2002) Functionalizable polycyclic aromatics through oxidative cyclization of pendant thiophenes. *J. Am. Chem. Soc.*, **124**, 7762–7769.
29. Boden, N. and Movaghar, B. (1998) in *Hand book of Liquid Crystals: Low Molecular Weight Liquid Crystals*, vol. 2b, Chapter IX (eds D. Demus, J., Goodby, G.W. Gray, H.W. Spies, and V. Will), Wiley-VCH, New York, p. 781.
30. Long, T.M. and Swager, T.M. (2001) Minimization of free volume: alignment of triptycenes in liquid crystals and stretched polymers. *Adv. Mater.*, **13**, 601–604.
31. Long, T.M. and Swager, T.M. (2002) Using 'internal free volume' to increase chromophore alignment. *J. Am. Chem. Soc.*, **124**, 3826–3827.
32. Zhu, Z. and Swager, T.M. (2002) Conjugated polymer liquid crystal solutions: control of conformation and alignment. *J. Am. Chem. Soc.*, **124**, 9670–9671.
33. Nesterov, E.E., Zhu, Z., and Swager, T.M. (2005) Conjugation enhancement of intramolecular exciton migration in poly(*p*-phenylene ethynylene)s. *J. Am. Chem. Soc.*, **127**, 10083–10088.
34. Ohira, A. and Swager, T.M. (2007) Ordering of poly(*p*-phenylene ethynylene)s in liquid crystals. *Macromolecules*, **40**, 19–25.
35. Bouffard, J. and Swager, T.M. (2008) Self-assembly of amphiphilic poly(phenylene ethynylene)s in water-potassium dodecanoate-decanol lyotropic liquid crystals. *Chem. Commun.*, 5387–5389.

Index

Charge and Exciton Transport through Molecular Wires. Edited by L.D.A. Siebbeles and F.C. Grozema

ISBN: 978-3-527-32501-6